Understanding Our Atmospheric Environment

Second Edition

Morris Neiburger
Professor Emeritus
University of California, Los Angeles

James G. Edinger
University of California, Los Angeles

William D. Bonner
National Oceanic
and Atmospheric Administration

W. H. Freeman and Company
San Francisco

Project Editor: Larry Olsen
Copy Editor: Julie Segedy
Designer: Sharon H. Smith
Production Coordinator: Linda Jupiter
Illustration Coordinator: Richard Quiñones
Artist: Evan Gillespie
Compositor: Graphic Typesetting Service
Printer and Binder: R. R. Donnelley & Sons Company

Cover: The photograph on the cover shows an unusual cloud formation over the San Jacinto Wilderness Area, San Bernardino National Forest, looking east toward Palm Springs. The photograph was taken between 2 AM and 4 AM in August, 1970. The cloud is rendered visible at night by the electric discharges within it; its pink color is caused by air pollution from the Los Angeles basin and dust particles from the nearby desert. (Photograph by John Deeks.)

Library of Congress Cataloging in Publication Data
Neiburger, Morris.
    Understanding our atmospheric environment.

    Bibliography: p.
    Includes index.
    1. Meteorology. 2. Weather. I. Edinger, James G.
II. Bonner, William D. III. Title
QC861.2.N45 1982   551.5   81-15160
ISBN 0-7167-1348-9   AACR2

Printed in the United States of America

9 8 7 6 5 4 3 2 1 DO 0 8 9 8 7 6 5 4 3 2

Contents

Preface

The old saying "Everybody talks about the weather . . ." is an expression of the pervasive interest people have in our atmospheric environment. This interest arises from concerns of individuals, communities, nations, and society as a whole. For example, the daily weather affects such personal decisions as whether to carry an umbrella or to seek safety from the destructive winds of a tornado or a hurricane. The food supply of entire nations is affected by droughts or floods. Among worldwide aspects, consideration must be given to possible unintentional deleterious effects of human activities, such as the depletion by SST flights or aerosol sprays of the high-altitude layer of ozone that screens out harmful ultraviolet radiation and the rise in world temperature due to accumulation in the atmosphere of carbon dioxide from the burning of fossil fuels. The concluding clause of the saying, "but nobody does anything about it," never completely true, is contradicted by the expansion during recent years of efforts in the field of intentional weather modification.

The scope of interest in atmospheric science has increased due to the extension of our observational resources to higher and higher altitudes by rockets and satellites and to the atmospheres of other planets by space vehicles.

This book, like its previous edition, attempts to answer some of the questions arising from this interest. Its emphasis is on *understanding*, that is, on explaining why various atmospheric phenomena take place in the way they do. It aims to give the reader a deeper appreciation of this part of the physical universe, which he or she experiences so intimately, and attempts to impart knowledge of the laws of physics that govern it.

Meteorology—the science of the atmosphere—is too extensive for all its phases to be treated with any degree of completeness in an introductory textbook. Most aspects of it are mentioned, but only those that are experienced by an observer on the ground or in an airplane are dealt with in detail. Thus we give only a brief account of the way corpuscular radiation from the sun (the solar wind) interacts with the rarefied outermost reaches of the atmosphere to produce the magnetosphere, but we treat extensively the interaction of electromagnetic radiation from the sun with the lower layers of the atmosphere. We discuss how the absorption of this radiation results in the observed horizontal and vertical temperature distribution and wind circulation. Water plays a role in the atmosphere principally in the troposphere, the layer of the atmosphere next to the ground, and we describe the processes by which fog, clouds, rain, and snow form there. We describe the way air masses and the frontal boundaries between them form as a consequence of the temperature and wind distribution, and we present the structure of storms of various scales and intensities, from local thunderstorms to continent-sized extratropical cyclones, with emphasis on the destructive storms, such as tornadoes and hurricanes. We discuss the problem of weather forecasting and describe current procedures, including numerical prediction using large high-speed digital electronic computers and statistical methods. Finally, we treat air pollution and other ways that human activities affect the atmosphere, either inadvertently or intentionally.

In this edition we have included the results of new measurements, particularly satellite observations, and results of recent research, but most of the material continues to be based on long-established observational data and theoretical principles.

This book is intended for students who are majoring in other subjects than a physical science and for other readers who are interested in understanding atmospheric phenomena but who do not have a background in science. Although it contains some mathematics at the level usually required for admission to college, almost all the concepts treated mathematically are also explained in words, so that it is not necessary to follow the mathematics to understand the ideas.

In preparing this edition we have thoroughly revised much of the text and changed the order of some of the chapters, but we have retained the essential features of the first edition, since its favorable reception has encouraged us to believe that there continues to be a need for an elementary textbook that emphasizes physical principles with some degree of rigor. We have given careful consideration to the comments and suggestions we have received, and we have revised sections that gave some readers difficulty. The Appendices have been expanded, and a Bibliography, a Glossary, and a list of Abbreviations and Acronyms have been added.

The new Appendices include a description of weather instruments and their use, written by James G. Edinger, and tables of saturation vapor pressure and the Beaufort wind scale, abstracted from the *International Meteorological Tables*.

In addition to elementary and popular books, the Bibliography includes lists of advanced textbooks, books on special subjects, reference books, and periodicals. The advanced and specialized books may be too technical for most students using this text, but they will be useful for those who have an adequate background in mathematics and science. We have limited the lists to books in English but have included a few published outside the United States.

The Glossary includes definitions of all specialized terms used in the book. As a project proposed in the exercises at the ends of the chapters, students are requested to prepare lists of definitions of the new terms they encounter in each chapter. They can then compare their definitions with those in the Glossary.

It would be impossible to acknowledge all of the comments and suggestions we received from users, reviewers, and recipients of examination copies. As stated earlier, we have given careful consideration to all of them and have acted on many. We extend our thanks to all, even though we cannot list them by name. We wish to thank particularly several persons who responded to requests or questions from us: Arnold Court of the California State University, Northridge; Julius London of the University of Colorado; William M. Gray and Thomas H. Vonder Haar of the Colorado State University; Joanne Simpson and Charles Warner of the University of Virginia; Toby N. Carlson of the Pennsylvania State University; Robert C. Sheets and Ronald L. Holle of the National Hurricane Research Laboratory, NOAA; Neil L. Streten of the Australian Numerical Meteorology Research Centre; Vincent J. Oliver and David Wark of the National Earth Satellite Service; Frederick G. Schuman and Norman A. Phillips of the National Weather Service; Robert Bacastow and Charles D. Keeling of the Scripps Institution of Oceanography, University of California, San Diego; T. N. Krishnamurti of the Florida State University; and George L. Siscoe and Edward A. Doty of the University of California, Los Angeles. We also acknowledge with thanks the permission by various copyright holders to reproduce figures and illustrations. The source of each figure is indicated in its legend.

We appreciate the fine work of the editorial and production staff of W. H. Freeman and Company, which has resulted in the excellent physical appearance of the book. We hope its content will be judged by readers to be of equally high quality.

January 1982   Morris Neiburger

A

B

Photographs of the earth from GOES satellites stationed over the equator on July 19, 1978. (A) GOES West (135° W) photo taken at 2045 GMT (11:45 A.M. local time) in visible light. Note the hurricanes and the waves in the Intertropical Convergence Zone. (B) GOES East (75° W) photo taken at 0400 GMT (11:00 P.M. July 18 local time) in the infrared. The white areas show clouds with tops at great heights, frequently cumulonimbus clouds from which showers are falling. The grey areas represent lower, warmer cloud tops. (Courtesy of National Environmental Satellite Center, NOAA.)

The Drama of the Weather 1

1.1 The excitement of meteorology

There are many reasons for studying meteorology. For many students in the United States a course in meteorology seems a less painful way of satisfying the curriculum requirement for a science than others. But rather than this negative reason, we hope that intrinsic interest of the subject is the primary motivation. To know the "hows and whys" of this most intimate part of our environment, the atmosphere, which surrounds us and enters our bodies with every breath we take, enhances our ability to cope with it and enables us to take deeper pleasure in our daily observations of it.

The most readily observed atmospheric phenomena, which we call weather, range from the quiet, serene, indeed delightful periods of mild warmth and sunshine to dramatically exciting, alarming, and even disastrous stormy spells. Examples of the latter make news almost daily. Thus, the crash of lightning and the rumble of thunder on the night of July 31, 1976, announced a downpour that resulted in a devastating flash flood in the canyon of the Big Thompson River in northeastern Colorado, bringing death to 139 people trapped in the raging torrent and property damage totaling more than $35 million. An unusually heavy snow storm in Chicago on January 12–14, 1979, dropped more than 21 inches in 30 hours on top of an already unusual accumulation, prostrated all activities for several days, and was largely responsible for the defeat of the mayor in his campaign for renomination. Severe cold, record-breaking in many places, throughout the winters of 1977, 1978, and 1979 in the central and eastern United States led to exceptionally heavy imports of oil, which further aggravated the unfavorable balance of trade and caused the dollar to fall to new lows in the international exchanges.

These and many other impacts of the weather show what an important role atmospheric phenomena play in our lives. Understanding the nature of these phenomena will enable us to relate better to them and to make intelligent decisions regarding them. These decisions include daily personal ones: What kind of clothes shall I wear? Shall I carry an umbrella? Shall I go surfing or skiing? They also include political decisions: Shall we support or permit experiments to increase rain? What shall we do about smog? Should we shift from the use of fossil fuels that is increasing the concentration of carbon dioxide in the atmosphere? And they include decisions that affect our safety: Shall we camp in a canyon that may be subject to a flash flood? What shall we do if a tornado warning for our locality is issued?

In addition to the familiar weather occurrences, there are other atmospheric phenomena, just as interesting, that are not part of our conscious experience. Some of them, like the variation of atmospheric pressure, are related to the weather, although they require instruments for their detection. Prior to the invention of the barometer, no one was aware that the atmospheric pressure varies; yet its variations are so intimately connected with the movement of storms and fair-weather systems that storms are frequently identified with centers of low pressure, while centers of high pressure are usually associated with fair weather. The pressure also has diurnal and semidiurnal oscillations, similar to the tides in the oceans, but these are evident only in tropical and subtropical regions. In higher latitudes, these oscillations are obscured by the larger changes due to moving storms.

Other atmospheric phenomena that are not part of our direct experience include some that have indirect influences on our lives or activities. Examples include the ionosphere, which reflects radio waves and makes possible long-distance radio reception, and the stratospheric ozone layer, which filters out deadly ultraviolet radiation from sunlight. Other phenomena, such as the air glow, a chemiluminescence occurring in the upper atmosphere, have no obvious effects on humans.

In this book we shall be concerned mainly with the phenomena that affect us most, namely, those that take place in the lower part of the atmosphere and comprise the weather, and we shall discuss much more briefly some of those occurring in the high atmosphere. Our objective, however, will not be only to enable the student better to deal with his or her individual environmental problems and responsibilities as a citizen. We shall also attempt to convey the sense of satisfaction and excitement that the scientist experiences in penetrating the mysteries of nature and the pleasure associated with appreciation of the logic, order, and beauty in the laws it obeys. This excitement and pleasure motivate researchers in all fields: sociologists, psychologists, political scientists, and economists seek them by understanding aspects of human behavior; biologists by understanding living organisms; and physical scientists, among whom are numbered the meteorologists, by understanding the "hows and whys" of the inanimate world about us. While the exhilaration of discovering for the first time a law of nature is reserved to a very few, such as Galileo, Newton, Harvey, Pasteur, and Einstein, the gratification of understanding these laws and being able to explain the occurrences around us in terms of them can be experienced by everyone.

Physicists and chemists are mostly concerned with things that are small, such as molecules, atoms, or even subatomic particles such as neutrons, electrons, mesons, and neutrinos. Astronomers are concerned with things that are very large, such as stars and galaxies. Meteorologists are concerned with things that range from microscopic (e.g., condensation nuclei) to earth-sized (general wind systems). Are meteorological phenomena as dramatic as splitting the atom or as awe-inspiring as the expanding galaxies? The drama of a developing thunderstorm, the awesome fury of a hurricane, or the mystery surrounding the growth of a raindrop or snowflake all excite an equally great sense of wonder in those who become aware of the processes and forces that produce them.

The beauty in meteorology lies not only in abstract processes and laws of forces. The myriad cloud forms, the manifold colors of the sky at sunset or its blueness at noon, auroral displays at night, glistening dew on the meadow at daybreak, and the infinitely varied forms of snowflakes all have intrinsic beauty that is augmented by an understanding of their causes.

To this beauty are added the excitement and adventure of never-ending change and development; a drama with plot within plot, a poem with intricate rhythm and rhyme, a novel in which characters grow and interact. Indeed, the titles of many literary works—Shakespeare's *Tempest*, Conrad's *Typhoon*, George Stewart's *Storm*, Norman Douglas's *South Wind*, and countless others—reflect their authors' awareness of the colorful and dramatic qualities of meteorological phenomena.

1.2   Characters in the drama. Clouds

Among the most interesting characters in the weather drama are the clouds. Consider, first, the typical sequence of clouds on a quiet summer day in the south–central United States. The early morning is clear, but as the day progresses and the temperature rises, scattered small heap clouds begin to form. These heap

Figure 1.1   Cumulus. [Courtesy of National Oceanic and Atmospheric Administration.]

clouds are named *cumulus* (abbreviated Cu) (Figure 1.1). They have distinct flat bottoms all at about the same height. The tops are rounded, in the form of dense mounds or domes, and are brilliantly white in the sunlight. With time the tops grow higher and higher, the growth taking the form of protuberances or towers. At this stage they are called *cumulus congestus* (Figure 1.2). The vertical growth is accompanied by horizontal growth as well. Then, in the afternoon, the cumulus top suddenly changes its character. Its "cauliflower" appearance becomes smooth, and the top usually stretches out principally in one direction to form the anvil shape so frequently typical of the *thunderhead* or *cumulonimbus* (Cb) (Figure 1.3). With the transition to the cumulonimbus come the flash of lightning, the peals of thunder, and the surge of heavy rain.

The clouds just described are formed by the condensation of water vapor when air heated near the ground ascends and is cooled by expansion. The process of transferring heat upward by air movement is called *convection*. Figure 1.4 shows schematically how the vertical motions of the air heated at the ground lead to the formation of cumulus clouds. The reason why rising air cools will be given in

Figure 1.2 Cumulus congestus. [Courtesy of NOAA.]

Figure 1.3 Cumulonimbus. [Courtesy of NOAA.]

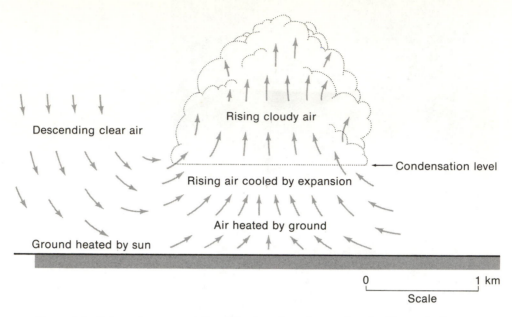

Figure 1.4   Schematic representation of the formation of a cumulus cloud by convection.

Chapter 4, and the details of the way in which cloud drops form from water vapor will be discussed in Chapter 5. The vertical extent of a convective cloud is usually about the same as its horizontal dimensions, ranging from a few hundred meters for small cumulus clouds to several kilometers for cumulonimbus.

In contrast to these clouds of vertical development are the layer, or *stratiform*, clouds. These clouds form sheets extending horizontally hundreds and even thousands of kilometers but may be as thin as a few tenths of a kilometer and are rarely thicker than about 5 kilometers.

Layer clouds frequently are associated with the approach of a storm in temperate latitudes. The first sign of its approach is the advance over the sky of a sheet of very high, thin clouds, feathery *cirrus* (Ci) (Figure 1.5) at first, thickening to a more even layer of *cirrostratus* (Cs) (Figure 1.6). The layer becomes progressively thicker and lower. When it has become sufficiently thick to obscure the outline of the sun's disk, it is called *altostratus* (As) (Figure 1.7). Gradually, the altostratus cloud becomes a uniform grey overcast, through which it is no longer possible to determine the position of the sun. Finally, precipitation begins to fall, at which time the cloud is called *nimbostratus* (Ns) (Figure 1.8); the rain or snow is usually continuous at this stage of the storm's development.

In winter storms over the northern United States, the transition from the first appearance of cirrus and cirrostratus in the blue sky to the sullen greyness and steady precipitation of the nimbostratus usually occurs in a period of 8–24 hours.

Figure 1.5 Cirrus. [Courtesy of NOAA.]

Figure 1.6 Cirrostratus with halo.

Figure 1.7   Altostratus. [Courtesy of NOAA.]

Figure 1.8   Nimbostratus. [Courtesy of NOAA.]

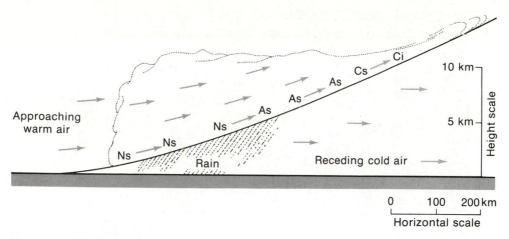

Figure 1.9   Schematic representation of clouds formed by air ascending a warm front.

Thus, if you recognize that the cirrostratus is of this type, you can forecast with some confidence that rain (or snow) will begin within a day.

This sequence of layer clouds is associated with the approach of the *warm-front* portion of a *frontal wave cyclone*. The details of this phenomenon will be discussed in Chapter 11, but a brief indication of its nature may be in order at this point.

When J. Bjerknes discovered, in 1918, that distinct boundaries occur in the atmosphere between masses of air that have different properties, and that much of the weather occurs as a result of "battles" between the air masses, he called the boundaries *fronts*. If cold air replaces warm air as a front moves, it is called a *cold front*. If warm air replaces cold air, the boundary is a *warm front*. At a warm front, however, the air doesn't merely move horizontally; the warm air moves faster than the receding cold air and climbs up along the frontal surface. The cold air doesn't get out of the way fast enough, and the warm air is forced to rise over it.

Figure 1.9 shows schematically the flow of warm air over the receding cold air at a warm front and the resulting cloud distribution. As the air masses move from left to right, the front, which separates them, moves in the same direction, and the cloud pattern moves with it.[1] An observer at the right edge of the diagram would thus experience the cloud sequence that has been described herein as the warm front and its associated pattern of clouds and precipitation would move past him.

---

1. The motion of the air masses referred to is only the component of motion perpendicular to the front. There is, in addition, air flow parallel to the front, and this is usually larger. The actual wind is the resultant (sum) of these two components. Furthermore, because clouds are forming and dissipating, the movement of the cloud pattern is usually slower than the movement of the warm air relative to the front.

Figure 1.10   Cirrocumulus. [Courtesy of NOAA.]

The battle of the air masses at a cold front is frequently more violent than at a warm front. The cold air pushes in under the warm air abruptly, forcing strong upward motions that are even more rapid than those occurring in a local thunderstorm. The frontal thunderstorms are more intense and are oriented in lines along (or ahead of) the cold front. These lines of heavy thunderstorms and strong gusty winds are called *squall lines*. Tornadoes occasionally develop in the severe thunderstorms accompanying squall lines. A tornado is an intense vortex, or rotating air column, rendered visible by the funnel cloud extending downward from the base of the cumulonimbus cloud and by the swirling mass of dust and debris lifted from the ground. It has winds up to 130 m s^{-1} (290 miles per hour)[2] in it and is extremely destructive. Sometimes erroneously called cyclones, tornadoes are among the most villainous characters of the meteorological drama.

The preceding discussion of clouds leads quite naturally to a classification system based on their physical characteristics, the levels at which they occur, and the processes by which they are formed. The classification now in use is an extension and modification of one introduced in 1803 by Luke Howard, an English amateur

---

2. m s^{-1} $\equiv$ meters per second. 1 m s^{-1} = 2.24 miles per hour. (See table of equivalents in Appendix A.)

Figure 1.11 Altocumulus.

naturalist. This classification divides them into mutually exclusive genera, according to their shape and height of occurrence; species, which are subdivisions of the genera, based on additional differences in shape and internal structure; and varieties based on special characteristics within a species. In this text the identification of clouds will be limited to the genus and, in a few instances, the species.

On the basis of height, clouds are classed in four groups—high level, middle level, low level, and clouds with marked vertical development—and within these groups according to whether they form layers (stratus), puffs (cumulus), or hairlike filaments (cirrus). The cirrus types are confined to high-level clouds and are always composed of ice crystals. If there is precipitation falling from the clouds, the word *nimbus* is combined with *stratus* or *cumulus*: nimbostratus or cumulonimbus. The upper parts of most nimbostratus and all cumulonimbus clouds contain ice crystals. Most other clouds are composed of tiny drops of liquid water, even at temperatures considerably lower than 0°C (273 K).

Table 1.1 outlines the basic cloud genera and their heights and forms. The full description of these genera and listing of species and varieties are given in the *International Cloud Atlas*, Volumes I and II, published by the World Meteorological Organization.

We have already discussed some of the ways in which these various clouds are

Figure 1.12 Stratocumulus.

Figure 1.13 Stratus. [Courtesy of NOAA.]

Table 1.1 Classification of Clouds

Genus	Height (of base above ground)	Shape and appearance
HIGH CLOUDS		
Cirrus (Ci)	6–18 km	Delicate streaks or patches
Cirrostratus (Cs)	6–18 km	Transparent thin white sheet or veil
Cirrocumulus (Cc)	6–18 km	Layer of small white puffs or ripples
MIDDLE CLOUDS		
Altocumulus (Ac)	2–6 km	White or grey puffs or waves in patches or layers
Altostratus (As)	2–6 km	Uniform white or grey sheet or layer
LOW CLOUDS		
Stratocumulus (Sc)	0–2 km	Patches or layer of large rolls or merged puffs
Stratus (St)	0–2 km	Uniform grey layer
Nimbostratus (Ns)	0–4 km	Uniform grey layer from which precipitation is falling
CLOUDS WITH VERTICAL DEVELOPMENT		
Cumulus (Cu)	0–3 km	Detached heaps or puffs with sharp outlines and flat bases, and slight or moderate vertical extent
Cumulonimbus (Cb)	0–3 km	Large puffy clouds of great vertical extent with smooth or flattened tops, frequently anvil shaped, from which showers fall, with thunder

formed. Further discussion will be given in connection with a more detailed study of the processes of condensation, precipitation, and convection, and the large-scale atmospheric systems, in Chapters 5, 6, 10, and 11. The study of the organization of clouds, precipitation, and other weather phenomena into systems is carried out by the use of weather maps. Simplified versions of the weather map are familiar to most people from television weathercasts and newspapers. In the next section a brief introduction to this important meteorological tool will be presented.

1.3   Weather systems and the weather map

Until the advent of the meteorological satellites, the only way to get an overall view of the state of the atmosphere was to enter on weather maps the conditions observed at weather stations throughout the world. Satellites provide an alternative view through the pictures of cloudiness as recorded from space by instruments measuring radiation reflected in the visible regions and emitted in the infrared regions.[3] These pictures give important information over areas of sparse conventional reports, such as the oceans, where data are available only from a few island stations and weather ships, and from vessels and airplanes traveling along established trade routes. In practice this information, together with more quantitative measurements that have recently begun to be obtained by satellite, is used in weather forecasting primarily by being entered on weather maps and used in analyzing them. The weather maps at various levels remain the fundamental tools for summarizing the present state of the atmosphere and predicting its future development.

By international agreement, the observations to be entered on the weather map are made simultaneously in all countries, using procedures that meet uniform standards of accuracy, and are transmitted by radio, teletype, or telegraph in a form or code that enables meteorologists of all countries to understand and use the information. Observations of the phenomena that can be evaluated by an observer at the ground without sending instruments aloft are called *surface observations*. Those obtained by instruments that are carried aloft by balloons, airplanes, or rockets are called *upper-air observations*. The upper-air observations that are carried out on a routine basis are made with *radiosondes* or *rawinsondes*, which are instruments that radio back the temperature, pressure, and humidity as they are carried upward by relatively small, unmanned balloons. The winds at upper levels are determined by tracking the radiosonde with radio direction-finding

---

3. Radiative spectral regions will be explained in Chapter 3.

equipment. Because the equipment used in upper-air observations is more expensive, upper-air observations are made at considerably fewer stations than are surface observations.

The principal elements observed and reported in surface observations are the amount and kinds of clouds, the wind speed and direction, the current and past state of weather (rain, fog, etc.), the pressure, the temperature, and the humidity. The routine upper-air observations give the pressure, temperature, humidity, and wind at various heights above the station. These data are collected at regional, national, and world data centers and are redistributed to various forecast centers either in their original form or in the form of *synoptic* maps and charts that summarize the weather conditions over a large portion of the earth. Until the development of high-speed electronic computers, the empirical study of synoptic weather maps formed the basis for the weather forecasts. Because of this, the branch of meteorology that is concerned with the interpretation of weather observations and weather maps is called *synoptic meteorology*.

A typical surface weather map is shown in Figure 1.14. It consists of groups of numbers and symbols at various places on the map, which represent the weather observations at those locations, and two sets of lines, the isobars (with numbers at the ends), which connect places having the same sea-level pressure, and the fronts (heavier lines with solid triangles or semicircles along them), which are the intersections with the ground of the surfaces separating different air masses. This map summarizes the weather conditions that were observed at 7:00 A.M., EST, December 15, 1971.

The complete explanation of the symbols on the weather map is given in Appendix E. Figure 1.15 is an abbreviated illustration of the interpretation of the observational data at each station. The circle, which is at the location of the station on the map, is shaded to show the fraction of the sky covered by clouds and serves as the "head" of an arrow, with a shaft extending from it in the direction from which the wind is blowing. Barbs or feathers at the other end of the shaft show the wind speed to the nearest 5 knots, a full barb (long) representing 10 knots, a half barb (short) representing 5 knots. To the upper right of the circle is a three-digit number, the pressure reduced to sea level, expressed in millibars (mb) and tenths, with the first one or two digits omitted. Since the pressure at sea level is always on one side or the other of 1000 mb and varies gradually for the most part, you can readily determine the missing digit(s).

To the left of the station circle at the same level as the pressure is the temperature, given in degrees Fahrenheit on surface maps in the United States. This procedure is a deviation from international practice. On the upper-air charts, the temperatures are given in degrees Celsius (centigrade) in the United States, as they are elsewhere in the world. To the left, a little lower than the station circle, is the dew-point temperature, also in degrees Fahrenheit, and immediately to the

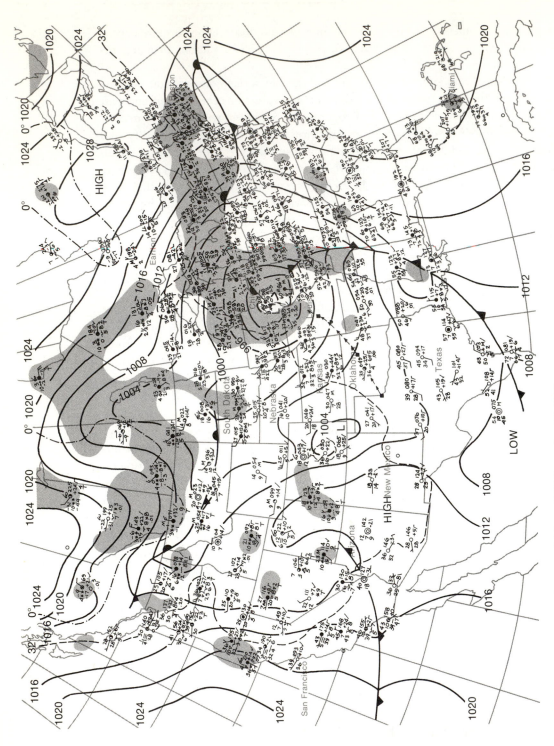

Figure 1.14 Surface weather map, 7:00 A.M. EST, December 15, 1971.

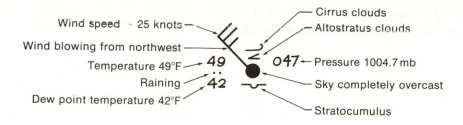

Figure 1.15   Example of interpretation of symbols on weather map.

left of the circle, between the two temperatures, is a symbol describing the character of the weather at the time of observation: groups of dots if it is raining, stars
if it is snowing, and so forth. (See the chart in Appendix E for the 100 different
symbols that can be used.)

Above and below the station circle are symbols that show what kinds of high,
middle, and low clouds are present. Other numbers and symbols, not shown in
Figure 1.15, give such additional information as the trend of pressure during the
past 3 hours, the state of weather and amount of precipitation during the 6 hours
previous to the observation time, the visibility, and the height of the base and
amount of the lowest cloud. Thus, the weather map enables us to see in considerable detail what the conditions were at the various stations.

The fact that pressure varies gradually and continuously from place to place
permits its distribution to be summarized by drawing *isobars*—lines of constant
pressure drawn so that the pressure on one side of the line is lower and on the
other side higher than the specified value. Thus, in Figure 1.14 the line going
southward across South Dakota and southeastward across Nebraska and northeastern Kansas has pressures lower than 1000 mb to the east of it and higher than
1000 mb to the west of it. Following this 1000-mb isobar, we see that it forms a
closed curve, enclosing a large area of lower pressures. The isobars are drawn at
4-mb intervals, and within the 1000-mb isobar there are successively smaller areas
enclosed by the 996-, 992-, 988-, and 984-mb isobars. A low-pressure area of this
sort is called a *cyclone*. The winds, as shown by the arrows, while tending to some
extent to blow across the isobars toward the low center, by and large show a
tendency to circle counterclockwise around the center—east of the center they
flow from the south, north of it from the east, west of it from the north, and south
of it from the west. The stations near the low center had overcast skies, and it was
raining at most of them at the time of observation. The winds were relatively
strong and the weather unpleasant near the center of this cyclonic storm.

In contrast, consider the area of high pressure over Arizona and New Mexico.
There the skies were clear and the winds light. High-pressure areas are frequently

regions of fair weather. On this map the high centers are not well enough defined to show the typical pattern of wind flow around them, which is clockwise (in the Northern Hemisphere). Having characteristics opposite to those of the cyclones, the high-pressure areas and associated wind systems are called *anticyclones*.

In the Southern Hemisphere, the air flow around lows and highs is the opposite of that in the Northern Hemisphere, namely clockwise around cyclones and counterclockwise around anticyclones. The flow around cyclones in both hemispheres is called *cyclonic*. Thus, cyclonic flow is counterclockwise in the Northern Hemisphere and clockwise in the Southern Hemisphere. Similarly, anticyclonic flow is clockwise in the Northern Hemisphere and counterclockwise in the Southern Hemisphere. The reason for this relationship will be explained in Chapter 8.

Cyclones and anticyclones move and their intensity changes with time, but, in general, they maintain their identity from day to day. Their movement and change are associated with the change in weather at any particular location. The problem of weather forecasting may be considered to be the prediction of the movement and development of these pressure systems and their associated weather patterns. The movement of the low center for the 24 hours previous to the map time for Figure 1.14 is shown by the string of arrows extending southwestward from it to the Texas–Oklahoma border. As a first guess we might assume that the movement in the next 24 hours would be the same distance in the same direction. This would place the low just east of Earlton, Ontario, at 7:00 A.M., EST, December 16. We would then guess that, as it moved, the area in which it had been would experience improving weather, with the rain stopping, the skies tending to clear, and the winds becoming lighter and shifting to the northwest (corresponding to the low being to the northeast of its previous position).

The surface weather map for 7:00 A.M., EST, December 16 is shown in Figure 1.16. While the low center has moved somewhat faster than it had during the previous 24 hours, its position is not very far from our estimate. As we guessed, the rain has stopped and the winds have shifted to the northwest in the region where the low passed; but contrary to expectations, a number of the stations there continue to have low clouds covering the sky. Thus, our first attempt at forecasting by using the weather maps has had some success, but also, as happens even when more sophisticated methods are used, some degree of failure.

The surface weather map gives a representation mainly of the conditions near the ground, that is, essentially in two dimensions. To give a picture of the atmospheric state in three dimensions, maps giving the conditions aloft are used. As will be discussed in Section 7.3, instead of using maps at constant elevations, maps are drawn representing the situations on surfaces on which the pressure is constant. The pressure decreases so rapidly with height that a constant-pressure map is close to a constant elevation. For instance, a 700-millibar chart gives conditions almost the same as a 3-kilometer elevation chart would. However, as will be explained in

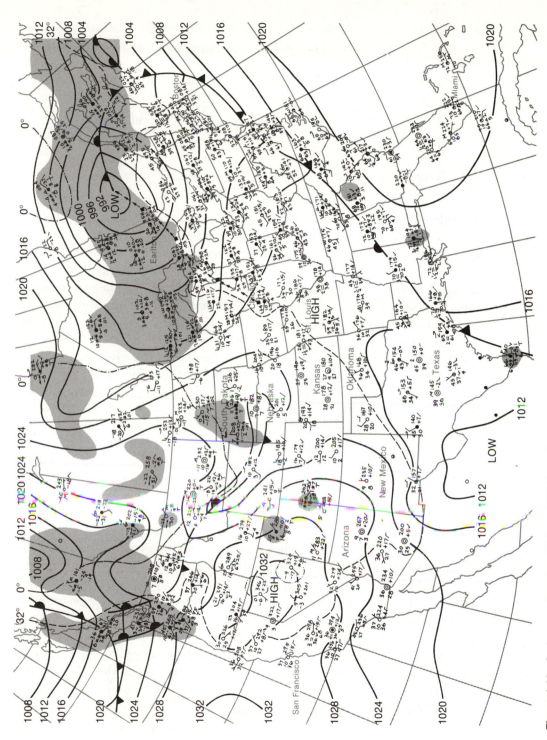

Figure 1.16 Surface weather map, 7:00 A.M. EST, December 16, 1971.

the chapters on atmospheric motions, the variation of height of constant-pressure surfaces, like the variation of pressure on a constant-level map, is a very important quantity.

In the next section a brief review of the instruments used for determining the state of the atmosphere will be presented. Details of the instruments used for measuring temperature, pressure, humidity, and wind are given in Appendix F.

1.4   The tools of meteorology

A single observer without instruments can do much toward characterizing the weather at his or her locality. He or she can determine the kinds of clouds overhead and the approximate fraction of the sky they cover, whether it is raining or snowing, whether it feels relatively warm or cold, and the direction and approximate strength of the wind. From such observations deductions regarding the relationships of various aspects of the weather can be made, and the observations of the ancients led to both accurate and inaccurate summations both in serious tracts and in weather proverbs that have been handed down and are quoted to this day.

From antiquity to the seventeenth century there was little advance in methods of observing the atmosphere or in explaining it. The invention of the barometer by Torricelli in 1643 and the thermometer by Galileo about 1598 introduced quantitative measurements, but it was not until the eighteenth century that instruments were standardized sufficiently for valid comparisons to be made between observations taken at different places. The first weather maps were made early in the nineteenth century, using observations collected by mail; however, it was not until the invention of the telegraph that preparation of daily weather maps on a current basis became possible.

Measurement of conditions at levels above the earth's surface began with the invention of the hot air balloon at the end of the eighteenth century. Manned balloon flights continued to provide information on temperature, pressure, and humidity in the upper atmosphere through the nineteenth century. Meteorographs, which recorded the temperature, humidity, and pressure on cylinders driven by clockwork, relieved the aeronauts of the need to read the instruments continuously. These meteorographs ultimately led to the introduction of unmanned sounding devices—first using large box kites that carried the meteorographs aloft, and later using small free balloons that ascended to great heights with very small meteorographs that descended to the ground on parachutes when the balloons burst. The balloon meteorographs had to be found and sent back before the record was available for analysis. The twentieth century saw the beginning of routine upper-air temperature soundings by airplane, enabling the use of the information in association with the weather map for forecasting.

The radiosonde, an instrument in which sensors control radio signals that give an immediate record at the ground of the temperature, pressure, and humidity as the instrument is carried upward by a balloon, was developed in the late 1920s and early 1930s. By the end of World War II there were enough radiosonde stations to enable construction of maps of conditions in the upper air over much of the Northern Hemisphere.

Upper-air winds were measured initially by following the motion of clouds, then by following small balloons of known ascension rates with theodolites. Using these means, the measurements were limited to at-cloud or below-cloud levels. Following the radiosonde balloons with radio-location devices or radar has eliminated this limitation.

Radar, which became available after World War II, has various meteorological applications in addition to obtaining upper-level winds by following radiosonde balloons. It enables location of precipitating clouds, including evaluation of cloud heights and approximate amounts of precipitation. The distribution of precipitation enables location of the centers of hurricanes, and the shape of the precipitation echoes sometimes makes possible the identification of tornadoes in severe thunderstorm situations. Doppler radar has been used to determine the details of wind-flow patterns in thunderstorms and other severe storms.

In radar, a pulse of shortwave radio signal is emitted and reflected back by the target, be it an enemy airplane or in our case a raindrop or snowflake. Cloud drops and dust particles suspended in the air (*aerosols*) are too small to reflect radio waves; *lidar*, which uses pulses of visible light or infrared waves emitted by a laser, is used to determine their height. Acoustic sounders make use of sound waves to determine irregularities in the thermal structure of the air that cause sound waves to be reflected. They are used to determine the thickness of the layer next to the ground through which rapid vertical mixing is taking place.

An important source of meteorological information is the reports of pilots of airplanes. These may be on commercial flights over established airways ("pireps") or they may be special reconnaissance flights to examine specific phenomena. Thus, the National Oceanic and Atmospheric Administration (NOAA) maintains a fleet of reconnaissance aircraft ("hurricane hunters") in southern Florida in the hurricane season (June–November) that fly out to determine the location and structure of tropical cyclones shown on satellite pictures.

Rockets and satellites provide information about the characteristics of the atmosphere at heights above those balloons and airplanes can reach. Before they were developed, indirect methods were used. For instance, studies of the reflection of radio waves demonstrated the existence of the ionosphere. The bending of sound waves from explosions showed that at great heights the air is about as warm as at the ground. The difference of absorption of the sun's ultraviolet radiation at different wavelengths by the atmosphere proved that there is a considerable amount of ozone in the air at high levels. Spectrographic analysis of the aurora enabled determination of some of the chemical constituents of the air hundreds of kilometers above the earth's surface.

The rocket and satellite measurements have corroborated the information obtained at the ground by indirect methods and have provided more details concerning it. They have extended our knowledge of the atmospheric structure out to where the atmosphere merges into interplanetary space. At present satellites are combined with indirect methods to provide many kinds of information about the earth and its atmosphere.

1.5   Meteorological satellites

With the launching of TIROS I, the first artificial satellite transmitting television pictures of clouds, on April 1, 1960, a new era in meteorology began. (TIROS is the acronym for Television InfraRed Observational Satellite; however, the cloud pictures were taken using visible light.) Previous to TIROS I the patterns of clouds and storms were inferred from the separate stations taking surface observations. The clouds observed and reported from these stations are those within a few tens of kilometers of the stations. Since the stations over continents are 200 km or more apart, and over oceans there are hundreds and sometimes thousands of kilometers without observations, the complete distribution of clouds cannot be recognized from them. The TIROS pictures gave a startlingly new perspective on cloud configurations.

Among the new patterns that were discovered in TIROS I pictures were the banded structure of clouds around large extratropical cyclones, similar to the rain bands that had been seen by radar to exist around tropical cyclones and hurricanes; the occurrence of eddies in the lee of islands; and the occurrence of large hollow convective cloud cells, indicating descending air in the center, the inverse of the usual cumulus cell with rising air in the center and descending clear air around them (see Figure 1.17). In addition to new discoveries, there were corroborations of older theories, such as those concerning the cloud distributions associated with frontal wave cyclones (Chapter 11), the jet stream, and the flow over mountains.

Perhaps more important than the new perspectives on cloud distributions in relation to the terrain and to features on the weather map was the contribution to improvement in forecasting. Particularly in Australia, where the weather systems come from the South Indian Ocean and the Southern Ocean, where few ships

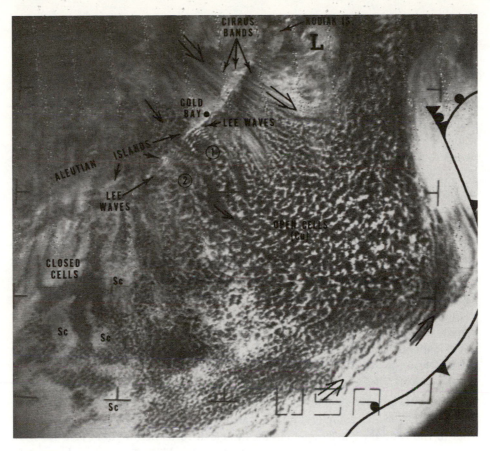

Figure 1.17 Examples of cloud forms viewed by satellite. [NOAA–NESS photo.]

travel, as soon as they became available the cloud pictures were put to use for forecasting the approach of fronts and storms that otherwise would not have been recognized until they were almost on shore. On the west coast of North America, the pictures were similarly useful in clarifying the structure and speed of the weather systems that were only vaguely indicated by the scanty observations over the Pacific Ocean. The characteristic cloud pattern of tropical storms enabled the location and tracking of hurricanes and typhoons, which otherwise might be located much later and tracked with much less precision.

These applications were greatly facilitated by the introduction of APT (Automatic Picture Transmission), which sent photographs of the local area directly to stations equipped with relatively inexpensive receiving equipment.

The TIROS satellites in the initial series, numbered 1 through 10 and launched from 1960 to 1965, were planned primarily as experimental vehicles to test various procedures for observing the atmosphere from space. They were followed by a series of TOS (TIROS Operational System) satellites, called ESSA 1 through ESSA 9, launched from 1966 to 1969. The ESSA (Environmental Survey Satellites) were operated in pairs. The odd-numbered members took and stored on tape the pictures on each orbit for transmission on command when in range of the command and data acquisition stations at Wallops, Virginia, and Gilmore Creek (near Fairbanks), Alaska. The even-numbered ESSA members were equipped with APT cameras for automatic transmission to ground stations within radio range. These were succeeded beginning in 1969 by the ITOS (Improved TIROS Operational System), a second-generation system in which each satellite (called NOAA 1 through NOAA 5) carried both the direct-readout APT cameras and the stored-data cameras. The ITOS had the additional feature of infrared sensors for providing pictures during the night as well as during periods of daylight. In 1972 the capability of getting vertical temperature soundings through cloud-free portions of the atmosphere was added. Finally, in late 1978, a third-generation satellite, TIROS N, was launched (see Figure 1.18), which provides higher-resolution visible and infrared cloud pictures, improved systems for observing the vertical temperature structure in clear and cloudy air, observations of the vertical distribution of water vapor, and a system for locating and getting data from free-floating buoys, constant-level balloons, and other observational platforms. NOAA 6, the second of the TIROS N series, was launched June 27, 1979.

All these satellites viewed the earth from relatively low levels, between about 650 and 1500 km (400–900 miles). After TIROS 8 they were launched in polar sun-synchronous orbits, going around the earth every 90–120 minutes in such a way that in the time they make one circuit the earth turns below them just the right amount so that they cross the equator at the same local (sun) time for each crossing. The current TIROS N and NOAA 6, in circular polar orbits every 102 minutes at about 850-km elevation, produce pictures in a strip about 1800 km wide from pole to pole. Thus, each completes approximately 14 orbits a day, but "sees" any particular place only twice a day—once in daylight and once in darkness. With two operational polar-orbiting satellites appropriately spaced (TIROS N and NOAA 6), the positions of a general storm system can be checked four times a day. Small systems such as local thunderstorms change rapidly, and it is desirable to view them more frequently—if possible, continuously. The Synchronous Meteorological Satellites (SMS) and Geostationary Operational Environmental Satellites (GOES) satisfy this need.

The higher a satellite is above the earth, the slower it goes. At an elevation of 35,850 km, its speed is just that required to go around the earth in 24 hours. When placed at this height in the plane of the equator, a satellite will move around the earth at the same rate the earth is rotating and thus will remain above the same spot on earth all the time; that is, its motion will be synchronous with that of the

Figure 1.18   Launching of Atlas rocket that placed TIROS N satellite into orbit October 13, 1978. [NASA/AF photo.]

earth's surface. At that height, also, the area viewed is much larger than that seen from the lower satellites. The ATS 1 (Applied Technology Satellite), launched in December 1966, and the ATS 3, launched a few months later, were very successful experimental prototypes of the present GOES satellites. Stationed at 150°W and at various positions between 45°W and 95°W along the equator, they covered most of the Atlantic and Pacific Oceans, with overlapping coverage of the Americas. ATS 1 and ATS 3 were equipped only for photography by visual light and were thus limited to daylight hours. SMS 1 and SMS 2 (developmental satellites launched in 1974 and 1975) and the operational GOES satellites that followed were

A

Figure 1.19   Pictures taken by GOES West geostationary satellite at 11:45 A.M. 135th Meridian Time. At that hour the entire disc is in daylight, enabling the photo by visible light (A) to be complete. The infrared (IR) photo (B) is complete both day and night, because it is made by

equipped with infrared as well as visual imagery. Maintained at 135°W (GOES West) and 75°W (GOES East), they have made pictures of the cloud distribution over most of the Western Hemisphere available every 30 minutes both day and night routinely, and more often when desired. The pictures from the GOES have become familiar to viewers of weather broadcasts on American television. In addition to the still pictures, the successive images for periods of 24 hours are combined into film loops that strikingly show the weather patterns in motion.

In the pictures taken with visible light, the clouds appear white because they reflect much more light than the ground, unless the ground is snow covered, or

B

infrared radiation emitted by the earth's surface and by the clouds. [NOAA–NESS photo. Courtesy of V. J. Oliver.]

the sea. In the infrared pictures what is really shown is the temperature of the material from which the radiation originates. Since clouds are at higher elevations where the temperature is lower, the areas of low temperature are interpreted as cloudy areas. To have the infrared pictures visually correspond to the visual light pictures, they are made with a grey scale in which the cold temperatures are shown in lighter shades and the warm areas in darker shades. Clouds that have their tops in the uppermost, coldest part of the troposphere are shown as white, while clouds near the ground, where the temperature is relatively high, are dark grey.

By measuring the movement of clouds that can be identified on two successive pictures, the wind velocity at cloud level can be determined. Systems have been developed at the National Earth Satellite Service headquarters to compute these winds automatically by computer.

Other kinds of data besides those used in weather forecasting—the cloud pictures and the vertical temperature profiles—are also collected by these operational satellites. These data include Space Environment Monitoring, in which high-energy particles and x rays coming into the atmosphere from the sun and from outer space are measured, as well as sea surface temperatures, which are useful to commercial fishing and other marine interests. Incidentally, just as the sea surface temperature measurements of meteorological satellites have provided valuable information to oceanography, the oceanographic research satellite SEASAT, which operated for 95 days in 1978, gave much interesting information to meteorologists, including the demonstration that winds at the ocean surface can be determined with a radar scatterometer, a device for measuring the microwave radiation sent back by the very small ocean waves (capillary waves) that respond almost instantaneously to changes in the wind.

In addition to the operational meteorological satellites, which have been developed cooperatively by the National Aeronautics and Space Administration (NASA) and the National Oceanic and Atmospheric Administration (NOAA), but operated by NOAA, a series of research meteorological satellites, the Nimbus series, has been operated by NASA. Besides testing some of the instruments and procedures subsequently put on the operational satellites, it has provided data on the energy received from the sun, the portion reflected by the earth and atmosphere, and the quantities of several trace gases and particles in the upper atmosphere. The latest in the series, Nimbus 7, launched in 1978, has an instrument called the Coastal Zone Color Scanner for mapping chlorophyll concentrations, sediments, and salinity of water along the coasts; it is expected to enable mapping of oil spills.

1.6   Computers in meteorology

Important as satellites are, even more important to meteorology has been the development of high-speed electronic digital computers, the use of which has completely revolutionized the whole process of weather prediction. Research into atmospheric processes has also been greatly facilitated by the use of these computers.

Before the introduction of the computer, the entire procedure of weather forecasting was empirical and highly subjective, not much more sophisticated than the exercise in prediction carried out at the end of Section 1.3. The reason for this is that the equations governing the behavior of the atmosphere are so complex that they cannot be solved by ordinary mathematical procedures. With the high-speed computers, methods have been devised for obtaining solutions by numerical approximation. The computers are used not only to solve the equations but to process the observations and plot the weather maps at the beginning of the forecast process, and to convert the predicted flow patterns that are the direct results of solving the fundamental equations into the quantities of specific interest to the man on the street, such as high or low temperatures and likelihood of rain or snow.

The keeping of weather records has likewise been revolutionized by the computer. Stored on tape or disc, the data are readily available at the touch of a few buttons on a keyboard.

Computers have also enabled improvement of ways of observing the weather. This is particularly true of satellite observations, in which computer techniques are involved at all stages.

Research studies of atmospheric phenomena of various scales, ranging from the microphysics of drop growth in clouds and the dynamics of the small-scale convection that produces cumulus clouds and local thunderstorms to the investigation of the general circulation of the atmosphere over the entire earth, have been carried out using computers to treat numerical models that simulate their behavior. These models have enabled clarification of such questions as the circumstances under which hail forms in thunderstorms, and what the likely effects would be of a doubling of the carbon dioxide content in the air. As bigger and faster computers are developed, it becomes possible to make the models represent more and more closely the complexity of the real atmosphere. Together with the development of better, more accurate methods of solving the complicated differential equations that represent atmospheric processes and better ways of observing the actual details of atmospheric processes, the introduction of improved computers should lead to more accurate weather forecasts and a more thorough understanding of what goes on in all parts of the air surrounding the earth.

1.7 The role of mathematics in science

The reference to equations in the preceding section brings up a subject that is a psychological block for many nonscience students. Having somehow surmounted the required mathematics courses in high school, they have put out of mind what they learned of it there and have made no attempt to use it in their daily lives. But much of our daily experience consists of quantitative relationships that are by their nature mathematical, and this is especially true of the scientific aspects of everyday life. While many of these relationships can be expressed in words, they are more easily and concisely expressed in numbers, symbols, and equations. In this sense, mathematics is the *language* of science. In addition to clarity and ease of expression, this language has the virtue of being universal. Scientists of practically all nationalities, however different their native tongues, use and understand it. Whatever the language and alphabet used in the rest of their writing, American, French, Russian, Chinese, or Japanese mathematicians, physicists, and meteorologists use the same systems of equations and symbols in the mathematical part.

In addition to serving as the language of science, mathematics provides a method of reasoning that enables the discovery of new relationships. Thus, in physics, mathematical rules are set up that correspond to processes occurring in the real physical world, and these rules enable deductions to be made concerning the consequences of these processes under prescribed conditions. We shall see how this is done in our analysis of some atmospheric processes.

In this book the quantitative relationships will be presented both in words and in mathematical symbols. The mathematics used will be that usually required for admission to college, namely, high school algebra or the equivalent. Both from the standpoint of appreciating the methods of atmospheric scientists in applying the laws of physics to atmospheric phenomena and from the standpoint of being better able to understand and remember the atmospheric relationships that will be discussed in the following chapters, it is desirable that the reader study the mathematical treatments presented. However, mastery of the equations is not necessary; the concepts can be obtained from the verbal presentation.

Finally, reference must be made to systems of units of measurement. To express relationships quantitatively, it is necessary to measure the pertinent aspects that are being related. In most of the world except the United States, the metric system is in common use, and in scientific circles everywhere a standardized version of it, the International System (SI), has been adopted. In anticipation of the eventual changeover from the English system (feet, pounds, etc.) to the metric system, elementary and high schools in the United States have included instruction in conversion from one system to the other. Therefore, some familiarity with metric units will be assumed in this book, and quantities will be expressed in them except when customary usage in meteorology retains other units, for instance, the knot as a unit of speed. Appendix A gives tables of the relationship between units in the metric and English systems and the factors for converting from one to the other.

Questions, Problems, and Projects

1. Write down for yourself, as frankly as you can, the reason why you are studying meteorology, and list the questions you expect to learn the answers to, or the topics you expect to find out about.

2. Keep a log of the cloud types you see every day. If convenient, choose two or three times of the day when you are out of doors regularly (e.g., on your way to class), and note the kinds of clouds in the sky at those times and your estimate of the fraction of the sky covered by each.

3. List the phenomena that you consider constitute *weather*. From these formulate a definition of weather.

4. Express the cloud heights given in Table 1.1 in feet.

5. From the data in Figure 1.14 write a description of the weather at 7:00 A.M., December 15, 1971, at Boston, Massachusetts; Chicago, Illinois; St. Louis, Missouri; Miami, Florida; and San Francisco, California.

6. What is meant by "fronts" in meteorology? Why do cloudiness and precipitation frequently occur in their vicinity?

7. What innovations have been introduced by modern technology in the study and prediction of the weather? What benefits have resulted from these innovations?

8. Make a glossary of definitions of all words you have encountered for the first time in this chapter. Arrange the glossary alphabetically and list after each term the page number on which it is introduced or defined. In some instances the complete definition is given in a later chapter, in which case it can be found by referring to the index. (Do the same for each chapter hereafter, as one of the ways of studying the material in it.)

2 The Composition and Thermal Structure of the Atmosphere

2.1 Nature and composition of the atmosphere

The atmosphere of the earth is a gaseous envelope that surrounds the solid and liquid surface of the earth. It extends upward for hundreds of kilometers, eventually meeting with the rarefied interplanetary medium of the solar system, which in turn may be considered an extension of the corona of the sun.

The phenomena that take place in the atmosphere are many and varied. In the lower layers winds blow, clouds form and dissipate, rain and snow fall, and warm and cold spells occur. These are among the phenomena that are commonly called *weather*. At higher levels the aurora borealis, zodiacal light, luminous meteors, and noctilucent clouds are seen, and radio signals are reflected or refracted by the ionosphere. Particles in the atmosphere cause visible effects, such as rainbows, halos, colors of the sky, and the changes in the appearance of the landscape due to haze. Invisible but equally important effects are caused by chemical (or photochemical) and electrical reactions between atmospheric constituents. The study of all these phenomena is the subject of meteorology.

The gas that constitutes the atmosphere is called *air*. Air is a mixture of several chemical elements and compounds. Some early Greek philosophers thought that air was the primary element, that is, the fundamental substance not further subdivisible into constituent components from which everything derived. Empedocles thought that the universe was composed of four such elements: air, fire, earth (soil), and water. We now know that there are many elements (105 have been discovered so far) and, further, that in addition to elements the air contains some compounds, that is, chemical combinations of elements. The most plentiful elements in air are nitrogen, oxygen, and argon. The compounds in it of greatest importance are water vapor and carbon dioxide. Another important constituent, even though it is present only in small amounts and mostly at high elevations, is ozone, the triatomic form of oxygen.

Water vapor is constantly being added to the atmosphere by evaporation from various bodies of water or damp ground and by transpiration from plants; it is being subtracted from it in other places by condensation into clouds or dew. For this reason the amount of water vapor present is quite variable. Over warm oceans or tropical jungles, it may be as much as 3 or (rarely) 4 percent of the air by volume; in cold regions, over deserts, or at great heights, it is a small fraction of 1 percent.[1]

Because of this variability, it is usual to consider separately the water-vapor content and the fractional composition of the remainder of the air (which for convenience we call *dry air*). It has been found that the composition of dry air is almost exactly the same all over the earth and to heights up to about 100 km. Table 2.1 gives the percentages of the major components, nitrogen, oxygen, argon, and carbon dioxide. The other constituents, which include the inert elements neon, helium, krypton, and xenon, as well as hydrogen and some compounds, such as methane, sulfur dioxide, and various oxides of nitrogen, that come from biological and industrial processes, are present in extremely small fractions, a few parts per million or less.

In addition to the gaseous components, air contains solid and liquid particles, most of them so small that the movements of the air offset their tendency to fall to the ground. These may originate as dust raised by the wind, as sea salt particles from the evaporation of ocean spray, or as products of combustion and industrial

---

1. It is shown in kinetic theory that, for a given pressure and temperature, the same number of molecules of any gas, whatever the mass or size of its molecules, occupies the same volume. When we speak of "percent by volume," we are thus identifying what percent of the total number of molecules are molecules of the particular constituent.

Table 2.1 Principal Components of Dry Air

Constituent	Chemical symbol	Content (% by volume)
Nitrogen	N_2	78.09
Oxygen	O_2	20.95
Argon	A	0.93
Carbon dioxide	CO_2	0.03
Total		100.00

processing. There are also particles that come into the atmosphere from extraterrestrial sources.

Our knowledge of the composition of air dates from the beginning of modern chemistry in the eighteenth century.[2] Prior to that time, John Mayow (1643–1679) conducted experiments that led him to conclude that air consists of two parts: one, called "fire-air," that supports combustion and sustains life, and another that does not. However, he did not isolate the separate constituents.

Mayow's discovery was overlooked, and, in fact, the first constituent of air to be isolated and studied in detail was not "fire air" (oxygen) or the more plentiful nitrogen, but carbon dioxide. This work was carried out by Joseph Black (1728–1799) in 1752, when he was a medical student at the University of Edinburgh. Black called the substance that he isolated "fixed air," because he found it to be somehow fastened to various substances (carbonates) from which it could be removed by heating.

The next advance, the discovery of nitrogen, was also made at Edinburgh, by a medical student named Daniel Rutherford (1749–1819). Working under Joseph Black in 1772, he studied the properties of the gas that remained after charcoal was burned in a closed volume for as long as possible and then the carbon dioxide was removed. He called this gas "mephitic air" and thought that it was a combination of ordinary air and "phlogiston," the material considered to be given off by substances in the process of burning.

Shortly after Rutherford isolated "mephitic air," Joseph Priestley (1733–1804) in England and Carl William Scheele (1742–1786) in Germany isolated oxygen. Priestley did so by heating mercury in the presence of air until it formed a red powder and then increasing the heat until a gas was given off that supported

---

2. The account given here is condensed from that published in *Ways of the Weather* by W. J. Humphreys (Lancaster, Pa.: Jaques Cattel Press, 1942).

combustion more strongly than ordinary air. He called this gas "dephlogistated air."

Subsequently, the nature of these components of air was clarified, and they were given their modern names. But it was not until 1894 that argon, which is almost 1 percent of dry air, was identified. Until that time it was lumped with nitrogen, but in that year Lord Rayleigh (1842–1919) recognized that "nitrogen" isolated from air had a heavier molecular weight than nitrogen obtained from the decomposition of chemical compounds. He and Sir William Ramsay (1852–1916) carried out experiments in which the nitrogen was removed from air. They found an inert residual gas, which they named argon.

The discovery that air contained a hitherto unknown constituent in such a large proportion stimulated more careful analysis to see whether other constituents were present. It was found that four additional inert gases—helium, neon, krypton, and xenon—are present in very small quantities, and that hydrogen, which is given off at the earth's surface and escapes upward to space, is present in a similarly small concentration.

Like water vapor, the gaseous constituents that enter the atmosphere from the biosphere and from industrial processes vary in concentration from place to place and from one time to another. Among the principal variable constituents are sulfur dioxide (SO_2), oxides of nitrogen (N_2O, NO, NO_2), ammonia (NH_3), methane (CH_4), carbon monoxide (CO), ozone (O_3), and various organic compounds. Some of these substances are serious air pollutants, having deleterious effects even when present in concentrations as low as one part per million or less. Others play important roles in chemical and photochemical processes that take place in the upper atmosphere.

Accurate measurements show that carbon dioxide also is a variable constituent. Not only is its concentration higher near urban and industrial complexes, but its average volume percentage has increased steadily from about 0.029 percent in 1900 to 0.0334 percent in 1979. It has been suggested that this increase has influenced the way in which the average temperature of the earth has changed since the beginning of the century and that further increase in its concentration may have catastrophic consequences by causing the thick ice sheets that cover Greenland and Antarctica to melt. This would raise the sea level enough to inundate coastal cities throughout the world. Carbon dioxide tends to raise the earth's temperature because it absorbs and radiates back some of the infrared radiation emitted by the ground. This additional heating makes it necessary for the ground to be warmer in order to emit enough radiation to balance the radiation received from the sun. Counteracting this effect of carbon dioxide, particles from volcanoes and pollution sources may significantly reduce the amount of solar radiation reaching the ground, thereby tending to decrease the earth's temperature. These possibilities of the world climate being changed by changes in the amounts of CO_2 and particles will be discussed in more detail in Chapter 15.

2.2   Vertical structure of the atmosphere

From common experience we know that the temperature of the air varies from time to time: warm afternoons follow cool mornings and warm summers follow cold winters in a regular cyclical fashion, and spells of cold days and warm days are interspersed in a single season in an irregular fashion. The temperature also varies from place to place at the same time; in general, regions at low latitudes are warmer than those at higher latitudes, and low lands are warmer than high mountains. The familiar Japanese woodprints showing trees in blossom with snow-capped Mt. Fuji towering in the background illustrate the effect of altitude on temperature.

If we consider the earth as a whole for the entire year, the average temperature near the ground at sea level (the surface temperature) is 15°C (288 K, 59°F).[3] This year-round overall average decreases with height, as indicated in the previous paragraph, but above about 12 km (40,000 ft) it stops decreasing and starts increasing. This surprising reversal was not believed when it was first discovered by sending up small balloons with recording instruments. At first it was thought to be an instrumental error, and only after several hundred balloons had reached these heights and the same result was obtained was the existence of this layer of inverse behavior of temperature accepted.

The bottom layer of the atmosphere, in which the temperature on the average decreases with height, is called the *troposphere*; the layer above, of constant or increasing temperature, is called the *stratosphere*; and the surface between them (sometimes a transition zone or layer rather than a surface) is the *tropopause*. In the troposphere the temperature on the average decreases at a fairly constant rate, 6.5°C/km (3.5°F/1000 ft). This rate is known as the *average lapse rate* of the troposphere; by *lapse rate* is meant the rate of *decrease* of temperature. Where the temperature decreases with height, as is usual in the troposphere, the lapse rate is positive; in the stratosphere, where the temperature is constant or increases with height, the lapse rate is zero or negative.

---

3. In this book we shall usually express temperatures on the Celsius (centigrade) scale or the Kelvin (absolute centigrade) scale rather than the Fahrenheit scale, which is commonly used in the U.S. Rules for converting from one of these scales to any other are given in Appendix A, and a discussion of them and the relationships for conversions is given in Chapter 4.

Early in the twentieth century, various pieces of indirect evidence, including the way in which some large explosions were heard both close to the places where the explosions occurred and at greater distances from them, with zones of silence between, led to the deduction that at a height of about 50 km the temperature is about the same as at the earth's surface. Other indirect evidence pointed to very low temperatures at about 80 km, and still other evidence indicated that above 80 km the temperature increases again. Thus, before the direct observations by rockets and satellites, the general atmospheric structure was known to considerable heights.

The use of rockets and satellites for the exploration of the high atmosphere has confirmed this knowledge and has provided precise measurements of the temperature to much greater heights. After the confirmation of the existence of the temperature maximum at about 50 km and the minimum at about 80 km, names were given to additional layers. The layer of decreasing temperature from 50 to 80 km is called the *mesosphere,* and the layer of increasing temperature above that is called the *thermosphere.* The boundary between the stratosphere and mesosphere is called the *stratopause;* that between the mesosphere and thermosphere is called the *mesopause.*

Other designations of atmospheric layers are based on electrical state and on composition. The *ionosphere* is a layer that contains ions (molecules and atoms that carry electric charges) and free electrons. It extends upward from about 80 km, and measurements of radio reflections indicate that it is composed of layers of varying ion density, the E- and F-layers. From the standpoint of composition, sometimes the layer of uniform composition with respect to major constituents, which extends from the earth's surface to 80 or 100 km, is called the *homosphere.* However, although within this layer the proportions of nitrogen, oxygen, and argon are almost constant, the concentration of ozone is much larger between 15 km and 50 km than elsewhere in the atmosphere (except for the effect of pollution over cities; see Chapter 14). This ozone-rich layer, called the *ozonosphere,* plays an important role in shielding humans and other forms of life from harmful ultraviolet radiation from the sun. Above 80 km, concentrations of the major chemical constituents start varying with height, and this part of the atmosphere is called the *heterosphere.*

Extending outward to as much as 10 earth radii (60,000 km), particles acted on by the earth's magnetic and gravitational fields move with the earth as it revolves around the sun and can therefore be considered part of the earth's atmosphere. Known as the *magnetosphere,* the region occupied by these particles has a complicated structure that includes layers such as the *plasmasphere* and the *radiation belts.* A brief description of this very tenuous outermost portion of the earth's atmosphere will be given in Chapter 3.

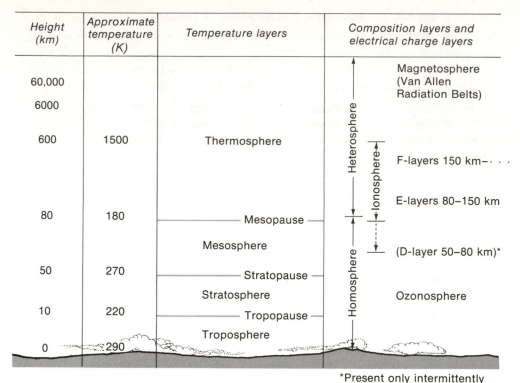

Height (km)	Approximate temperature (K)	Temperature layers	Composition layers and electrical charge layers
60,000			Magnetosphere (Van Allen Radiation Belts)
6000			
600	1500	Thermosphere	F-layers 150 km–· · ·
			E-layers 80–150 km
80	180	Mesopause	
		Mesosphere	(D-layer 50–80 km)*
50	270	Stratopause	
		Stratosphere	Ozonosphere
10	220	Tropopause	
		Troposphere	
0	290		

*Present only intermittently

Figure 2.1   Layers of the atmosphere.

2.3   Variation of average temperature with height

The vertical structure of the atmosphere, based on temperature, composition, and electrical state, is shown in Figure 2.1. The heights and temperatures given in Figure 2.1 are rough averages characteristic of middle latitudes, where most of the observations of temperatures aloft have been made. As an indication of the degree of approximation, the heights are given only to one significant figure,[4] and the temperature to two. To this degree of approximation, the data in the figure may be considered to represent the annual averages over the entire earth.

---

4. By "significant figures" is meant the digits (other than the zeros adjacent to the decimal point on either side) in a number that are known accurately. Thus, the statement that the height of 50 km in the figure is given only to one significant figure means that we know that it is closer to 50 km than to 40 km or 60 km, but that the accurate value might be anything between 45 and 55 km.

Table 2.2 Temperatures of the Standard Atmosphere

Height (km)	Temperature (K)	Height (km)	Temperature (K)
0.000	288.15	150	634.39
11.019	216.65	160	696.29
20.063	216.65	170	747.57
32.162	228.65	190	825.31
47.350	270.65	230	915.78
51.413	270.65	260	950.99
71.802	214.65	300	976.01
85.638	187.65	350	990.06
86.000	186.87	400	995.83
91.000	186.87	500	999.24
100	195.08	750	999.99
110	240.00	1000	1000.00
120	360.00		

Source: U.S. Standard Atmosphere 1976 (Washington, D.C.: U.S. Government Printing Office, October 1976).

Approximate average temperature conditions at various heights over the earth are given in the *U.S. Standard Atmosphere 1976*. This is an idealized representation of the conditions considered typical for middle latitudes, say 45°N, up to a height of 1000 km. It was arrived at because of the need to define a reference atmosphere for use in the design of aircraft and missiles and the instruments used on them. Up to 32 km, it is the same as the *ICAO Standard Atmosphere* adopted by the International Civil Aviation Organization, and the lowest 50 km was accepted by the International Standards Organization as the *ISO Standard Atmosphere*. For greater heights, the estimates of the appropriate values were arrived at by a committee of U.S. scientists using the observations obtained from rockets and satellites.

The temperatures of the *U.S. Standard Atmosphere 1976* are given in Table 2.2. In Figure 2.2, they are shown graphically for the lowest 120 km. In the standard atmosphere, the temperature in the troposphere decreases at a constant rate of 6.5°C/km to approximately 11 km. The stratosphere consists of three sublayers: isothermal from about 11 to 20 km, temperature increasing from about 20 to 47 km, and isothermal again from about 47 to 51 km. The mesosphere, with temperature decreasing again, extends from about 51 km to about 86 km. The thermosphere begins with an isothermal layer, above which the temperature rises with height at an increasing rate up to about 130 km, then rises more and more

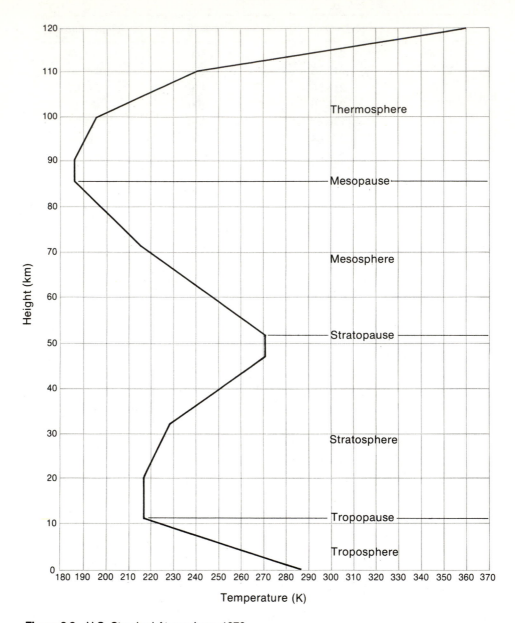

Figure 2.2 U.S. Standard Atmosphere, 1976.

slowly until it becomes practically isothermal above 500 km. Thus, the tropopause in the standard atmosphere is at 11 km, the stratopause is at 51 km, and the mesopause is at 86 km.

It should be remembered that the standard atmosphere is representative of year-round average conditions at 45° latitude. The average temperatures at all heights are different at other latitudes, and the actual temperatures vary with the time of year and, in fact, from day to day with the fluctuations of the weather. In the next section, examples of the vertical temperature structure at various latitudes and different seasons will be given.

The accumulation of our present knowledge of the thermal structure of the upper atmosphere was a gradual process. Apart from measurements at a few mountain-top observatories, it began with the raising of thermometers on kites in the middle of the eighteenth century and the invention of the lighter-than-air balloon at the end of that century. J. L. Gay-Lussac (1778–1850), the French scientist who later gained fame for his research in physics and chemistry, made two balloon ascents in 1804, the second of which reached 7 km. He measured the temperature and humidity up to this height and collected samples of air at different heights. In his analysis of the air samples he found no detectable variation of composition with height.

There were further measurements using manned balloons, particularly during the latter half of the nineteenth century. A record height estimated to be 11.2 km was attained by an English meteorologist, James Glaisher (1809–1903). The accuracy of the temperature and humidity measurements was improved by the introduction by Richard Assmann (1845–1918) of the aspirated psychrometer. The psychrometer (see Appendix F) consists of two thermometers fastened to a common backing, one of which has its bulb covered with muslin or linen, which is kept moist (the wet bulb thermometer). When properly ventilated, the dry bulb will give accurate values of the temperature, and the wet bulb thermometer will be cooled by evaporation an amount that depends on the humidity. The lower the relative humidity, the larger is the wet-bulb depression. The accuracy of the wet- and dry-bulb readings depends on an adequate movement of air past the thermometers; but a balloon is carried with the wind, and thus instruments on it have no air moving relative to them. In Assmann's psychrometer (see Figure 2.3), air is drawn past the thermometer bulbs by a built-in fan that eliminates the errors to which most measurements of temperature and humidity on manned balloons had been subject.

The next important advance was made in 1892, when Gustave Hermite (1822–1901) and Georges Besançon developed meteorographs that could be sent aloft on unmanned balloons. In some meteorographs, the pressure, temperature, and humidity were automatically recorded on a drum rotated by clockwork. In others the temperature and humidity were recorded on a slide moved by a pressure-sensing unit. The balloons burst at great heights and the meteorographs

Figure 2.3 Assmann ventilated psychrometer. A spring-driven motor, wound by the key at the bottom, operates a fan that draws air across the bulbs of two thermometers. The bulb of one of the thermometers is covered with a muslin wick, which is moistened with distilled water. The wet-bulb thermometer is cooled by evaporation below the temperature shown by the dry-bulb thermometer. The humidity is evaluated by computation or taken from a table, using the readings of the two thermometers and the pressure. [Courtesy of Science Associates, Inc., Princeton, New Jersey.]

descended by parachute to the ground. When the instruments were recovered, the records were evaluated. Using unmanned balloons, L. P. Teisserenc de Bort (1855–1913) discovered in 1898, and Assmann subsequently confirmed, that the temperature stopped decreasing upward above a height ranging from 8 to 12 km in different situations. The region above this, now called the *stratosphere,* was termed by them the "isothermal region."

In the early part of the twentieth century, upper-air observations were begun on a routine basis to obtain data for use in weather forecasting. For this purpose, recording meteorographs were flown on huge box kites, captive balloons, and airplanes. This enabled the immediate evaluation of the record at the end of the flight, eliminating the wait until the parachuted records from the unmanned balloon flights were found. However, the heights reached were limited to the lowest five kilometers.

During the period 1928–1937, the radio-meteorograph[5] was introduced. With this device, the temperature, pressure, and humidity at each level is immediately transmitted by radio to the observer at the ground, so that the sounding from an unmanned balloon can be evaluated while it is ascending, making the recovery of the meteorograph unnecessary. Since its invention, systematic daily soundings

---

5. Now called the *radiosonde*.

have been made at an increasing number of points, and the balloons and instruments have been improved so that heights above 30 km are reached regularly. As these observations accumulated, it became clear that the apparent isothermal region always merges into a region of increasing temperature. At low latitudes, where the troposphere extends to 16 or 17 km, there is no isothermal layer, the temperature rise beginning immediately at the tropopause. At first, the term stratosphere was limited by some to the isothermal layer, and the layer in which the temperature increases with height and then decreases to the temperature minimum at 80 km was called the mesosphere. The view of those who advocated calling the entire layer through which the temperature does not decrease the stratosphere (including the isothermal regions and the region of increasing temperature) has prevailed, but in some articles, books, and reference works (dictionaries and encyclopedias) you may still find the older definition.

As stated earlier, the existence of high temperatures at about 50 km, a temperature minimum at about 80 km, and higher temperatures in the ionosphere had already been determined on the basis of indirect measurements before the means for carrying instruments to these great heights became available. This took place principally in the 1920s and 1930s. The use of rockets after World War II and artificial satellites since 1958 has made possible the evaluation of the properties of air at great heights from data obtained at those levels.

2.4   Latitudinal and seasonal variations of temperature

The general characteristics of temperature near the earth's surface are familiar to everyone: cold in polar regions, particularly in winter; seasonal changes in middle latitudes, from hot summers to cold winters; and warm in equatorial regions throughout the year. Is the variation with latitude and season the same at higher levels in the atmosphere? To examine this question, the variation of temperature with height at different latitudes and seasons are graphed in Figures 2.4, 2.5, and 2.6.

The curves in Figure 2.4 show how the average temperature in January varies with height at several latitudes. Up to about 11 km, it is warm at low latitudes and gets colder the farther north one goes, the same as near the ground. While the lapse rate at high latitudes changes to a low value or zero at or below 10 km, from 30°N southward to the equator the temperature continues to decrease rapidly up to about 17 km. This difference results in the temperature between 12 and 20 km being lower at low latitudes than at high latitudes. Above the level of the tropical tropopause the temperature increases with height at a greater rate in low latitudes than nearer the pole, and the difference between the temperatures at low latitudes and at high latitudes decreases. At sufficient heights this difference changes sign

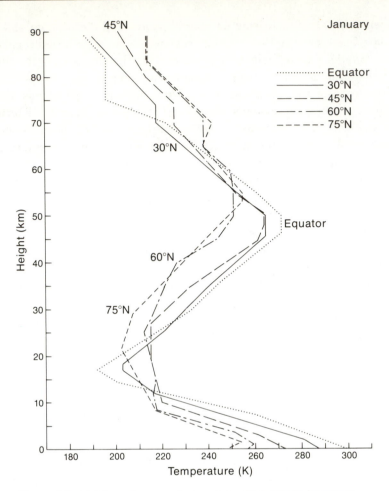

Figure 2.4 Variation of average temperature with height in January at equator, 30°N, 45°N, 60°N, and 75°N. [Data from *Air Force Reference Atmosphere* by A. E. Cole and A. J. Kantor, Air Force Surveys in Geophysics No. 382.]

again. In the upper part of the stratosphere the pattern in January, typical of the winter season, is similar to that in the lower troposphere.

In Figure 2.5, the curves for the same latitudes in July are shown. There is very little difference in temperature at all levels between the equator and 30°N, except for the equatorial tropopause being colder and slightly higher than at 30°N. North of 30°, however, the situation is similar to January: Up to about 11 km it is colder the farther north you go, and above about 12 km, the relationship is reversed. But in July the temperature is higher at high latitudes than at low latitudes throughout the upper part of the stratosphere instead of becoming colder again above 22 km, as in January.

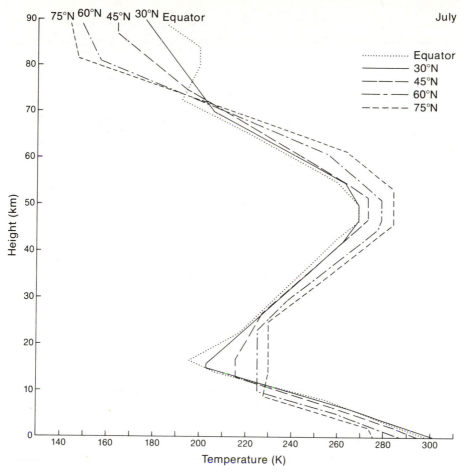

Figure 2.5   Variation of average temperature with height in July at equator, 30°N, 45°N, 60°N, and 75°N. [Data from *Air Force Reference Atmosphere* by A. E. Cole and A. J. Kantor, Air Force Surveys in Geophysics No. 382.]

It should be noted that in the lowest kilometer or two at 60°N and 75°N in January, the average temperature increases with height, instead of the decrease with height that characterizes the troposphere in general. This is due to the cooling of the air by the cold ground in the Arctic night. Layers in which the usual decrease of temperature with height is reversed are called *inversion layers* or *inversions*. Inversions within the troposphere are far from rare; however, at most latitudes they occur sporadically and not at all longitudes, so that they don't show up in the averages taken all around the earth over many months (such as all the Januaries for which records are available). In winter the inversion is present most of the time over the snow-covered continents of the far north and the frozen Arctic Ocean, so

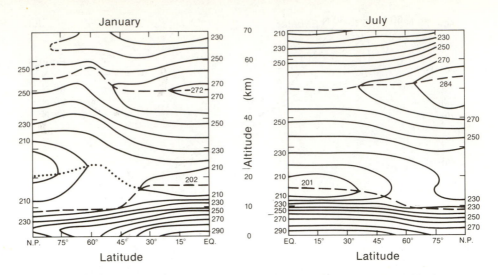

Figure 2.6 Variation with latitude and height of temperature averaged around the Northern Hemisphere in January and July. [Data from *Air Force Reference Atmosphere* by A. E. Cole and A. J. Kantor, Air Force Surveys in Geophysics No. 382.]

that it shows up in the average. This is even more true of the Antarctic continent in the Southern Hemisphere winter, for instance in July.

Of course, much of the stratosphere is a permanent inversion layer, as is the thermosphere.

A further aspect to be noticed is that, whereas in the July curves, the tropopause is well defined at levels that are progressively lower as one goes to higher latitudes, if one uses the strict definition of the stratosphere as a region of constant or increasing temperature with height, the tropopause in the January curves for 45°N and 75°N would be above 20 km. Actually there is no special reason for picking zero lapse rate as the criterion; the breaks in the curves from large values to much smaller positive ones at heights of 10 km and 8 km, respectively, as well as the one at 8.5 km in the 60°N curve, are more significant. They represent the actual upper limit to overturning and cloud formation, for reasons that will be discussed subsequently, and as such identify the effective tropopause, the place where turning (tropo) stops. By defining the tropopause in this way, its height in January varies with latitude in a fashion similar to that shown by the July curves.

The variation of tropopause height with latitude is shown more clearly in Figure 2.6. In this diagram, the horizontal coordinate is latitude, the vertical coordinate is height, and the temperature distribution is shown by isotherms—continuous lines drawn at 10-degree intervals of temperature, separating temperatures higher than the given value on one side of the line from lower temperatures on the other.

Thus, the diagram shows the variation of temperature with height at all latitudes, separately for January and July, as representative of the Northern Hemisphere winter and summer averages around the earth. The tropopause and stratopause are shown by dashed lines, and the position where the lapse rate changes from positive to negative values in the January diagram is shown by a dotted line. As discussed above, this latter position cannot be regarded as the true location of the tropopause in the sense that it is the upper boundary of the troposphere, the region in which rapid vertical mixing and overturning of the atmosphere takes place. However, for want of a better name for it, we shall call this position the *secondary winter tropopause*. In both seasons, the tropopause is almost horizontal south of 30°N and north of 60°N, but it slopes downward toward the pole between these latitudes, with a more abrupt slope in January than in July. The isotherms have a similar slope in the troposphere, corresponding to the lower temperatures toward the north, particularly in the winter month of January. Near the ground north of about 70° in the January diagram the isotherms bend back, showing the inversion layer, in which the air is warmer at 2 km than at the ground.

The reversal of the horizontal temperature gradient at heights that are in the troposphere at low latitudes and in the stratosphere at higher latitudes is readily seen in both parts of Figure 2.6. The temperature at the tropopause near the equator is the lowest anywhere up to that height in both seasons, but in January the temperature near the North Pole at the secondary tropopause is even lower.

The only place at the earth's surface where temperatures as low as the equatorial tropopause are observed is on the Antarctic continent, near the South Pole, in the southern winter. The lowest temperature ever observed in the air at the earth's surface was $-88.3°C$ (184.8 K, $-127°F$) at Vostok, a Russian meteorological station in Antarctica, and the average August temperature there is $-70°C$ (203 K, $-94°F$). The highest temperatures on earth occur not at the equator but in the deserts of subtropical latitudes in summer. The record high temperature appears to be $54°C$ ($129°F$) at Furnace Creek, Death Valley. Slightly higher temperatures have been reported both in Death Valley and in Tripoli, Libya, but analysis has shown them to be unreliable.

The long-term average surface air temperatures are also higher over subtropical portions of continents in summer and lower over the high-latitude interior of continents in winter than elsewhere. Figure 2.7, in which the distribution of average temperatures near the ground over the earth in January and July is shown by isotherms, illustrates these effects. Near the equator and over the oceans generally, the difference between the average temperatures in January and July is much less than over continents at middle and high latitudes. The moderating influence of the ocean affects the temperatures of the adjoining coastal areas.

Naturally, scientists have been concerned not only with determining the structure of the atmosphere, but also with the question *why?* Why does the temperature

Average January Temperature

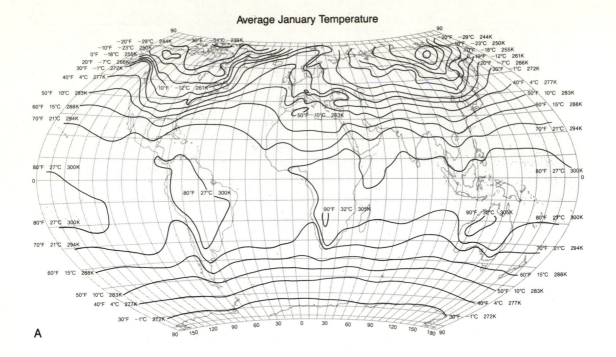

A

Average July Temperature

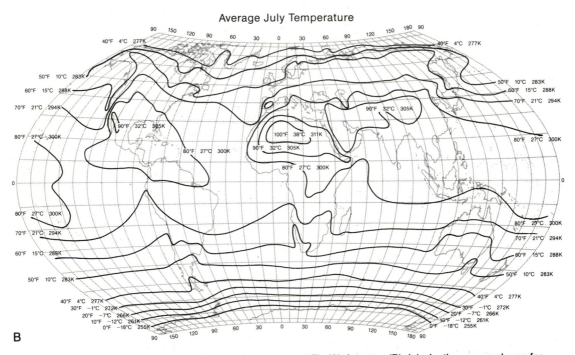

B

Figure 2.7 Average surface air temperatures over the earth (°F): (A) January; (B) July. Isotherms are drawn for every 10°F, with the corresponding values in the Celsius and Kelvin scales shown. [After *Climates of the World*, Washington, D.C.: U.S. Department of Commerce, Environmental Science Service Administration, Environmental Data Service, 1969 (reprinted 1977).]

vary as it does in the vertical and horizontal? Why is the mean temperature at sea level 288 K? Why does the temperature decrease with height in the troposphere, and why at the particular rate that is observed? Why does it increase in the stratosphere, decrease in the mesosphere, and increase in the thermosphere? Why is the temperature higher in summer at subtropical latitudes than at the equator? Why do the extreme values occur over continents rather than oceans?

Other questions concern the ionosphere. Why does the atmosphere above 100 km contain charged particles, and below that level, almost entirely neutral ones? And with respect to composition, why does dry air to heights up to about 100 km contain the same proportions of carbon dioxide, argon, oxygen, and nitrogen, even though the molecular weights are 44, 40, 32, and 28, respectively, and we should expect the heavier gases to be separated from the lighter gases by gravity? Within this uniformity, why is there a layer in which oxygen takes the form O_3 (ozone) rather than O_2 (molecular oxygen)? And why is the layer of a relatively high (but still very low in absolute magnitude) concentration of ozone located in the upper stratosphere, in spite of the fact that its molecular weight (48) is the highest of common atmospheric gases?

The answers to all these questions lie entirely or in large part with the disposition of radiant energy from the sun as it passes through the atmosphere to the ground. The energy that heats the atmosphere and drives its circulations originates in the sun. Small amounts of energy from other stars reach the earth, and minute quantities are released by radioactive transformations of matter on earth, but these are extremely small indeed compared with the large steady flow of energy from the sun.

The next chapter deals with the nature of radiation, the properties of solar radiation, and the interaction of solar radiation with the atmosphere. In it are presented explanations of some of the characteristics of the atmospheric structure described in this chapter.

Questions, Problems, and Projects

1. Define the following terms:

   a. Atmosphere
   b. Troposphere
   c. Stratosphere
   d. Mesosphere
   e. Thermosphere
   f. Atom
   g. Molecule
   h. Element
   i. Compound
   j. Ion
   k. Standard atmosphere
   l. Lapse rate
   m. Inversion

   Include these definitions in the glossary you prepare in response to question 8 of Chapter 1.

2. List the constituents of the atmosphere that affect the weather in a major way in the order of their relative importance. Describe the role of each.

3. From Figure 2.6 make graphs of temperature versus height at the equator, 30°N, 45°N, 60°N, and 75°N in summer and winter, and compare the resulting curves with those in Figures 2.4 and 2.5.

4. From Figure 2.7 make graphs of surface air temperature versus latitude at the following longitudes: 0°, 80°W, 180°, and 100°E in January and July.

5. Radio waves, like light, normally travel in straight lines. In view of the curvature of the earth's surface, why can radio signals be received at great distances from the place at which they are transmitted?

6. From Figure 2.5 find the ratio between the average rate of change of temperature with distance in winter from the equator to 75°N at the ground, and the average rate of change of temperature with height from sea level to 10 km at 45°N. (Note that one degree of latitude is 111 km.)

7. Draw isotherms for every 10°F on Figures 1.14 and 1.16. Explain why they do not follow parallels of latitude. Compare them with those in Figure 2.7A and give reasons for the differences.

Radiation Through the Atmosphere   3

3.1 The sun and the earth

The earth revolves around the sun at an average distance of $149.6 \cdot 10^6$ km (about 150 million kilometers or 93 million miles).[1] Its orbit is an ellipse, but when it is closest to the sun (the perihelion), on about January 4, it is $147 \cdot 10^6$ km away, only 1.7 percent nearer the sun, and at its aphelion, about July 5, it is the same percentage farther away. Thus, its path around the sun departs only a little from a circle. At all times of the year the energy that controls the temperature and motions of the atmosphere must travel about 150 million kilometers, mostly through empty space, to reach the earth. How does it do it?

---

1. See Appendix A for an explanation of the "power of ten" notation.

Another puzzling question concerns the amount of energy the sun is sending out. The sun is not "beaming" its energy only toward the earth; it sends it out equally in all directions. The earth intercepts only a small fraction of it, although this amount is $174 \cdot 10^{12}$ kilowatts, or more than 500,000 times the capacity of all the electric-generating plants in the United States. This huge amount of energy, which keeps the earth warm, provides for the growth of plants and the life of animals and humans, and creates the circulation of the atmosphere, is an almost negligible part of the solar energy. The total energy emitted by the sun is more than 2.2 billion times as much as the earth receives. If this tremendous amount of energy were produced by combustion, the sun would have burned up long ago; an amount of coal equal to the mass of the sun could produce energy at this rate for less than five thousand years. Where does all this energy come from?

The answer to the second question is that the sun is a huge nuclear reactor. At the extremely high temperatures and pressures in the interior of the sun, a hydrogen-helium fusion process takes place by which mass is transformed into energy. The amount of mass used in this process is so small that it will take about $14 \cdot 10^{12}$ years for the sun's mass to be exhausted. The estimated age of the solar system is $4.6 \cdot 10^9$ years; at the present rate of emission of energy, the sun can be expected to last more than 3000 times as long as it has existed so far.

If a method could be devised to operate on earth fusion reactors of this type, in which the nuclei of small atoms (hydrogen) are combined or fused, to yield larger atoms, with conversion of some of their mass into energy, instead of the present nuclear reactors that use the fission or splitting of large atoms, such as uranium, the energy problem for human activities might well be solved. Not only is hydrogen plentiful, while uranium is scarce, but the radioactive wastes from the fusion reaction promise to be less difficult to dispose of than those from fission. Intensive research continues on the problems of harnessing thermonuclear energy. One major problem is finding a commercially feasible way of containing gases at the extremely high temperatures required for fusion. Solutions to these problems seem remote, and it appears that the only fusion-generated energy that will be available for use on earth in the foreseeable future is that which comes from the sun.

How this energy is transferred from the sun through empty space is different for the two forms that the energy takes. A very small fraction, less than one-millionth of the total energy, is in the form of particles, sometimes called *corpuscular radiation*, streaming out from the sun to form what is known as the *solar wind*. The rest (almost all of it) is in the form of *electromagnetic radiation*. While the radiation has a dual character, behaving like particles in certain respects, for most purposes it can be considered to be composed of electromagnetic waves, waves which propagate through empty space and also through transparent materials, even solids. Thus, glass is transparent to visible light, brick walls to radio waves, and flesh to x rays. All three are forms of electromagnetic radiation having different wavelengths.

3.2   Corpuscular radiation: The solar wind and the magnetosphere

While the corpuscular radiation accounts for less than one-millionth of the total solar energy incident on the earth and its atmosphere, it is responsible for important effects in a region extending from a few hundred kilometers to several earth's radii distance above the earth's surface. It is composed of a variety of particles, predominantly protons and electrons in approximately equal numbers, that result from the ionization of material in the high solar atmosphere. Since solar wind particles bear electric charges, they interact with the earth's magnetic field as they approach the earth.

The details of solar wind particles' behavior are too complicated to be discussed here, but some indication of its complexity may be gathered from Figure 3.1, which delineates the various layers that arise as the particles contained in the solar wind are deflected by the earth's magnetic forces and collide with the rarefied portion of the earth's atmosphere. By *plasma* is meant a gas consisting primarily of charged particles. The entire configuration of charged particles moving with the earth that arises in response to the interaction between the solar wind and the earth's magnetism is called the *magnetosphere*. It extends about 10 earth's radii in the direction of the sun, but trails outward much farther from the dark side of the earth, forming a comet-like magnetospheric tail.

Among the layers in the magnetosphere are the *radiation belts*, discovered early in the period of unmanned space exploration by James A. Van Allen (1914–   ). Radiation belts are regions of high radioactivity from which astronauts and cosmonauts must be shielded as they pass through.

Among the effects associated with the incoming corpuscular radiation are the spectacular visual phenomena of polar regions, the *aurora borealis* or Northern Lights and the *aurora australis* or Southern Lights. These phenomena are caused by the focusing effect of the earth's magnetism on solar wind particles that are stored and energized in the geomagnetic tail, and then are precipitated along the lines of magnetic force into the atmosphere. Auroras formed this way are almost always visible at night in the *auroral zones*, the regions between 65 and 70 degrees latitude in both hemispheres. During periods of geomagnetic storms, the magnetospheric reservoir of aurora-creating particles moves closer to the earth, and

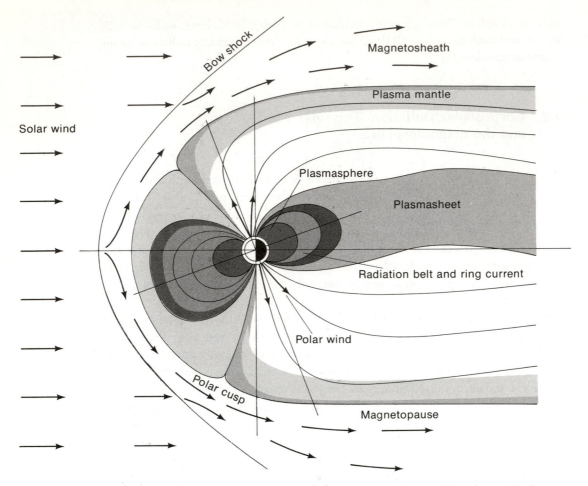

Figure 3.1   Schematic representation of the earth's magnetosphere, which is created by the solar wind approaching from the left and being deflected by the earth's magnetic field. [From Heos 2 Plasma Observations in the Distant Polar Magnetosphere: The Plasma Mantle, by H. Rosenbauer, H. Grünwaldt, M. D. Montgomery, G. Paschmann, and N. Sckopke, *Journal of Geophysical Research* 80, 19 (July 1, 1975), p. 2734. Copyright © 1975 by the American Geophysical Union.]

the resulting auroras occur at lower latitudes. (Other effects of geomagnetic storms include disturbances of the ionosphere that interfere with long-distance radio communication.) As the precipitating charged particles follow the magnetic lines of force poleward and descend to levels where the air density is higher, they collide with neutral atoms or molecules and either raise them to an excited state, in which their electrons are at higher energy levels, or else remove one of their orbital electrons, thereby ionizing them. The subsequent decay of the excited air mole-

Figure 3.2   The aurora borealis. [Courtesy of NOAA.]

cules or atoms to their normal states involves emission of light (quanta of electromagnetic radiation) that is responsible for the characteristic red, green, and violet colors of the aurora.

The corpuscular radiation is deflected or stopped by collision with air particles in the very high atmosphere. The energy from the sun that is responsible for the variation of temperature with height from the ground up to the thermosphere is in the form of electromagnetic waves.

3.3   The nature and laws of electromagnetic radiation

The waves we are most directly familiar with are those occurring at the surface of a body of water, for instance, ocean waves. In general, these waves are very irregular and complex, but their main properties are readily observed, particularly near shore where the waves are simpler. They consist of a series of moving troughs

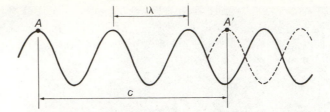

Figure 3.3 Schematic representation of the characteristics of a moving wave: the distance from one crest to the next is the wavelength λ; the dashed curve at A' is the position of the wave crest initially at A after unit time (for instance, one second); and c is the distance traveled in unit time, that is, the wave speed.

and ridges in the water surface (Figure 3.3). The horizontal movement of the waves is not produced by bodily horizontal movement of the water, but primarily by alternating up and down movements, up ahead of the crest and down behind the crest, and ahead of the trough. The *wave speed* is thus different from the speed of flow of the water. It is the speed with which an identifiable part of the wave, for instance, the crest, moves. In Figure 3.3 the dashed curve marked A' represents the wave crest originally shown by the solid curve at A after a unit time interval. The wave speed is thus equal to the distance AA' per unit time. The distance from corresponding parts of successive waves at a given time (e.g., from crest to crest) is called the *wavelength*. The number of waves that go by a point in unit time is called the *wave frequency*, and the time that it takes for a complete single wave to pass is called the *wave period*. Since the waves between A and A' in the solid curve in the diagram would pass A' in unit time, their number (approximately 2.5) is the frequency. If this number is represented by ν, the time τ for one of them to pass is $1/\nu$, that is, the period is the reciprocal of the frequency ν:[2]

$$\tau = 1/\nu \qquad (3.1)$$

The wavelength and wave speed are also interrelated. If we represent the wave speed by c and its length by λ,

$$c = \lambda/\tau = \lambda\nu \qquad (3.2)$$

The units of wavelength and wave speed in the SI (International System) are meters (m) and meters per second (m s^{-1}). The unit of frequency (cycles) per

---

2. See Appendix C for names of Greek letters.

second, or s^{-1}, has a special name, Hertz, abbreviated Hz, and the period is given in units of seconds. A typical ocean swell may have a wavelength of 100 m and a period of 10 seconds. Since $\lambda = 100$ m and $\tau = 10$ s, we have from equation (3.2)

$$c = 100/10 = 10 \text{ m s}^{-1}$$

and from equation (3.1) we get

$$\nu = 1/\tau = 1/10 = 0.1 \text{ Hz}$$

As a wave moves it transfers energy (the ability to do work) from place to place. The lifting of a boat by the crest of a water wave as it moves by demonstrates this ability to do work, and the lifting in succession of a group of small boats at various distances from a passing ship demonstrates the transfer of energy by the bow wave of the ship.

Electromagnetic waves differ from water waves in that they do not require the presence of matter to transport energy. At one time physicists were puzzled by the propagation of waves through a vacuum. Since it was difficult to conceive of waves without some medium doing the waving, they postulated the existence of an all-pervading substance, the "luminescent ether." As the nature of forces operating at a distance, such as the force of gravity and electrostatic and electromagnetic forces, became better understood, the idea of the oscillation of a force at a point in space owing to the vibrating movement of a distant electric charge became more acceptable, even if there is no matter at the position where the oscillation takes place.

Electromagnetic waves move with a speed of $3 \cdot 10^8$ m s^{-1} (186,000 miles per second) in empty space and only slightly slower in space that is occupied by air. Their wavelengths range from several kilometers for long radio waves to nanometers for x rays and gamma rays. Figure 3.4 shows the ranges of wavelengths and the corresponding frequencies for various kinds of electromagnetic radiation. The wavelengths of the radiation that carry most of the energy from the sun to the earth's atmosphere, and from the earth and its atmosphere to space, are in the vicinity of those of visible light. These wavelengths are so short that it is convenient to express them in micrometers, abbreviated μm, 1 μm $= 10^{-6}$ m; in nanometers, nm, 1 nm $= 10^{-9}$ m; or in Ångstrom units, Å, 1 Å $= 10^{-10}$ m.

Visible light ranges from 0.4 μm to 0.7μm (400 nm to 700 nm, 4000 Å to 7000 Å),[3] where the shortest visible wavelengths give the sensation of violet color

3. The human eye is actually sensitive to light of wavelengths slightly shorter and longer than this range, but so slightly that these limits are commonly used.

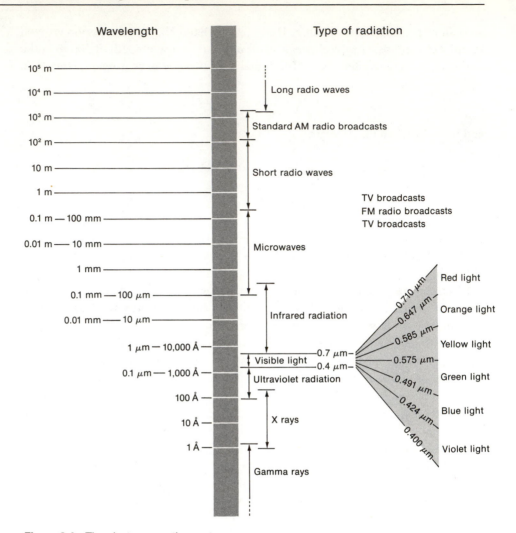

Figure 3.4 The electromagnetic spectrum.

and the longest give the sensation of red. Radiation with wavelengths shorter than 0.4 μm is called *ultraviolet* (UV); radiation with wavelengths longer than 0.7 μm is called *infrared* (IR).

The behavior of electromagnetic radiation is described by certain physical laws. The following is a simplified summary of them.

1. All matter that is not at absolute zero can emit radiation in amounts that depend on its temperature.

2. Some substances emit radiation only in certain wavelengths. This is particularly true of gases.

3. Substances that emit the maximum amount for their temperature in all wavelengths are called *black bodies*. The amount of energy radiated by a black body is proportional to the fourth power of its absolute temperature.

4. Substances absorb only radiation of wavelengths that they can emit.

5. The hotter the substance, the shorter the wavelengths at which most of the radiant energy is emitted.

6. Since the total radiation from a point source passes through the surface of successively larger spheres as it radiates outward, the amount passing through a unit area is inversely proportional to the square of the distance of the area from the source.

7. If a gas absorbs radiation of any wavelength, the amount absorbed will be proportional to (a) the number of molecules of gas and (b) the intensity of radiation of that wavelength.

Some of these laws can be expressed mathematically very simply. Thus, law number 3, which is known as the Stefan-Boltzmann law, is represented by the equation

$$E = \sigma T^4 \tag{3.3}$$

where E is the energy emitted per second from the unit area of a black body whose temperature is T and σ is a constant that has the value $5.67 \cdot 10^{-8}$ watts/m^{2}K^4 or $5.67 \cdot 10^{-5}$ ergs/cm^{2}K^4 sec. Law number 5, called Wien's displacement law (after Wilhelm Wien, 1864–1928), is

$$\lambda_M = a/T \tag{3.4}$$

where λ_M is the wavelength at which the peak occurs in the spectrum of the radiation from a black body whose absolute temperature is T and a is a constant that has the value 2898 if λ_M is expressed in micrometers.

Both of these laws are consequences of a more general law, Planck's law, which gives the amount of radiant energy emitted at each wavelength (or frequency) by a black body at any particular temperature. Planck's law is of special importance because, in his effort to find a formula that gave the distribution that was observed, Max Planck (1858–1947) found it necessary to postulate that radiant energy was emitted in discrete packets, which he called *quanta*. Previously, Wien had derived an empirical expression on continuous radiation theory that fit the experimental data for short wavelengths and gave the shift of the maximum toward shorter wavelengths when the temperature goes up, as shown in his displacement law. However, it did not correspond to experimental results at longer wavelengths. Lord Rayleigh and Sir James Jeans (1877–1946) derived an equation based on the theory of statistical mechanics that, while apparently based on sound theoretical considerations, fit the observed data only at very long wavelengths. In evolving a theory that matched the experimental results for all wavelengths, Planck introduced the *quantum theory*, which has completely revolutionized physics.

It is with respect to quanta that electromagnetic radiation behaves as though it were composed of particles rather than waves. The particles or quanta of radiant energy, called *photons*, have sizes given by the equation

$$E = h\nu \tag{3.5}$$

That is, the higher the frequency or shorter the wavelength, the more energetic the photon. But there is no such thing as a fraction of a photon. Each time radiant energy is emitted, it comes out as a whole photon of the appropriate size for that wavelength. Radiation of sufficiently high frequency or short wavelength will be composed of photons having enough energy to break apart molecules or atoms that absorb them, thus producing molecular dissociation or ionization. Photons of longer wavelength do not have sufficient energy to do this, and serve only to raise the temperature when they are absorbed.

Figure 3.5 shows the radiation spectra for the emission from a black body at (A) 6000 K, approximately the effective temperature of the sun's surface, and (B) some temperatures representative of the range occurring at the surface of the earth. It is important to notice that the scale of radiation intensity in diagram A is in units 1 million times as large as in diagram B. Note that the wavelength of the energy maximum for 6000 K is near 0.5 μm, in the visible range, with magnitude about 100 MW/m^2 μm, while for terrestrial temperature it is in the far infrared, at about 10 μm or longer wavelengths, with magnitude 32 W/m^2 μm or less. The peak intensity at 6000 K is thus more than 3 million times as great as that at temperatures that occur on earth.

The wavelength of the peaks shown in the curves representing Planck's law in Figure 3.5 are in accord with Wien's law, equation 3.4. Thus, for 6000 K, equation (3.4) gives

$$\lambda_M = 2898/6000 = 0.483 \ \mu m$$

and for 300 K it gives

$$\lambda_M = 2898/300 = 9.66 \ \mu m$$

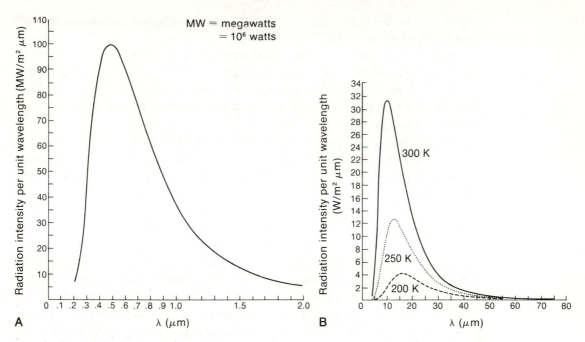

Figure 3.5 Variation of intensity of black-body radiation with wavelength: (A) T = 6000 K; (B) T = 200 K, 250 K, and 300 K.

If we apply equation (3.3) for these temperatures, we get for 6000 K

$$5.67 \cdot 10^{-8} (6000)^4 = 7.35 \cdot 10^8 \ \text{W/m}^2 = 735 \ \text{MW/m}^2$$

and for 300 K it gives

$$5.67 \cdot 10^{-8} (300)^4 = 4592.7 \ \text{W/m}^2 = 4.6 \ \text{kW/m}^2$$

The total radiation per square meter at the approximate temperature of the sun's surface is 160,000 times that emitted by the earth's surface on a warm day. It is the reduction due to distance that brings the sun's radiation intensity down so that there is on the average a balance between the energy received from the sun and the energy emitted by the earth and the atmosphere.

The effect of distance is expressed by law number 6. Written as an equation, it states

$$E = C/d^2$$

or

$$E_2/E_1 = d_1^{\,2}/d_2^{\,2} \tag{3.6}$$

where E is the flux of radiation (i.e., the amount of radiant energy passing through unit area per second) at a distance d from the source; E_1 is the energy flux at distance d_1; E_2 is the energy flux at a distance d_2; and C is a constant depending on the intensity of the source and represents the radiant energy at unit distance. In this treatment the radiation is regarded as originating at a point. We shall see how this law is applied when we discuss the measured intensity of the radiation from the sun in Section 3.5.

The relationships between emitted radiation and temperature, given by equations (3.3) and (3.4) and Planck's laws, are used in converting the measurements of infrared radiation received from earth by the NOAA and GOES satellites into pictures that show areas of cloudiness, snow cover, and so forth. (See Figure 1.19B as an example.) Actually the pictures are representations of the temperature distribution. Regions of small amounts of radiation received, corresponding to low temperatures, are represented by white and light shades of grey; large amounts of radiation, corresponding to high temperatures, are shown as black or dark grey. Since the temperature decreases upward in the troposphere, the result is that clouds having tops at high elevations emit small amounts of radiation and appear white or light grey in the pictures, while the earth's surface, being warmer, appears dark grey or black, and low clouds have intermediate shades of grey. This enables the satellites to give pictures of the cloud distribution at night, in addition to the ones using reflected visible light that can be taken during the day. Comparison of the visible and infrared pictures during the day gives additional information about the nature and height of the clouds, as well as characteristics of the earth's surface in cloudfree areas.

3.4   Energy, heat, and work

The word "energy" is part of our everyday language, and in a general way it is commonly understood. Unless you have studied its nature in a course in physical science, however, its precise nature and, in particular, the units in which its measured values are expressed may not be familiar to you.

Energy is defined as the capacity for doing work. This capacity may take various forms, some of which follow:

Potential energy: Energy due to position, such as that produced by lifting a weight. Potential energy can be released by letting go of the weight and allowing it to fall.

Kinetic energy: Energy due to motion, for instance, the motion of a hammer swinging toward the head of a nail.

Heat: Energy associated with the ability of one body to raise the temperature of a cooler one. The capacity of heat to perform work is illustrated, for instance, in the steam engine.

Chemical energy: Energy stored by the combination of molecules into chemical compounds. Examples of the release of chemical energy are the combustion of gasoline that propels automobiles and the chemical reactions inside dry cells (batteries) that produce electrical energy.

Nuclear energy: Energy stored in the nucleus of an atom.

Radiant energy: Energy associated with electromagnetic waves propagating through space, whether empty or occupied by a substance. If radiant energy can pass through a substance without change, the substance is said to be *transparent* for radiation of that wavelength.

One form of energy can be transformed into another. Thus, the release of a suspended weight results in the transformation of its potential energy into kinetic energy. The burning of gasoline in the cylinders of an automobile results, first, in the conversion of chemical energy into heat and then of part of that heat into the kinetic energy of the automobile's motion. The absorption of radiant energy passing through air may heat it; the part that passes through the air and reaches the ground heats the ground.

The equivalence of heat and work, or mechanical energy, was first demonstrated by Benjamin Thompson, Count Rumford (1753–1814), toward the end of the eighteenth century. Rumford, who was born near Boston but sympathized with the British and went to England at the time of the American Revolution, had a very impressive career in England and in Austria and Bavaria. He was noted for his introduction of important social reforms in Bavaria, but his most important contribution was the demonstration that the then current view that heat was a material fluid, called caloric, was wrong. He showed this in a series of experiments he was led to by considering the heat arising from friction during the boring of cannons at the military arsenal in Munich of which he was in charge.

Even now different units are frequently used to express heat energy—the calorie in the metric system and the BTU (British Thermal Unit) in the English system of units—from those used for other forms of energy. One calorie is the amount of energy required to raise the temperature of one gram of water one degree Celsius (from 14.5°C to 15.5°C). One BTU is the amount of energy required to raise the temperature of one pound of water one degree Fahrenheit (from 62°F to 63°F). [The calorie in which energy produced by food in the body is expressed is the "large calorie" or kilogram calorie, 1000 times as large as the (gram) calorie.]

Since all forms of energy represent the ability to do work, the natural way in which to express quantity of energy is in work units. The unit of energy in the International System (SI, the meter–kilogram–second, or mks system) is the *joule,* abbreviated J, which is the work done when unit force in this system, one newton, acts through one meter. Many scientists still use the cgs (centimeter–gram–second) system some of the time. In this system the unit of energy, the *erg,* is similarly the amount of work done when a force of unit magnitude in this system, one dyne, acts through unit distance, one centimeter. One joule is equal to 10^7 ergs. One calorie is equal to 4.186 J or $4.186 \cdot 10^7$ ergs.

In the SI the unit of *power,* the rate at which energy is released or converted, which is 1 joule per second, is called the *watt,* abbreviated W. Thus, a 100-W electric bulb is one that converts 100 J of electric energy per second into heat and light.

In the English system the unit of work is the *foot–pound,* and the power, or rate of doing work, is expressed in horsepower, where one horsepower equals 550 foot–pounds per second. One foot–pound is equal to 1.36 J, and one horsepower is equal to 746 W.

In discussions of radiation, we are frequently concerned with the amount of energy falling on or passing through a unit area. In the SI and cgs units, this is expressed in J/m^2 and $ergs/cm^2$, where $1 \ J/m^2 = 10^3 \ ergs/cm^2$. The unit for this quantity in heat units has been given a special name—the langley (ly), named after Samuel P. Langley (1834–1906), the American scientist who, in addition to his experiments in heavier-than-air flight, carried on important studies of radiation. One langley is one calorie per square centimeter; thus $1 \ ly = 4.186 \cdot 10^4 \ J/m^2 = 4.186 \cdot 10^7 \ ergs/cm^2$.

The rate at which the energy falls on or passes through unit area is called *radiant flux density* or *irradiance.* It is usually expressed in watts per square meter (sometimes per square centimeter) or langleys per minute, where $1 \ ly/min = 697.7 \ W/m^2$.

3.5   Solar radiation

Except for the effect of distance, as stated in law number 6 of Section 3.3, the radiation from the sun reaches the outer limit of the earth's atmosphere undepleted, but in passing through the air it is scattered and absorbed by the molecules and the dust, haze, and cloud particles, so that only a part of it reaches the earth's surface. Consequently, previous to the development of rockets and artificial satellites, the amount of radiation coming from the sun could only be estimated from the measurements made at observatories on high mountains. The best values until very recently were those obtained many years ago by C. G. Abbott (1872–1972) and other staff members of the Smithsonian Institution from a long series of measurements at Mt. Wilson, California. These measurements were adjusted to allow for the absorption of the radiation by the atmosphere above the mountain, using the differences in readings in adjacent parts of the spectrum to estimate the amount of absorption. They obtained an average value of 1.94 ly/min, corresponding to 1354 W/m² for the "solar constant," the radiant flux at mean solar distance before depletion by the earth's atmosphere. They estimated that its value varies about 1 percent during the 11-year solar cycle, that is, between sunspot maximum and sunspot minimum. However, the uncertainty of their measurements was estimated to be greater than the 1 percent variation with time.

In the past few years measurements have been made from high balloons, rockets, and satellites. These measurements are subject to some uncertainties because of instrumental difficulties. Observations made in June 1976 by rocket with an improved instrument gave a value of 1367 W/m² or 1.959 ly/min. This value, made near the time of minimum solar activity, is regarded to have an accuracy better than 0.5 percent. A measurement of 1373 W/m² was made on a rocket with similar instrumentation in 1978, when the solar activity had increased somewhat. From November 15, 1978 to May 15, 1979, measurements made with refined instruments aboard the Nimbus 7 satellite ranged within 0.1 percent of 1376 W/m². This progressive increase with solar activity suggests that the solar "constant" does vary, but because the change measured is only slightly greater than the instrumental uncertainty, the question of whether the variation is real remains in doubt.

It is from the measurement of the solar constant that the rate of radiant energy emitted by the surface of the sun is computed. The mean distance from the sun to the earth is 149.6 · 10⁶ km; the radius of the surface of the sun is 6.95 · 10⁵ km. Let these distances from the sun's center be d_2 and d_1, respectively, in equation (3.6). In this case, the solar constant, for which we use the value 1370 W/m², is E_2, and the intensity at the sun's surface, E_1, has the value

$$E_1 = \frac{(149.6 \cdot 10^6)^2}{(6.95 \cdot 10^5)^2} \cdot 1370 = 6.348 \cdot 10^7 \text{ W/m}^2$$

The radiation leaving the sun is more than 46,000 times that reaching the earth.

We can now apply equation (3.3) to obtain the effective temperature of the sun's surface. We have

$$5.67 \cdot 10^{-8} T^4 = 6.348 \cdot 10^7$$
$$T^4 = 6.348 \cdot 10^7 / 5.67 \cdot 10^{-8}$$
$$T = \sqrt[4]{1119.58 \cdot 10^{12}}$$

The computation of the fourth root of a number can be carried out by taking the square root twice, by the use of logarithms, or more conveniently on an electronic hand computer (scientific) to yield

$$T = 5784 \, \text{K}$$

Thus, the total amount of radiation emitted from the sun is approximately that of a black body having an absolute temperature of 5784 K.

The wavelength distribution of solar radiation is also similar to that of a black body at 5784 K. Figure 3.6 shows the measured intensity of the sun's radiation at various wavelengths reduced to "zero air mass," meaning the value evaluated for outside the atmosphere, and the curve for the radiation of a black body reduced to the mean distance from the sun. (We used 6000 K rather than 5784 K because the data for 6000 K are available in published tables.) The peaks in the two curves are at about the same wavelength (0.47 μm for the observed, and 0.48 μm for the 6000-K curve), and the general shape is much the same. The curve for 5784 K would, of course, be lower than the 6000-K curve at all wavelengths. Up to about 1.0 μm it would come closer to the observed, but at longer wavelengths where the measured solar irradiance is greater than the 6000-K values, the difference would be larger.

On the short end of the spectrum, the observed curve falls almost to zero at 0.2 μm, and on the long wavelength side it reaches very low values by 2.0 μm. Practically all the solar energy reaching the earth's atmosphere is in wavelengths between 0.15 μm and 4.0 μm. Referring back to Figure 3.5(B), we see that radiation emitted at temperatures typical of the earth and its atmosphere is almost all at wavelengths greater than 4.0 μm. For practical purposes there is no overlap, and it is convenient to speak of two different kinds of radiation passing through the atmosphere, short-wave or solar radiation, and long-wave (far infrared) or terrestrial radiation. These types of radiation are affected differently as they pass through the atmosphere.

About 9 percent of the solar radiation is in the ultraviolet, 41 percent is in the visible, and 50 percent is in the near infrared parts of the spectrum.

The solar constant is defined as the radiation that falls on unit area of surface normal to the line from the sun per unit time at the outside of the atmosphere and at *mean solar distance*. The deviation from this value due to varying distance from the sun is slight. The 1.7 percent departure in distance leads to about 3.4 percent difference in radiation. That is, on January 4 there is 3.4 percent more, or 1.41 kW/m^2, falling on a unit area normal to the line from the sun at the outside of the atmosphere, and about July 5 the amount is 1.32 kW/m^2. The big variations in energy from the sun reaching the ground at various times of the year are not due

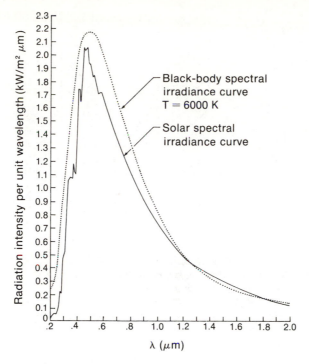

Figure 3.6 Solar irradiance spectrum and spectrum of 6000 K black-body radiation reduced to mean solar distance.

to this effect, but are due to the inclination of the earth's surface to the sun and the depletion of the energy from the sun in passing through the atmosphere.

3.6 Solar radiation at the earth's surface

The solar radiation reaching the earth's surface consists of one part that comes directly from the sun and another portion that comes from all parts of the sky. The radiation from the sky is due to scattering of solar radiation by air molecules and other particles. (If it weren't for this scattering, the sky would be black during the day as well as at night.) The direct radiation, expressed as the amount of radiant energy falling on unit horizontal area in unit time (W/m² or ly/min), is called *insolation*. The indirect radiation is called *sky radiation,* and the total direct and diffuse is called *global radiation.* Sometimes the word "insolation" is taken to include the indirect as well as direct, and thus to be equivalent to global radiation.

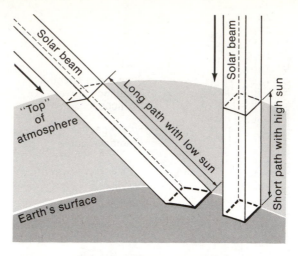

Figure 3.7 Effect of sun's altitude on insolation, showing that, when the sun is low, the same amount of radiation is spread over a larger area, and there is a longer path through the atmosphere in which the radiation can be depleted by scattering and absorption than when the sun is near the zenith.

The insolation varies because of the variation of solar altitude for two reasons: (1) When the sun is high, the radiation falls almost perpendicularly on the ground, whereas when it is low, the radiation passing through unit area normal to the line from the sun is spread over a large area (see Figure 3.7). (2) When the sun is low, the radiation traverses a longer path through the atmosphere along which a greater amount of scattering and absorption can take place. Both effects reduce the insolation when the sun is low, in accord with our common experience during any clear day. At midday the sun's radiation is much more intense than in the early morning and late afternoon.

The total insolation for a whole day is affected by the length of time from sunrise to sunset, in addition to the above two factors.

The insolation varies diurnally, with time of year, and with latitude because of these factors. The diurnal variation is an obvious consequence of the changing height of the sun during the day. The seasonal variation is a consequence of the inclination of the earth's axis to the plane of the ecliptic (the plane in which its orbit around the sun lies). The direction of the earth's axis of rotation stays practically the same throughout the motion of the earth around the sun, maintaining an angle of 66½° with the orbital plane. The consequence in terms of the length of day and height of the sun at noon may be seen in Figures 3.8 and 3.9. As shown

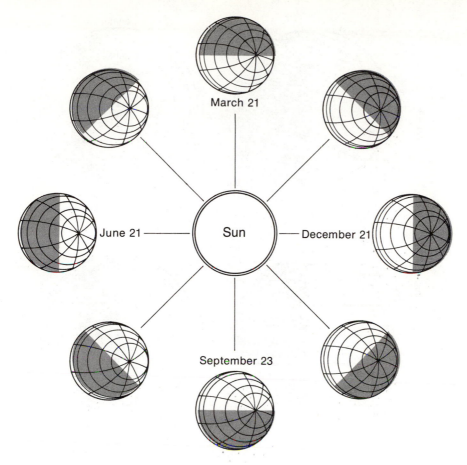

Figure 3.8 Earth's orbit around the sun.

in Figure 3.8, while the earth goes around the sun with the axis pointing in a constant direction, the area that is illuminated by the sun changes; it includes the North Pole at the June solstice and excludes it in December. Figure 3.9 shows this effect in more detail. The fraction of the entire day during which the sun shines at each latitude is shown by the portion of the parallel of latitude that is in the illuminated hemisphere. At the June solstice, the sun shines throughout the entire period of rotation of the earth at all latitudes north of the Arctic Circle, while south of the Antarctic Circle there are 24 hours of darkness. In the latitudes between the Arctic and Antarctic Circles there are progressively fewer hours of daylight as one goes from north to south. At the equator there are 12 hours of

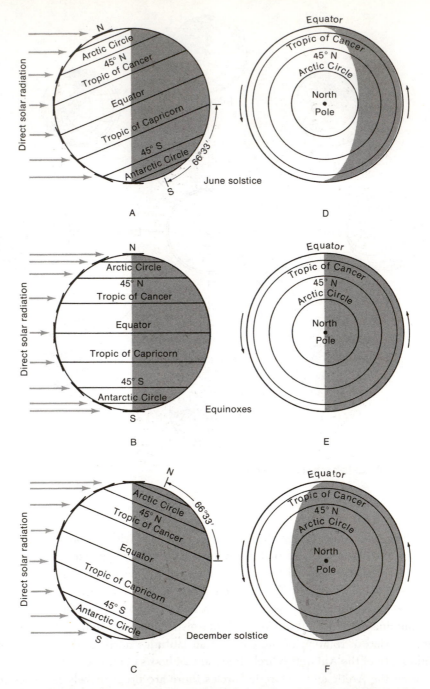

Figure 3.9 Exposure of the earth to the sun's radiation: A, B, and C show the altitude of the midday sun; D, E, and F, the relative length of daily periods of sunshine and darkness. [From "Seasons," by John B. Leighly, *Encyclopaedia Britannica*, © 1962.]

Table 3.1   Time from Sunrise to Sunset

Latitude	Winter solstice	Vernal or autumnal equinox	Summer solstice
90°	0	12 hr 0 min	6 months
80°	0	12 hr 0 min	4 months
70°	0	12 hr 0 min	2 months
60°	5 hr 33 min	12 hr 0 min	18 hr 27 min
50°	7 hr 42 min	12 hr 0 min	16 hr 18 min
40°	9 hr 8 min	12 hr 0 min	14 hr 52 min
30°	10 hr 4 min	12 hr 0 min	13 hr 56 min
20°	10 hr 48 min	12 hr 0 min	13 hr 12 min
10°	11 hr 25 min	12 hr 0 min	12 hr 38 min
0°	12 hr 0 min	12 hr 0 min	12 hr 0 min

daylight and 12 hours of darkness. At the December solstice, the duration of daylight is just reversed, with the region south of the Antarctic Circle having 24 hours of daylight. At the equinoxes there are 12 hours of daylight at all latitudes.[4]

The variation of length of day with season and latitude is shown in Table 3.1. At the equator it is 12 hours throughout the year, but at 60° latitude it ranges from less than 6 hours in winter to more than 18 in summer, and of course at the poles the sun is below the horizon for six months and above the horizon for the other six months.

Figure 3.9 also shows how the height of the sun at noon varies with latitude and season. At the June solstice the noon sun is at the zenith (exactly overhead) at the Tropic of Cancer; farther north it is lower, but even at the North Pole it makes an angle of 23½° with the horizontal plane (the solar altitude) and 66½° with the vertical (its zenith angle). As one goes south, the noon sun is lower, and at the Antarctic Circle it is at the horizon. Conditions are just reversed at the December solstice, with the noon sun overhead at the Tropic of Capricorn and on the horizon at the Arctic Circle. At the equator the noon sun is at the zenith at the equinoxes and departs at most 23½° from it at the solstices, so that the sun's radiation is always almost perpendicular to the horizontal at noon. At the poles the highest

---

4. Actually, the time from sunrise to sunset is slightly longer than indicated by these astronomical considerations because the sun's rays are refracted (bent) by the atmosphere, making the sun appear to be at the horizon when it is slightly below it; daylight is also increased by twilight, which is produced by the scattering of sunlight by the upper part of the atmosphere, which remains illuminated after the sun has set for an observer at the ground.

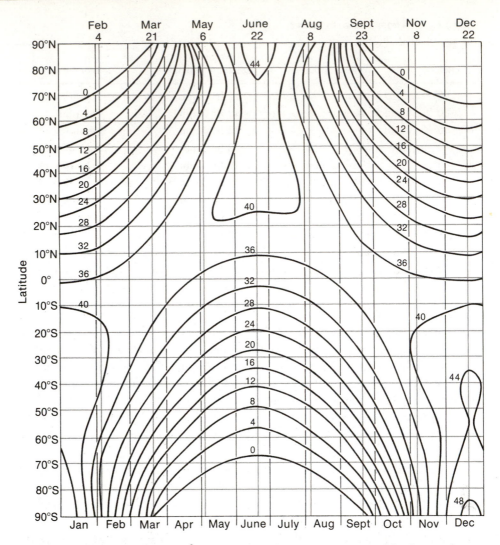

Figure 3.10   Solar radiation (MJ/m² day) falling on horizontal surface at outside of atmosphere.

the sun gets above the horizon is 23½°, so that even when it is above the horizon 24 hours a day the radiation spreads over a large horizontal area all day.

Figure 3.10 shows the distribution of daily insolation that would be received at the earth's surface at various latitudes and seasons if there were no atmosphere. At the equator, where the noon sun is never far from the zenith and the sun is above the horizon 12 hours every day, the total insolation before depletion by the atmosphere varies only about 12 percent, from a high of about 38 MJ/m² day (megajoules/m² day) in March to a low of about 32 MJ/m² day in June. At 45°N latitude it ranges from about 12 MJ/m² day at the winter solstice to about 41 MJ/m² day at the summer solstice, and poleward of 66½° it varies from zero at the winter solstice to more than 42 MJ/m² day at the summer solstice.

The small added effect of the varying distance from the sun gives the South Pole the largest amount of radiation on a horizontal surface outside of the atmosphere in a day anywhere on earth, 48 MJ/m² on December 22. The maximum at the North Pole is 45 MJ/m² on June 21. It should be noted that the low angle of the sun tends to offset the duration of sunlight at the poles, but, even so, in the absence of the atmosphere the sun would heat the poles more than the equator at the summer solstice. That the poles are cooler than lower latitudes in summer is due to the depletion of the radiation as it passes through the air, as well as such effects as the reflection of radiation by snow and ice and the utilization of heat in melting the snow and ice.

In the next section we shall discuss how radiation of different wavelengths is depleted differently by the constituents of the atmosphere. To get a rough idea of how the insolation is affected by the differences in path length through the air at different latitudes and seasons, Milankovich computed the radiation on the assumption that, if the sun were in the vertical, 70 percent of the incident radiation would be transmitted to the ground everywhere. The results of his computations, adjusted for the presently accepted value of the solar constant, are shown in Figure 3.11.

In Figure 3.11, we see that the maximum insolation is near the latitude at which the sun is vertical at noon. At the summer solstice in each hemisphere, the amount varies little from pole to equator. The maximum then is at about 30° latitude, totaling more than 25 MJ/m² day in the Southern Hemisphere and a little less in the Northern Hemisphere. The equator receives about 20 MJ/m² day, and the poles both receive somewhat more than 16 MJ/m² day at their respective summer solstices. At the winter solstice the variation of daily insolation with latitude is large, from zero in the region of polar night to about 20 MJ/m² day at the equator.

The assumed transmissivity of 70 percent is frequently exceeded, and the sky radiation is added to the direct radiation, so that total daily global radiation amounts in excess of 30 MJ/m² (720 ly) are not rare even at low-altitude stations, and mountain stations may have even higher values. When the sun is behind clouds, the global radiation consists only of diffuse sky radiation, and on days with overcast skies the total global radiation is much less than on clear days. However, when there are scattered cumulus clouds, the reflection of sunlight to the ground from the sides of the clouds augments the direct radiation so that the global radiation part of the time may exceed that with clear skies. In fact, once in a while under

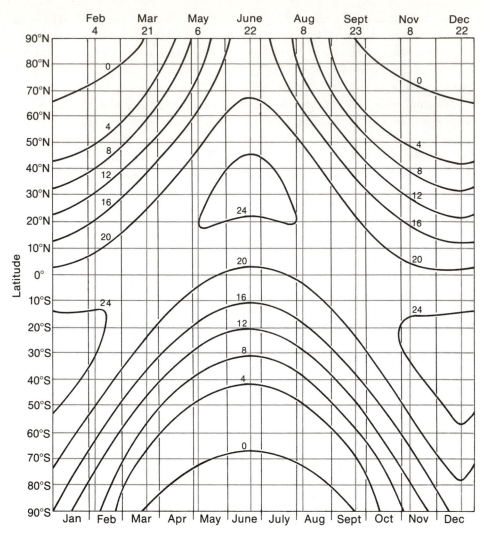

Figure 3.11 Solar radiation (MJ/m² day) falling on horizontal surface at sea level, assuming atmospheric transmission of 0.7 for vertical sun and cloudless skies. [Milankovich's computed values, adjusted to presently accepted value of solar constant.]

these circumstances the global radiation may momentarily be greater than the solar radiation reaching the outside of the atmosphere (the solar constant adjusted for solar distance). This happens when the sun is near the zenith, the air is dust-free and very dry except near the ground, so that the direct radiation comes through with very little depletion, and there are some cumulus clouds. When the

Table 3.2   Albedo of Various Surfaces

Surface	%
Fresh snow, high sun	80–85
Fresh snow, low sun	90–95
Old snow	50–60
Sand	20–30
Grass	20–25
Dry earth	15–25
Wet earth	10
Forest	5–10
Water (sun near horizon)	50–80
Water (sun near zenith)	3–5
Thick cloud	70–80
Thin cloud	25–50
Planetary albedo	30

ground is not shaded by the clouds, it receives radiation reflected by the sides of the clouds. This reflected radiation is added to the direct radiation, already intense because of the clarity of the air, and the scattered radiation from the blue sky. The total solar radiation falling on the ground may then be raised for a short time above the amount entering the atmosphere. Of course, the global radiation in the shadow of the clouds is correspondingly reduced, and the total over a longer time, an hour or an entire day, is always less than the solar radiation at the "top" of the atmosphere.

Of the global radiation falling on the earth's surface, some is absorbed and some is reflected back toward the sky. The fraction of the solar radiation that is reflected by a surface is called the *albedo* of the surface. The albedo of the ground depends on its character and on the kind of vegetation growing on it. At liquid water, ice, and snow surfaces, the albedo depends also on the angle of the sun above the horizon. The fraction of the solar radiation entering the atmosphere that is reflected back to space by the earth and atmosphere as a whole is called the *planetary albedo*. Table 3.2 gives approximate values of the albedo for various types of surfaces and for clouds, and the average value of the planetary albedo.

The planetary albedo combines the radiation reflected by the ground and by clouds with that scattered upward by the cloudfree atmosphere. Because of the variation of cloudiness and snow cover, it varies with season. The fraction reflected to space from earth and atmosphere is different at different places, as well, due to the varying character of the underlying surface and the varying amounts of cloud-

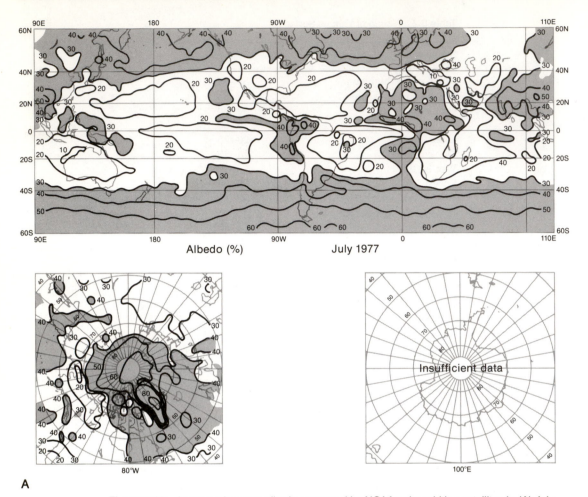

Figure 3.12   Average planetary albedo measured by NOAA polar-orbiting satellites in (A) July 1977 and (B) January 1978. [From *Earth–Atmosphere Radiation Budget Analyses Derived from NOAA Satellite Data, June 1974–February 1978,* Washington, D.C., NOAA–NESS, 1979.]

iness. This is illustrated in Figure 3.12, which shows maps of the albedo of the earth and atmosphere observed by the NOAA polar-orbiting satellites in July 1977 and January 1978. In the regions of polar darkness (Antarctica in July, the Arctic Ocean in January), the albedo could not be measured, of course. Elsewhere it has high values over regions of frequent cloudiness and areas covered by ice and snow, and low values over cloudfree and snowless places, particularly in the subtropics. The variation with latitude is shown in Figure 3.13, in which the average values around the earth along parallels of latitude for the year June 1, 1976 to May 31,

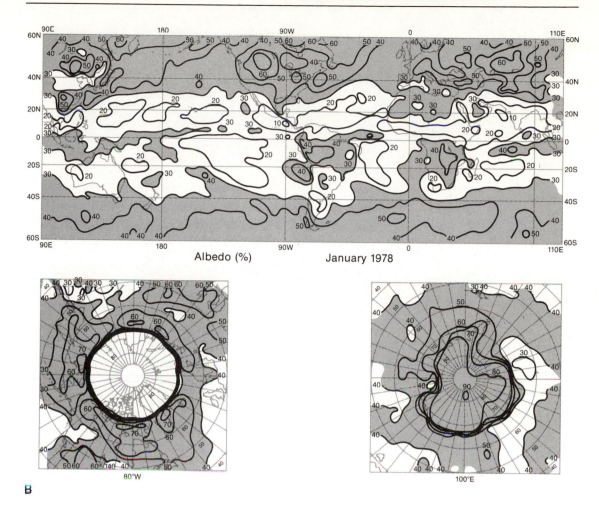

Albedo (%)   January 1978

B

1977 is presented. The high values in polar latitudes is due to the snow cover there. In middle latitudes, intermediate values correspond to the relatively large amounts of cloudiness in the region of storminess associated with the polar front. The low values in low latitudes are due to the infrequency of clouds in the belt of subtropical high pressure. The slightly higher values just north of the equator correspond to the greater cloudiness in the intertropical convergence zone, where the northeast trade winds and the southeast trade winds meet. These features of the atmospheric circulation will be discussed in Chapter 9.

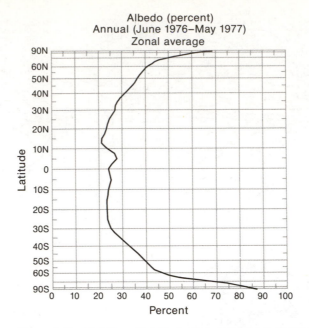

Figure 3.13 Variation with latitude of average planetary albedo for the year June 1976–May 1977, as measured by NOAA polar-orbiting satellites. [From *Earth–Atmosphere Radiation Budget Analyses Derived from NOAA Satellite Data, June 1974–February 1978*, Washington, D.C., NOAA–NESS, 1979.]

3.7   Depletion of solar radiation by the atmosphere

Radiation passing through the air is subjected to two types of attenuation: *scattering* and *absorption*.

When radiation is scattered by a particle—say, an air molecule—it is as though the radiation were momentarily captured and then sent out unchanged in amount and wavelength, but dispersed in all directions, not just in the direction from which it came. Consequently, of the radiation from the sun thus captured, only about one-half continues downward, the other half being sent back into space. Radiation scattered by air molecules reaches an observer from all parts of the sky, not just from the direction of the sun. For this reason the daytime sky appears bright to someone near the bottom of the atmosphere. At high levels, where relatively few molecules are above one, the sky is black. Small particles, such as individual molecules or clusters containing a few molecules, scatter a larger proportion of the short-wave radiation, the blue and violet light, than the longer-wave yellow and red light. That is the reason the sky is blue in the absence of haze or smog. Haze, fog, and smog contain larger particles, which scatter more nearly equally in all wavelengths. When they are present, the sky tends to be white, particularly near the horizon, except when smog with absorptive properties gives it a yellow or brownish color.

In absorption, air molecules actually take up part of the radiant energy and convert it into internal energy, whereby the motion of the molecules or their component atoms and electrons is changed. This energy increase manifests itself as a temperature change. Layers in which radiation is absorbed may be expected to be the warm layers of the atmosphere. In addition, when very short UV radiation, having quanta with high energy, in accord with Planck's equation [3.5], is absorbed by air molecules, the molecules may be "knocked" apart (dissociated) into separate atoms, or electrons may be "knocked" out of them, creating ions, which are particles carrying electric charges.

Let us now consider what happens to solar radiation as it passes downward through the atmosphere. The components of air, like all gases, absorb radiation of certain wavelengths and are transparent to others. Figure 3.14 shows the absorptivity, that is, the relative ability to absorb, of several atmospheric gases and of the atmosphere as a whole at various wavelengths. In the wavelengths in which solar radiation reaches the earth with significant intensity, oxygen and ozone are very absorptive in the UV and slightly absorptive in the visible; water vapor absorbs slightly in the visible and somewhat more strongly in the near IR; and carbon dioxide absorbs slightly in the near IR. Nitrogen (not shown in the figure) absorbs only in the very short UV.

As the radiation from the sun impinges on the rarefied outer atmosphere, oxygen and nitrogen absorb the small amount of very short-wave UV radiation, and are heated, dissociated, and ionized, thereby producing the thermosphere and the ionosphere. By the time the radiation reaches 100 km, all of the far UV has been absorbed. Below this level is a region in which there is little absorption because none of the radiation that is strongly absorbed reaches it, and the density (number of molecules) is insufficient for weakly absorbed radiation to be important. This region extends for some distance above and below 80 km. The lack of absorption of radiation causes the minimum of temperature there.

At still lower levels an interesting development takes place. The weak absorption by oxygen of the UV radiation that reaches these levels produces some atoms, but now the density of molecules is great enough for the atoms to collide with the ordinary diatomic molecules and produce triatomic molecular oxygen (ozone). The ozone absorbs UV radiation of longer wavelengths (up to about 0.3 μm), and since

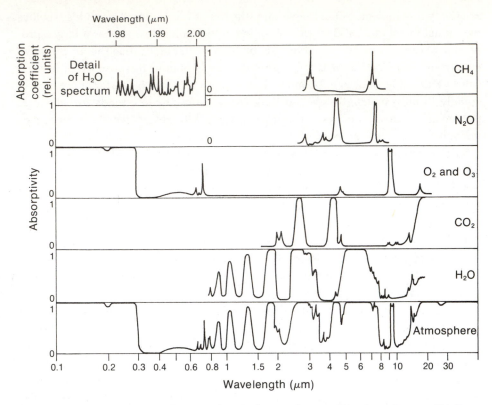

Figure 3.14 Absorptivity at various wavelengths by constituents of the atmosphere and by the atmosphere as a whole. [From R. G. Fleagle and J. A. Businger, *An Introduction to Atmospheric Physics.* Copyright 1963 by Academic Press.]

there is more of this radiation, both because the intensity at these wavelengths incident on the atmosphere is greater and because it has not been subject to absorption higher up, it heats up the layer to quite high temperatures, producing the temperature maximum at 50 km. At levels below 50 km the UV radiation that ozone can absorb gets used up, so that with the increased density of air to heat and the decreased absorption the temperature is lower.

The UV radiation of wavelengths greater than 0.3 μm and the visible radiation are absorbed only weakly—in some wavelengths not at all—by the atmosphere. Effectively, the atmosphere may be regarded as transparent to visible light. At some near-IR wavelengths there is considerable absorption by carbon dioxide and water vapor, the latter particularly in the lower part of the troposphere where it is plentiful. Between the far UV, which is absorbed in and above the stratosphere, and the IR, which is absorbed by CO_2 and water vapor, about 25 percent of the

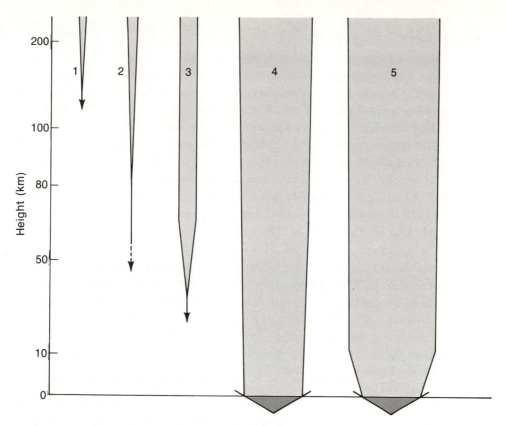

Figure 3.15 Transfer and absorption of solar radiation.
1. UV, $\lambda < 0.12$ μm, absorbed by N_2 and O_2.
2. UV, 0.12 μm $\leqslant \lambda < 0.21$ μm, absorbed by O_2.
3. UV, 0.21 μm $\leqslant \lambda < 0.34$ μm, absorbed by O_3.
4. Near UV and visible, 0.04 μm $\leqslant \lambda < 0.7$ μm, transmitted nearly undiminished except for scattering.
5. Near IR, 0.7 μm $\leqslant \lambda < 4$ μm, absorbed slightly by O_2 and CO_2, and in troposphere by H_2O vapor.

incoming radiation is absorbed by the atmosphere. About 30 percent is scattered and reflected to space by the atmosphere, clouds, and the ground. The remaining 45 percent comes through the atmosphere and is absorbed by the ground. The temperature of the ground is raised by absorbing this radiation.

The disposition of solar radiation in passing through the atmosphere is illustrated schematically in Figure 3.15. The various wavelengths of incoming radiation are represented by bands terminating in arrowheads, with the width of the bands representing crudely the amount of incident radiation in those wavelengths. The depletion by absorption and scattering is represented by the narrowing of the bands, and the arrowhead shows the depth in the atmosphere to which the radiation of those wavelengths penetrates. Thus, the narrowness of band number 1, at the left side of the diagram, is intended to show that the amount of extremely short-wave UV radiation, with wavelengths so short it can ionize nitrogen and oxygen, is very small (so small that it is not shown in Figure 3.6). Its narrowing and ending before reaching 100 km represent the absorption of the radiation at the greater heights, producing the F-layers of the ionosphere. Band 2 shows that the slightly longer-wave UV, in the range 0.12 μm to 0.21 μm, is absorbed somewhat less strongly by oxygen and consequently contributes to producing atomic oxygen and heating the air farther down. By 80 km most of it has been absorbed, but a little of the longer-wave part of this band penetrates below 50 km. So little of it reaches 80 km that there is little heating at that height, which accounts for the low temperature at that level. However, the part that goes lower produces oxygen atoms there, and with the higher density there are enough oxygen molecules present for collisions between the atoms and molecules to produce ozone.

Band 3 in Figure 3.15 represents the UV radiation in wavelengths that are absorbed by ozone. As was shown in Figure 3.6, there is somewhat more solar radiation in these wavelengths; hence, the greater width of the band. This radiation passes through the atmosphere undiminished except for scattering until it starts being absorbed somewhat below 80 km, where ozone is present. The temperature maximum at 50 km is due to heating produced by absorption of this radiation by ozone. By 20 km practically all of the radiation in these wavelengths is gone.

The near-UV and visible radiation represented by band 4 almost all gets through to the ground, while some of the solar IR radiation, band 5, is absorbed in the troposphere. As indicated by the width of these bands, most of the solar energy is in these wavelengths. A small amount of the visible radiation is absorbed by oxygen and ozone, and the infrared is absorbed slightly by oxygen and CO_2 and somewhat more strongly in the troposphere by water vapor.

The minimum temperature on either side of the tropopause is due to the small amount of absorption of solar radiation there. The higher temperature at lower levels in the troposphere is due to other processes.

When the solar radiation that gets through the atmosphere strikes solid ground, the part that is not reflected is converted to heat right at the surface, and this heat raises the temperature of a very thin layer of soil or rock; as a consequence, the temperature increase is large. Where the earth's surface consists of water (oceans or lakes), the radiation can penetrate and be absorbed through a considerable thickness, so that the same amount of heat is spread through a larger mass, resulting in a much smaller temperature increase. Other factors also contribute to reducing the temperature rise when radiation is absorbed by a water surface:

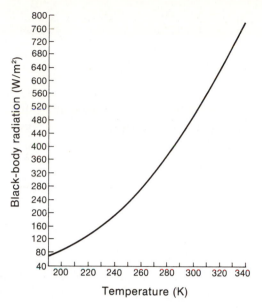

Figure 3.16   Black-body radiation at terrestrial temperatures.

evaporation uses some of the heat, and the motions of the water may spread it through even deeper layers than the radiation penetrates.

The ground heated by radiation tends to heat the air above it in two ways: by conduction and convection—the way in which the burner of a stove heats a pot of water—and by radiation. The ground radiates approximately as a black body at its own temperature. This radiation passes upward through the overlying atmosphere, and some of it is absorbed by the layers of air it passes through and some escapes into space. As shown in Figure 3.5B, at the temperatures that occur at the ground the emitted radiation is in the far infrared, with maximum energy at about 10μm. [Applying equation (3.4) to the average temperature of the earth's surface, estimated to be 286 K, we have λ_M = 2898/286 = 10.1 μm.] At many wavelengths in this region, water vapor and carbon dioxide are good absorbers, and the air layers near the ground absorb a large part of the radiation. It is this heating from below by contact and by radiation that is responsible for the warmth of the lower layers of the atmosphere and the decrease of temperature upward through the troposphere.

Figure 3.16 is a graph of the total radiation emitted from a black body for temperatures that occur in the earth's atmosphere. The energy emitted by a black body at the mean temperature of the earth's surface is seen to be approximately 380 W/m².

In the next section we shall examine in more detail the exchange of the long-wave (infrared) radiation.

3.8   Transfer of terrestrial radiation through the atmosphere

We have seen that black body radiation at temperatures that occur at the earth's surface is long-wave radiation, mostly longer than 4 μm, with maximum intensity in the vicinity of 10 μm, whereas radiation reaching the earth from the sun is short-wave radiation, almost entirely shorter than 4 μm. Since the atmosphere has temperatures in the same general range as the earth's surface, the radiation from it is also long-wave radiation. Consisting of gases, however, the atmosphere does not radiate in all wavelengths, like a black body, but only in certain wavelengths, that is, the wavelengths at which it can absorb.

Figure 3.14 shows that the atmosphere is much more transparent for short-wave than for long-wave radiation, particularly in the visible. For long-wave radiation there is a band between 8 and 11 μm in which the atmosphere absorbs very little. The absorption that shows in the middle of this band, centered at 9.6 μm, is due to ozone in the stratosphere, so that the troposphere is practically completely transparent in this band.

The result of the atmosphere being transparent to solar radiation and more absorptive to long-wave radiation from the earth is that the earth's surface is kept at a higher temperature on the average than it would have if there were no atmosphere. The radiant energy absorbed by the atmosphere is partially radiated back to the earth's surface, increasing the total energy received there. The raising of the surface temperature because of the back-radiation from the atmosphere is known as the *greenhouse effect*. This is because it was thought that the glass roof of a greenhouse, being transparent to solar radiation and relatively opaque to long-wave radiation, plays the same role as the atmosphere in keeping the temperature in the greenhouse high enough for plants to grow in winter. However, it has been shown that in greenhouses the prevention of air movement is the more important factor; in small greenhouses in which the glass panes in the roofs were replaced by rock salt, which allows long-wave radiation to escape, the temperature was about as high. For this reason, it has been suggested that the increase in temperature due to trapping of radiation by the air be called the *atmospheric effect* instead of the greenhouse effect. However, we shall use the customary term, while recognizing that it is based on a misconception.

Figure 3.14 shows that the principal absorbers of long-wave radiation are water vapor and carbon dioxide. Water vapor absorbs strongly at wavelengths between 5 and 7 μm and longer than 12 μm, and carbon dioxide between 4 and 5 μm and longer than 14 μm. The band between 8 and 11 μm, which is practically transparent except for the absorption at 9.6 μm by stratospheric ozone, is called the *atmospheric window*. Radiation from the ground or cloud tops in these wavelengths passes directly through the troposphere, and all but the portion absorbed by ozone at 9.6 μm goes out to space unimpeded. In the other wavelengths, radiation from the ground is absorbed at various levels in the atmosphere, and in turn the atmosphere at those levels radiates up and down in amounts that depend on its temperature and water vapor content.

Figure 3.17 illustrates schematically the way in which the various layers of air gain or lose energy by radiative exchange. The troposphere is assumed to be divided into three layers of equal absorptivity due to water vapor and carbon dioxide. The double arrows pointing into the layers and into the ground represent the amount of radiation absorbed there and not, as in Figure 3.15, the amount passing through. The single arrows represent radiation emitted by the layers or by the ground, or the amount of radiation passing through the tropopause level at the top of layer 3.

It will be convenient to use a general designation for the various quantities pertaining to the layers, so that, for instance, when we write T_i ($i = 1, 2,$ or 3), we mean the temperature of any one of the layers. Each layer emits radiation E_i both up and down in an amount that depends on its temperature and its water vapor and carbon dioxide content. Since we have assumed equal amounts of these gases, the emissions of the layer will depend on the temperature only. The figure is drawn assuming a normal decrease of temperature with height ($T_3 < T_2 < T_1$), so that the amount of radiation emitted, indicated by the length of the arrows, is correspondingly smaller for the higher layers, that is, $E_3 < E_2 < E_1$. Note that the total amount of energy emitted by each layer is $2E_i$.

The emission from the ground, designated by E_s, is approximately black body radiation. Since it includes radiation in wavelengths (between 8 and 11 μm) in which the air layers do not emit, E_s will be considerably greater than E_1 for this reason, in addition to T_s being greater, during the day at least, than T_1. Part of the radiation from the ground will be absorbed by layer 1. Since much of E_s in the wavelengths absorbed by water vapor and carbon dioxide will be removed by the time it has passed through layer 1, a smaller part will be absorbed by layer 2, and, similarly, the amount absorbed by layer 3 will be still smaller. If we designate the part of E_s that is absorbed by the ith layer A_{si}, we have seen that $A_{s1} > A_{s2} > A_{s3}$. But the part of E_s that is between 8 and 11 μm will pass through all the layers practically unabsorbed; thus, an amount E_{sT} is emitted through the tropopause from the ground.

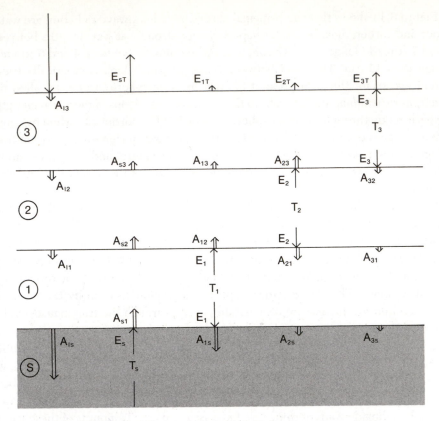

Figure 3.17   Illustration of radiative transfer through the atmosphere: single arrows—radiation emitted by ith layer, E_i (or solar radiation I); double arrows—radiation A_{ij} absorbed by jth layer from E_i. (See text for explanation.)

The emission up or down from each layer may similarly be traced, as shown in Figure 3.17. For instance, of the emission upward from layer 2, a part A_{23} is absorbed by layer 3, and the rest, E_{2T}, passes through the tropopause, while of its downward emission, layer 1 absorbs A_{21} and the ground receives A_{2s}.

We may similarly treat the solar radiation passing downward through the layers. Of the amount I passing through the tropopause, A_{I3} is absorbed by layer 3, A_{I2} by layer 2, and A_{I1} by layer 1. Because of the small absorptivity of air for short wavelengths (once the radiation at wavelengths shorter than 0.3 μm has been removed), these quantities are small; the remainder, A_{Is}, is the amount absorbed at the ground. Of course, I varies with time of day—maximum at noon and zero at night—and A_{Is} undergoes a similar variation. At any one time the absorption of radiation at the ground is

$$A_s = A_{Is} + A_{1s} + A_{2s} + A_{3s}$$

The net amount of energy available for heating the ground is

$$H_s = A_s - E_s$$

During the day, when the sun is high, A_s is larger than E_s, and the temperature at the ground rises. During the night, when $A_{Is} = 0$, A_s is smaller than E_s, and the surface temperature decreases.

We can likewise discuss the radiation budget for each of the layers. For instance, the net amount of energy added to layer 2 by radiation is

$$H_2 = A_{I2} + A_{s2} + A_{12} + A_{32} - 2E_2$$

It turns out that usually the layers of air away from immediate contact with the ground are nearly in radiative equilibrium; that is, H_i is very small in these layers. During the day there may be a slight surplus of radiation, but on the average through the day and night there is a net deficiency that must be made up for by other processes. The processes by which the deficit in the troposphere is made up are conduction, convection, and transfer of latent heat by evaporation from the earth's surface and condensation to form clouds.

Of the long-wave radiation passing upward through the tropopause (E_{sT} + $E_{1T} + E_{2T} + E_{3T}$), some is absorbed by gases in the stratosphere and higher layers of the atmosphere, but most of it passes upward into space. This loss of energy to space is augmented by the relatively small upward radiation from the absorbing gases in the high atmosphere. Figure 3.18 shows the total outgoing long-wave radiation measured by the NOAA polar-orbiting satellites, averaged for the year June 1, 1976 to May 31, 1977. The warm regions at low latitudes radiate more energy to space than the colder regions near the poles, but in areas near the equator where convective cloudiness extends to great heights, the outgoing radiation is smaller than in subtropical areas where clouds tend to be absent or are present only at low levels.

In the long run these processes of absorption and emission at the ground and in various layers of the atmosphere result in a balance between the incoming and outgoing energy for the earth and atmosphere and for each separately. Figure 3.19 shows the estimates of the component streams of radiation and the other processes that produce this balance. The radiation from the sun on the earth, averaged for

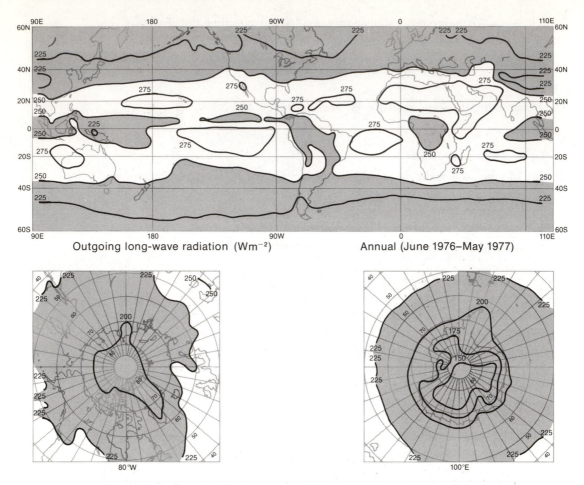

Figure 3.18 Average daily long-wave radiation passing from earth and atmosphere to space during the year June 1976–May 1977, as measured by NOAA polar-orbiting satellites. [From *Earth–Atmosphere Radiation Budget Analyses Derived from NOAA Satellite Data, June 1974–February 1978,* Washington, D.C., NOAA-NESS, 1979.]

the entire year, is represented by 100 units, of which less than half is absorbed at the ground. Due to the greenhouse effect, which produces a higher surface temperature than would occur in the absence of the atmosphere, the earth's surface emits 114.5 units, most of which is absorbed by the atmosphere. The atmosphere in turn radiates back to the ground 96.5 units, more than twice that absorbed at the ground from the sun, providing the excess that heats the atmosphere from below by conduction, convection, and transfer of latent heat and thus contributing to the decrease in temperature with height through the troposphere.

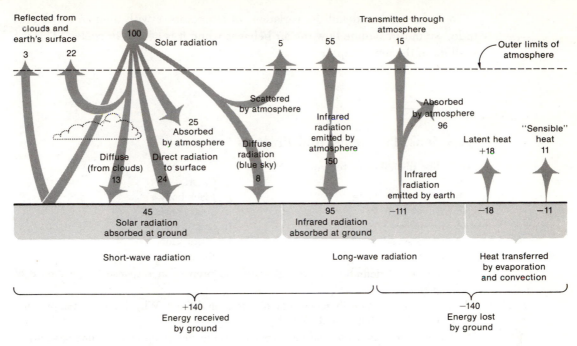

Figure 3.19   Budget of radiation from the sun, the atmosphere, and the ground.

We have seen that the high atmosphere, above 100 km, is heated, dissociated, and ionized by the very short ultraviolet radiation that is strongly absorbed by atomic and molecular oxygen, that the layer around 80 km is shielded from the radiation it could absorb by the layers above it, and that below this, where the air gets dense enough for ozone to form by collisions, there is a layer that is heated by the absorption of ultraviolet radiation by ozone. The ultraviolet radiation at wavelengths below 0.3 µm is completely absorbed by the ozone layer. The longer-wave radiation passes through this layer and is absorbed only slightly in the lower stratosphere and the troposphere on its way to the ground. The warmth of the lower part of the troposphere is due to heating from below by long-wave radiational exchange, transfer of latent heat, conduction, and convection.

Quantitative studies have shown that these processes explain the observed average temperature structure in the vertical. Similarly, computations of the ionizing effects of the radiation result in ion density distributions corresponding to that which is observed. These computations demonstrate that the way in which solar radiation is absorbed and transmitted through the atmosphere provides the explanation of the remarkable variation of temperature and composition with height.

To understand in detail the variations of temperature with time and with latitude, we must examine how the air behaves when it is heated or cooled. This we shall do in the next chapter.

Questions, Problems, and Projects

1. Define the following terms:

   a. Radiation   g. Calorie
   b. Wave   h. Black body
   c. Wavelength   i. Solar constant
   d. Frequency   j. Scattering
   e. Wave speed   k. Absorption
   f. Energy   l. Reflection

   Include these definitions in the glossary you prepare in response to question 8 of Chapter 1.

2. The earth is continually receiving radiation from the sun. Why doesn't its temperature continually get higher and higher?

3. Discuss the reason why the time from sunrise to sunset varies with time of year and with latitude. Why is it always 12 hours at the equator?

4. What would be the consequence of decreasing the amount of ozone in the stratosphere? It has been shown that chlorofluoromethanes (Freons) that are released at the ground get up into the stratosphere unchanged and then react with and reduce the amount of ozone. What are the implications of this for their use in commercial applications?

5. How does the amount and dominant wavelength of radiant energy from a black body vary with temperature? Discuss the radiation from the sun and the earth's surface in these terms.

6. Why is the diurnal variation of temperature over land larger than over the sea?

7. Discuss what happens to solar radiation in passing through the atmosphere, from the "top" of the atmosphere to the ground.

8. What constituents of the atmosphere absorb long-wave radiation? How would changes in the amount of carbon dioxide affect the temperature near the earth's surface?

9.a. Calculate the frequency of radiation having a wavelength of 0.5 μm.
  b. Calculate the wavelength of a radio signal having a frequency of 1000 kHz.
  c. Calculate the frequency of a radar having a wavelength of 10 cm.
  d. Calculate the wavelength of a weather radio broadcast having a frequency of 162.4 MHz.

10.a. Calculate the temperature of a red-hot iron bar that has its peak in energy emitted at 0.65 μm.
  b. Calculate the rate of emission of energy from it.
     [*Hint:* Calculate its temperature using equation 3.4, and then the rate of emission using equation 3.3. Remember that the fourth root of a number can be obtained by taking the square root twice.]

11.a. Compute the temperature which the earth would have in the absence of the atmosphere in order to emit exactly the amount of radiation received on the average from the sun. Use 1.370 kW/m^2 for the solar constant, and 6370 km for the radius R of the earth. Remember that the radiation received is that intercepted by the cross section of the earth, πR^2, but the radiation emitted comes from its entire surface, $4\pi R^2$.

b. If the albedo would be exactly what the average planetary albedo is at present, namely 0.30, but the atmosphere was absent, what would the temperature have to be in order that the energy emitted by the earth be exactly equal to the amount absorbed at the ground?

[*Answers:* (a) 278.8 K; (b) 255.0 K.]

12. What would the temperatures be in question 11 when the earth is at its closest to the sun, about January 4, and when it is at its farthest point, July 5? What would the answer to 11b be if the albedo were 0.1?

[*Note:* Since the earth must on the average emit as much energy as it receives, it behaves as though it is a black body with average temperature of the value given in question 11b. This temperature is known as the *effective temperature* of the earth. The fact that the actual average temperature of the earth's surface is much higher is due to the presence of the atmosphere, and largely to the greenhouse effect.]

4 The Gas Laws. Heat and Temperature Changes

4.1 The behavior of gases

The gas phase of a substance differs from the other two phases, liquid and solid, in that a gas tends to occupy all the space available to it. If a small amount of gas is placed in a closed container, it does not form a free surface, as a liquid does, but spreads throughout the entire volume. The molecules of a gas are not bound to one another, but move about freely, except for bouncing against each other or the walls of the container. When one molecule collides with another or with the wall, it rebounds like an elastic ball, and in so doing it exerts *pressure*. Until the molecules are uniformly spread through the container, the pressure will be uneven and the molecules will be pushed more toward the places where there are fewer of them pushing back. Very quickly an equilibrium state is reached, in which the internal pressure of the gas either is constant or varies in such a way that it just balances any external forces, such as gravity, that are acting on the gas.

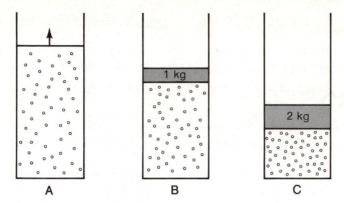

Figure 4.1 Illustration of the effect of pressure on volume of gas when the temperature is kept constant: (A) weightless piston moves upward because of pressure of gas; (B and C) piston weighing 2 kg compresses gas to one-half the volume and twice the density of that compressed by a piston weighing 1 kg at the same temperature.

Figure 4.1 illustrates another kind of experiment we might carry out. In each of the diagrams, the situation is illustrated in which a particular quantity of gas has been placed in a container that has a movable top, that is, a cylinder with a piston at the top. (For the purposes of this illustration, we assume the space above the piston to contain no matter.) If the piston were weightless, the pressure on it due to the bombardment of it by the gas molecules would push it upward indefinitely, as indicated in diagram A. Since a real piston has weight, it will be pushed to the position where the upward force due to the gas pressure inside the cylinder is just equal to the weight, as shown in diagram B. If we increase the weight, the piston will be pulled down by the force of gravity, squeezing the molecules of gas together and increasing the number colliding with it until the increase in pressure force so produced is just equal to the increase in weight. This downward displacement of the piston corresponds to a decrease in volume and an increase in density of the gas. If the weight on the piston is doubled and at the same time the temperature is kept constant (diagram C), the volume will be halved and the density will be doubled.

The average speed of the molecules, and therefore the force they exert when they collide, vary with the temperature. If the temperature is raised in a rigid container (constant volume), the pressure will rise because of the increased speed of the molecules. If the temperature is raised in a cylinder with a piston of fixed weight (constant pressure), the piston will be pushed upward and thus the volume increased and density decreased because of the increased speed of the gas molecules.

The behavior of gases can thus be expressed in the following rules:

1. The density of a gas is proportional to the pressure, and the volume is inversely proportional to the pressure, if the temperature is constant.
2. The pressure of a gas is proportional to its absolute temperature, if the volume is constant.
3. The volume of a gas is proportional to the absolute temperature, and the density is inversely proportional to it, if the pressure is constant.

These rules can be expressed quantitatively in a single equation, which is called *the equation of state of a perfect gas,* or *the gas law:*

$$p = \rho RT \qquad (4.1)$$

or

$$p\alpha = RT$$

where the letters have the following meanings:

p Pressure
ρ Density
T Absolute temperature
α Specific volume
R A constant of proportionality having a particular value for each gas in any particular system of units. R is called the *gas constant*. In the SI, its value for dry air is 287 J/kg K; in the cgs system, $2.87 \cdot 10^6$ ergs/gram K. For all gases, the gas constant multiplied by the molecular weight, the weight of the molecules of the gas relative to a scale in which the most plentiful isotope of carbon is taken to have a molecular weight of exactly 12, is a constant:

$$Rm = R*$$

$R*$ is the universal gas constant, having the value $8.31432 \cdot 10^3$ J/kmol K, and m is the molecular weight.

We have used the terms "pressure," "density," and "temperature" without defining them, and while they have meanings for us gained from common experience, it is desirable to make these meanings precise:

Pressure is the force per unit area acting perpendicular to the area. On the walls of the container, it is the force pushing outward on the wall because of the bombardment by the molecules of gas. In the interior of the gas, it is the force that the molecules on one side of an (imaginary) unit area exert on those on the other side,

Figure 4.2 Pressure-measuring instruments. (A) Mercury barometer, in which the atmospheric pressure is balanced by the mercury in an evacuated tube. The length of the mercury in the tube is a measure of the pressure; consequently, pressure is frequently expressed in terms of length. Thus, standard atmospheric pressure (1013.25 mb) is 760 mm of mercury. (B) Aneroid barometer, in which the atmospheric pressure compresses a partially evacuated flexible metal box. The changes in the size of the box with pressure are converted by a mechanical linkage to the position of the pointer on the dial. In the example shown, the scale indicates the pressure in millibars (inner scale) and inches of mercury (outer scale). [Courtesy of Science Associates, Inc., Princeton, New Jersey.]

and it is equal and opposite to the force that is exerted by those on the other side. Since we can have such imaginary unit areas constructed in all directions through a point, *the pressure at a point is the same in all directions*.

The unit of force in the SI or mks system is the *newton* (N), where one newton is the force that, acting on a mass of one kilogram, would produce an acceleration of one meter per second per second, and the unit of pressure is the newton per square meter, to which the name *pascal* (Pa) has recently been given. Similarly, in the cgs system the unit of force is the *dyne*, defined as the force that would accelerate one gram of mass one centimeter per second per second, and the unit of pressure is the dyne per square centimeter. These units are so small that normal or standard pressure at sea level is 101,325 Pa or 1,013,250 dynes/cm^2. To avoid such big numbers in meteorological practice, a special unit, the *millibar* (mb) is commonly used, where 1 mb = 100 Pa = 1000 dynes/cm^2. Atmospheric pressures are also expressed in kilopascals, kPa. Normal sea-level pressure is 1013.25 mb or 101.325 kPa.

Temperature is the property of a body[1] that determines whether heat will flow to or from it when it is placed in contact with another body. If there is no flow of heat between the two bodies, they are at the same temperature; otherwise the body that receives heat from the other is at the lower temperature. Absolute temperature can be measured by the expansion of a gas at constant pressure or the change in its pressure in a constant volume. More conveniently, for most purposes temperature is measured by the expansion of a liquid in a glass container (relative to the expansion of the container).

The Celsius (centigrade) scale of temperature has 0° for the melting point of ice and 100° for the boiling point of water at normal sea-level atmospheric pressure. It was named for Anders Celsius, a Swedish astronomer, who in 1742 introduced a scale with 100 degrees between ice point and boiling point, but with the 0 and 100 reversed. The Celsius scale has come into general use in science, and, except in English-speaking countries, it is used in everyday affairs as well. England is in the process of going over to the Celsius scale; weather reports are broadcast there with temperatures in Celsius, with the Fahrenheit equivalents given parenthetically.

Gabriel Daniel Fahrenheit, who was born in Germany, spent most of his life in Holland as a manufacturer of and experimenter with meteorological instruments. He introduced the mercury thermometer and his temperature scale in 1714. The first thermometer, introduced a century earlier, used the expansion of air and was subject to many sources of error. Subsequently, sealed tubes containing alcohol

---

1. By "body" we mean any delimited or circumscribed quantity of matter. Thus, we might refer to a book as a body, or the gas in a particular volume as a body.

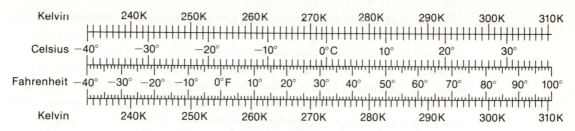

Figure 4.3 Relationship among Fahrenheit, Celsius (centigrade), and Kelvin (absolute) temperature scales.

were used. Individually, these gave good readings, but the readings from one to another could not be compared for lack of a reference scale. Fahrenheit showed that the freezing point of pure water in the presence of pure ice is always the same, and he invented the mercury thermometer to test whether the boiling point of water is also constant. For zero on his scale of temperature he chose the lowest temperature he could attain by a mixture of ice and common salt, and for the other fixed point on his scale he used body temperature, which he arbitrarily called 96°. On this scale he found the freezing point of water to be 32° and the boiling point 212°, and these became the standard reference points for the Fahrenheit scale. With improved thermometers, the average human blood temperature was shown to be 98.6°F instead of 96°F.

In the SI, the unit of temperature is the Celsius degree, one one-hundredth of the temperature change from the melting point of pure ice to the boiling point of pure water. This unit is called the *kelvin* (represented by K without a degree symbol) after William Thompson, Lord Kelvin (1824–1907), who in 1848 showed the thermodynamic significance of absolute temperature. The absolute temperature scale, with zero based on equation (4.1), is called the Kelvin scale. Experiments have determined that 0 K $= -273.15°$C or 0°C $= 273.15$ K.

In meteorology in the United States all three temperature scales are used, and it is important to be familiar with the equivalents in the different scales. The conversion relationships are given in Appendix A.

It facilitates computations to note that $1.8 = 2 - 0.2$ and $5/9 = 1/2 + 1/20 + 1/200 + \ldots$. For example, to convert 20°C to °F, we simply multiply by two, giving 40°, subtract one-tenth of that, or 4°, giving 36°, and add 32°, giving 68°F. To convert -28°F to °C, we first subtract 32, giving -60°; take one-half, -30°, and add one-tenth of that, -3, one one-hundredth, -0.3, and so forth, obtaining $-33.333 \ldots$ °C. For approximate conversions the scales in Figure 4.3 may be used.

Density is the mass of unit volume of a substance. Mass is the measure of the amount of matter; it is the property that provides inertia, that is, resists changes in amount of motion, and provides gravitational attraction of other bodies. From the latter characteristic we have, for bodies on earth, the convenient method of measuring the mass of a body by weighing it. If the body's mass is M, its weight is Mg, where g is the attracting force that the earth exerts on unit mass. The quantity g is called the *acceleration of gravity*. If the density varies from point to point, we must define it as the limit of the ratio of the mass of a very small volume of the substance to the volume as the volume gets smaller and smaller.

The unit of mass in the SI is the *kilogram,* equal to 1000 grams, where the *gram,* the unit of mass in the cgs system, was originally defined as the mass of one cubic centimeter of water. The SI unit of density is the *kilogram per cubic meter* (kg/m³), and in the cgs system it is the *gram per cubic centimeter*. The numerical value of density in SI units is 1000 times the value in the cgs units. Thus, the density of water in the cgs system is 1 g/cm³; in the SI it is 1000 kg/m³.

We can compute the density of dry air by using equation (4.1). Solved for the density, we have

$$\rho = p/RT$$

As an example we shall do so for "NTP" conditions (normal temperature and pressure), which is a temperature of 0°C or 273.15 K, and standard sea-level pressure, 1013.25 mb or 101,325 Pa. Substituting the values in SI units in the equation,

$$\rho = \frac{101325}{287 \cdot 273.15} = 1.293 \, \text{kg/m}^3$$

The density of dry air at sea level is thus less than 1/800 the density of water. In cgs units, it is $1.293 \cdot 10^{-3}$ g/cm³. One cubic meter of air near sea level weighs somewhat more than a kilogram.

Specific volume is the volume occupied by unit mass of a substance. It is the reciprocal of the density:

$$\alpha = 1/\rho$$

In the SI its units are cubic meters per kilogram; in the cgs system, cubic centimeters per gram. At standard conditions air has a specific volume of 0.773 m³/kg or 773 cm³/g.

When using any physical equation, such as the equation of state, all quantities must be expressed in the same system of units, for example, either SI units or cgs units.

4.2   Variation of pressure with height

We are now in a position to explain why the pressure (and the density) decreases with height in the atmosphere. Consider an air column of *unit* cross section from the ground up to the "top" of the atmosphere (Figure 4.4). The pressure at any arbitrary height h in the air column is equal to the weight of the part of the air column above it, in the same way that the pressure inside the cylinder considered in the previous section is equal to the weight of the piston.

Suppose that the weight of the air column above height h is W. If we designate the pressure at that height p_A, we have $p_A = W$. Now consider a position B a distance Δh above A.[2] The weight of the air column above B will be less than W by the weight of the part of the air column between A and B. Since the column has unit cross-sectional area, the volume of this portion is equal to Δh, and its weight is therefore $\rho\, g\, \Delta h$. The pressure at B is therefore

$$p_B = W - \rho g \Delta h = p_A - \rho g \Delta h$$

The change in pressure in going up a distance Δh is thus

$$\Delta p = p_B - p_A = -\rho g \Delta h \qquad (4.3)$$

This equation is called the *hydrostatic equation*.

If we now substitute the value of ρ from the gas law, equation (4.1), we have

$$\Delta p = -\frac{pg}{RT}\Delta h \qquad (4.4)$$

From equation (4.3) we see that if we go up a distance Δh, we will experience a decrease in pressure that is proportional to the pressure and inversely proportional to the temperature. Near sea level, where the pressure is high, the decrease in pressure for a given increase in height is large, and at great heights, where the pressure is small, the rate of decrease in pressure with height is much smaller.

The temperature also appears in equation (4.4). However, temperature changes much less rapidly with height than pressure, and to a first approximation its variation can be ignored. The change of pressure for a given change of height is thus approximately a constant fraction times the pressure itself. A quantity that changes

---

2. The notation Δh means that we are considering a small increase or increment in the quantity h. The symbol Δ (Greek letter delta) before a quantity means a small increment in it. It is important to distinguish this notation, in which Δh is a single quantity, from the usual algebraic notation, in which ab means the product of quantity a times quantity b.

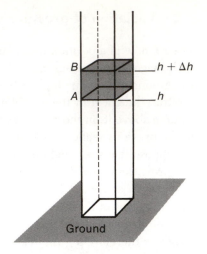

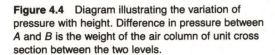

Figure 4.4 Diagram illustrating the variation of pressure with height. Difference in pressure between A and B is the weight of the air column of unit cross section between the two levels.

a fixed fraction of itself for each change of a fixed amount in another variable is said to vary *exponentially*; thus, pressure decreases approximately exponentially with height. It is reduced to one-half its sea-level value at a height of about 5.5 km, one-fourth at about 10 km, one-eighth at about 15 km, and one-sixteenth at about 20 km.

As an example of the application of equation (4.4), let us compute the approximate change of pressure in going upward 10 m if the pressure at sea level has its standard value, 1013.25 mb, and the temperature is 15°C. In SI units, we have

$$p = 101325 \text{ Pa}$$
$$R = 287 \text{ J/kg } K$$
$$T = 288.15 \text{ K}$$
$$g = 9.8 \text{ m/sec}^2$$
$$\Delta h = 10 \text{ m}$$

$$\Delta p = -\frac{1.01325 \cdot 10^5 \cdot 9.8 \cdot 10}{287 \cdot 288.15} = -120 \text{ Pa} = -1.2 \text{ mb}$$

From this we see that, near sea level, for typical terrestrial temperatures, the pressure decreases approximately one millibar for every ten meters we go upward.

The U.S. Standard Atmosphere is defined in terms of temperatures and height, as was given in Table 2.2. The corresponding pressures at various heights up to the stratopause are given in Table 4.1.

Equation (4.4) is also used in evaluating elevations determined by altimeters and in reducing the pressure observed at weather stations to sea level. The usual altimeter consists of an aneroid barometer with the pressure scale replaced by a

Table 4.1 U.S. Standard Atmosphere to the Stratopause

Height (km)	Temperature (K)	Pressure (mb)	Height (km)	Temperature (K)	Pressure (mb)
0.0	288.2	1013.2	11.0	216.8	227.0
1.0	281.7	898.8	12.0	216.6	194.0
2.0	275.2	795.0	14.0	216.6	141.7
3.0	268.7	701.2	16.0	216.6	103.5
4.0	262.2	616.6	18.0	216.6	75.65
5.0	255.7	540.5	20.0	216.6	55.29
6.0	249.2	472.2	25.0	221.6	25.49
7.0	242.6	411.0	30.0	226.5	11.97
8.0	236.2	356.5	35.0	236.5	5.746
9.0	229.7	308.0	40.0	250.4	2.871
10.0	223.3	265.0	50.0	270.6	0.798

height scale that gives the elevation in the standard atmosphere at which the pressure occurs. In most altimeters the zero of the altitude scale can be shifted to allow both for the departure of the sea-level pressure from its standard value and for an additional correction to compensate for the departure of the actual temperature from that in the standard atmosphere. For aircraft, radio altimeters are available that determine the height above the terrain by measuring the time it takes for a radio signal sent from the plane to be reflected back from the ground below.

To enable comparison, the pressure measured at various weather stations, which are at different elevations, is "reduced to sea level." In effect, the pressure difference Δp of an imaginary air column with thickness Δh equal to the altitude of the station and an appropriate average temperature T is added to the station pressure. The temperature used is an estimate of the average temperature an air column from the station level to sea level would have if the solid earth between were absent, based on the observed temperature at the station.

4.3 Heat and temperature change

Heat is a form of energy. It is associated with the random motions of the molecules of which matter is composed. In solids and liquids these motions consist largely of vibrations of the molecules; in gases they are the free movements of the molecules between collisions. The temperature is a measure of the kinetic energy of the molecules. If heat is added to a quantity of matter, the speed of the molecules, and thus the temperature, will be increased. This can be expressed by the equation

$$H = cM(T_2 - T_1)$$

where H is the amount of heat added, M is the mass (amount of matter), T_1 and T_2 are the temperatures before and after the heat was added, and c is a constant of proportionality that is called the *specific heat of the substance*.

If we are dealing with a gas, the amount of heat required to produce a given temperature increase will be different for different circumstances under which the heating takes place. If the gas is in a closed container, so that its volume is kept constant, less heat will be needed than if the gas is in a container with a movable piston. In the latter case, in which the pressure remains constant, the piston rises, increasing the volume. Some of the heat is used to do the work of pushing the piston upward. Therefore, gases are considered to have two specific heats, c_v and c_p, where c_v, the specific heat at constant volume, is less than c_p, the specific heat at constant pressure.

We shall be dealing with phenomena in the open air where, on the one hand, the air is free to expand, so that the volume of a given quantity (say, unit mass) of air will not remain constant; but at the same time the air may rise or fall, so that the pressure will change also. For convenience we shall consider what happens to a unit mass (1 kg or 1 g) of air. It can be shown that the heat required per unit mass to produce a change from T_1 to T_2, when both the volume and the pressure change, may be written in terms of either c_v or c_p as follows:

$$H = c_v(T_2 - T_1) + p(\alpha_2 - \alpha_1)$$
$$H = c_p(T_2 - T_1) - \alpha(p_2 - p_1)$$

Properly these equations can be applied only for small (infinitesimal) changes in pressure and specific volume, since they are both changing, and otherwise one would not know what value of the pressure to use in the first equation, or of specific volume to use in the second. To show this we let

$$T_2 - T_1 = \Delta T$$
$$p_2 - p_1 = \Delta p$$
$$\alpha_2 - \alpha_1 = \Delta \alpha$$

and write ΔH instead of H, where, as explained in footnote 2 (Section 4.2), Δ is understood to represent a small change in the quantity represented by the letters that follow. The equations then become

$$\Delta H = c_v \Delta T + p \, \Delta \alpha \tag{4.5}$$

$$\Delta H = c_p \Delta T - \alpha \, \Delta p \tag{4.6}$$

Except in the layers close to the ground and in the high atmosphere, heat is added to (or subtracted from) the air very slowly and the air is usually moving fast, so that if we are considering changes over a short time (say, less than one day) in the upper part of the troposphere and lower stratosphere, the exchange of heat between an air parcel and its environment can be neglected. A process in which no heat exchange between an air parcel and its surroundings occurs is called *adiabatic*, and thus motions of the air are approximately adiabatic, except near the ground.

For processes in which the pressure changes adiabatically, we set $\Delta H = 0$ in equation (4.6). Solving for ΔT, we get

$$\Delta T = \frac{\alpha}{c_p} \Delta p \tag{4.7}$$

If we substitute for Δp from equation (4.4) we get

$$\Delta T = - \frac{g}{c_p} \Delta h$$

or

$$\frac{\Delta T}{\Delta h} = - \frac{g}{c_p} \tag{4.8}$$

Since g, the acceleration of gravity, and c_p, the specific heat at constant pressure, are nearly constant through the troposphere and lower stratosphere, we see that the change in temperature for a given change in height in an adiabatic process is practically a constant.

We represent this adiabatic rate of temperature decrease with height by Γ (Greek capital letter gamma)

$$\Gamma = \frac{g}{c_p} \tag{4.9}$$

The value of Γ depends on the choice of units for Δh. It is convenient to remember its value for Δh expressed in hundreds of meters:

$$\Gamma = 0.98°C/100 \, m$$

For practical purposes the value can be taken as one degree per hundred meters. This rate applies only to cloudless air; when clouds form or are present, other processes affect the rate of temperature change. We may therefore state the following rule:

If unsaturated air rises (falls) without receiving or losing heat, it will cool (warm) one Celsius degree for each 100-meter change in height.

This suggests part of the reason why the temperature decreases in the troposphere, for in moving over the rough and unevenly heated surface of the earth it is bound to rise and fall. The displaced air will mix with the surroundings at its new position, lowering the temperature aloft and raising it at low levels.

4.4   Thermodynamic diagrams. Process curves and sounding curves

Let us consider a small mass of air (which we refer to as an *air parcel*). As the parcel moves (or even when it remains stationary), its pressure and temperature will vary. For instance, it may lose heat by emitting radiation faster than it regains it by absorbing radiation, without the pressure changing; in this case we say it undergoes cooling during an *isobaric process*. Or it may move horizontally across isobars or up or down, with associated changes in pressure. If it does not receive or lose any heat while moving, its temperature will change at the adiabatic process rate, or if heat is added or subtracted during its motion, its temperature will undergo some other variation. To study these processes it is convenient to represent the changes in the state of an air parcel graphically. A diagram on which variations of the thermal state of a system are shown is called a *thermodynamic diagram*.

Thermodynamic diagrams may have any two of the quantities in the equation of state, equation (4.1), as coordinates. For instance, in one diagram that is used when studying heat engines in physics or engineering, the pressure is plotted against volume (the cylinder volume or the specific volume of the gas it contains). This choice of coordinates is made because, on the pressure-volume (P-V) diagram, work done in a cyclic process is proportional to the area enclosed by the curve representing the process. In meteorology it is convenient to use the directly measured quantities, pressure and temperature, as coordinates. By choosing appropriate scales for the coordinates, the work-area property of the P-V diagram can be retained. One such combination of scales is the pressure on a logarithmic scale and the temperature on a linear scale.

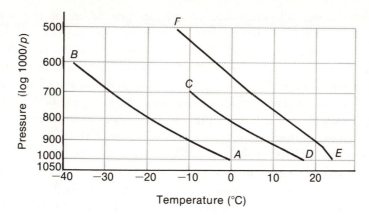

Figure 4.5 Thermodynamic diagram: *AB* and *CD* are process curves representing the change in temperature with adiabatic change in pressure; *EF* is the sounding curve representing the variation of temperature with pressure in the air above Washington, D.C., 7:00 P.M., July 1, 1957.

Figure 4.5 shows a thermodynamic diagram with temperature (T) as the abscissa, evenly spaced, and the logarithm of the pressure (really, log p_0/p, where p_0 is a constant, for instance, 1000 mb) as ordinate. Note that the logarithmic scale results in larger spacing for the same pressure difference as the pressure decreases. A consequence of this choice is that the ordinate corresponds to height; distances upward on the diagram are approximately proportional to distances upward in the atmosphere.

The state of the air parcel (its temperature and pressure) is represented by a point on this diagram. Isobaric processes are represented by horizontal lines (isobars). Isothermal processes in which pressure varies but heat is added (or subtracted) at exactly the right rate to keep the temperature constant would be represented by vertical lines (isotherms). Lines representing adiabatic processes are called *adiabats*—in this case *dry adiabats,* since we are dealing with cloudless air. Dry adiabats for two examples of adiabatic processes are drawn on the diagram. Line *AB* represents the variation of temperature for a parcel of air having an initial temperature of 0°C as its pressure changes adiabatically from 1000 mb to 600 mb. At 600 mb its temperature would be -37.3°C.

The temperature a parcel of air would attain if it were to undergo an adiabatic compression or expansion until the pressure reaches 1000 mb is called the *potential temperature*. Line *CD* in Figure 4.5 represents the adiabatic process a parcel having an initial temperature of -10°C and pressure of 700 mb would follow in being compressed. It is seen that its potential temperature is 18.5°C, or 291.6 K.

For convenience, adiabatic process curves are printed on the thermodynamic diagrams used routinely in meteorology, in addition to the grid of coordinate lines. The potential temperature of any parcel whose representative point falls on an adiabat is the same, namely, the temperature at which the adiabat intersects the 1000-mb isobar. On the diagrams printed for routine use, the adiabats are labeled with the corresponding potential temperature, in the same way that the isotherms are labeled with the actual temperature and the isobars with the pressure. Thus, the thermodynamic diagrams used in meteorology have three basic sets of lines printed on them: isobars, isotherms, and adiabats.

So far we have discussed the representation of processes on a thermodynamic diagram. These processes are changes experienced by an individual air parcel with the passage of time. In addition, curves can be drawn to represent the state of all the parcels in a vertical air column at a particular time. This type of curve is called a *sounding curve*.

We have discussed in Chapter 2 the way in which temperature varies with height in the atmosphere, and in Table 4.1 we presented the variation of pressure and temperature with height in the Standard Atmosphere. The Standard Atmosphere represents average conditions at middle latitudes. At any particular time and place the temperature at various heights (and therefore pressures) differs from these standard values. The actual values are measured routinely at hundreds of places all over the earth. The radiosonde measures the temperature and pressure (and the humidity) as the balloon ascends, and sends back the values by radio. The measurement of the properties of the atmosphere in the vertical is called a *sounding*; hence the term *sounding curve* for the line representing the observed temperatures and pressures along the vertical. Curve *EF* in Figure 4.5 is the sounding curve showing the way in which the temperature and pressure varied over Washington, D.C., at approximately 7:00 P.M. on July 1, 1957.

It is important to keep clear the distinction between process curves and sounding curves. A process curve represents the variation with time of temperature and pressure of a single parcel; a sounding curve represents the measured temperatures and pressures of different parcels of air at various heights in the atmosphere at one particular time.

Since height and pressure are related through the hydrostatic equation [equations (4.3) and (4.4)], process curves and sounding curves can also be represented on a diagram with height and temperature as coordinates. On such a diagram the adiabatic process curves are straight lines.

Usually the sounding curve in a particular instance will not lie along the adiabatic process curve. The variation of temperature with height in an air column thus is generally different from the adiabatic rate of cooling Γ. The rate of decrease with height of observed temperature as one goes upward in an air column is called the *lapse rate*, usually represented by γ, the lowercase Greek letter gamma. Thus, we may write symbolically

$$-\left(\frac{\Delta T}{\Delta h}\right)_{\text{observed}} = \gamma$$

$$-\left(\frac{\Delta T}{\Delta h}\right)_{\text{adiabatic process}} = \Gamma$$

If the lapse rate in an air column is equal to the rate of adiabatic cooling Γ, the air column is said to have an *adiabatic lapse rate*.

4.5   Atmospheric stability

The ease with which air parcels move up and down relative to the surrounding air depends on a property called *stability*. The stability of air columns or layers of air is of importance in connection with the diffusion of pollution, the occurrence of turbulence, which may make the flight of airplanes uncomfortable or dangerous, and the development of showers and thunderstorms. To find the criteria for stability, we shall consider what happens when one parcel of air in an air column is moved up or down relative to the rest of the column.

We assume that initially all of the parcels in the column are in equilibrium, that is, that the forces acting on them are in balance. (This is the assumption that was made in deriving the hydrostatic equation.) When a parcel is not in equilibrium, the unbalanced forces produce an acceleration.

The equilibrium of an object is *stable*, *neutral*, or *unstable*, depending on what would happen if the object were moved slightly. If the displacement of the object gives rise to forces that tend to bring it back to its original equilibrium position, the equilibrium is said to be *stable*. If the displacement leads to forces that tend to increase the displacement from the equilibrium position, the equilibrium is called *unstable*. If no forces arise from the displacement, the equilibrium is *neutral*.

These types of equilibrium can be illustrated by simple experiments with a marble and a hemispherical bowl. If the bowl is placed on a table with the hollow side up and the marble is placed inside, it will come to equilibrium at the bottom, as in Figure 4.6A. If the marble is pushed a small distance away from this position and released, it will roll back under the force of gravity. (Actually, in falling it will gain momentum that will carry it past the equilibrium position. It will oscillate back and forth until friction damps out its motion.) Thus, at the bottom of the bowl the marble is in stable equilibrium. Consider next the situation illustrated in

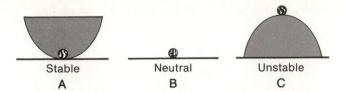

Figure 4.6 Illustration of types of stability: (A) marble in stable equilibrium inside bowl; (B) marble in neutral equilibrium on horizontal table top; (C) marble in unstable equilibrium on top of inverted bowl.

Figure 4.6C, with the bowl inverted and the marble balanced at the top. In principle there is such an equilibrium position, although in practice it may be hard to place the marble exactly at the highest point of the bowl, where it will stay. If the marble is moved the least amount from this position, it will continue to roll off the bowl. This situation is thus one of unstable equilibrium. Neutral equilibrium occurs if the marble is at rest on a flat horizontal table (Figure 4.6B). If it is moved a short distance on the table, it will remain there; the displacement does not give rise to any force either toward or away from the original position.

The corresponding conditions with respect to the equilibrium of the parcels of air composing an air column are shown in Figure 4.7. In this figure, the solid curves represent the sounding curves for four air columns, each of which has a temperature of 14.0°C at a height of 300 m. The column represented by diagram A has a temperature of 13.4°C at 400 m, so that its lapse rate, γ, is 0.6°C/100 m ($\Delta T = 13.4 - 14.0 = -0.6°C$; $\Delta h = 400 - 300 = 100$ m; $\gamma = -\Delta T/\Delta h = 0.6°C/100$ m). Similarly, for the column represented by diagram B, $\gamma = 1.0°C/100$ m, and for the one represented by diagram C, $\gamma = 1.1°C/100$ m. Thus, the lapse rates of the columns represented by diagrams A, B, and C are, respectively, subadiabatic (normal), adiabatic, and superadiabatic.

Let us consider the temperature of a parcel in each column that is lifted adiabatically from 300 to 400 m in each case. In each instance the parcel would cool from 14.0°C to 13.0°C, the process rate of cooling being shown in the diagrams by the dotted lines. In the case represented by diagram A, the displaced parcel would be 0.4°C colder than the surroundings. Being cooler and thus heavier, it would tend to fall back to its original equilibrium position. Under these conditions the air column is in stable equilibrium.

In the case represented by diagram B, the displaced parcel would have the same temperature as the surroundings, and thus subject to no force, neither one to restore it to its original position nor one tending to move it farther away. In its new position it would be in equilibrium, just as it was initially. Under these circumstances the air column is in neutral equilibrium.

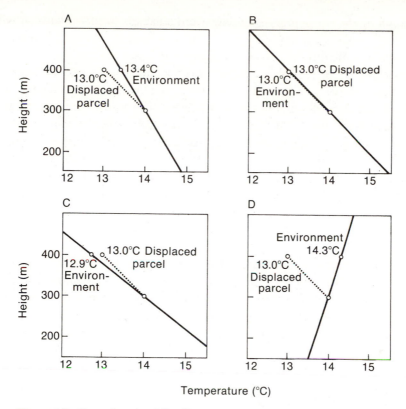

Figure 4.7 Examples of stability. The solid lines are sounding curves and the dotted lines process curves for four schematic cases: (A) normal lapse rate (slightly stable); (B) adiabatic lapse rate (neutral equilibrium); (C) superadiabatic lapse rate (unstable); (D) inversion (very stable).

In the case represented by diagram C, the temperature of the displaced parcel, 13.0°C, is 0.1°C warmer than the undisplaced air column at that level. The displaced parcel of air would thus be lighter than the surrounding air and would experience an upward buoyant force that would tend to displace it farther from its original equilibrium position. The original equilibrium of the air column is thus unstable; any small disturbance of a parcel of it would give rise to forces that increase the departure from the equilibrium position.

It is readily seen that the following criteria apply:

$\gamma < \Gamma$ (subadiabatic lapse rate): stable equilibrium
$\gamma = \Gamma$ (adiabatic lapse rate): neutral equilibrium
$\gamma > \Gamma$ (superadiabatic lapse rate): unstable equilibrium

By examining the consequences of downward adiabatic displacements, it will be found that these criteria apply for the stability of the equilibrium of air columns independent of the direction of the displacement.

During the day, when solar radiation is received at the ground and converted to heat there, the heated ground becomes warmer than the air in contact with it, and heat is transferred from the ground to the air. This process leads to the lowest layers becoming unstable, as in diagram C. Any little irregular movement, such as that taking place over uneven ground, causes large vertical displacements that mix the heated air below with the potentially cooler air above. This vertical mixing tends to change the lapse rate from the unstable value toward the adiabatic value. The mixing tends to establish a neutral equilibrium, that is, an adiabatic lapse rate, throughout a layer immediately above the ground in the daytime. The thickness of the adiabatic layer that develops depends on the amount of insolation and the initial stability of the air column. On clear days in summer and at low latitudes the adiabatic layer may extend upward 3 km or more.

The sounding curve in diagram D of Figure 4.7 shows a situation in which the temperature increases with height instead of the normal decrease. A layer of air with this reversal of the normal situation is called an *inversion*. If a parcel were displaced adiabatically up (or down) in an inversion, it would be much cooler (or warmer) than the surrounding air and subjected to a very strong restoring force. Because of this, vertical motions are effectively prevented by an inversion. When an inversion is present, it suppresses the upward transfer of water vapor or pollutants that are introduced at or near the earth's surface.

At night, with solar radiation absent, the ground loses more energy by its outgoing radiation than it receives from the atmosphere. This is both because it is warmer than the bulk of the water vapor and carbon dioxide that are the principal radiating substances in the atmosphere and because it radiates in all wavelengths, whereas the water vapor and carbon dioxide radiate only in certain wavelength intervals. When the ground cools by radiation, the air next to it is cooled by conduction, and later, when the ground is sufficiently colder, by radiation also. The cooling of the air is greatest at the ground and decreases upward. The result is that an inversion is produced next to the ground. This type of inversion, called a *radiation inversion* or *ground inversion,* occurs over land every clear night that the wind is not too strong. In high latitudes in winter, the daytime solar radiation is weak or absent and the ground inversion becomes deeper and more intense day after day.

Another type of inversion is the *trade-wind inversion* or *subtropical inversion.* This type of inversion is present over the eastern part of oceans and the adjoining coasts at subtropical latitudes (equatorward of about 45° latitude) both day and night every day during the warmest one-third of the year and frequently in other seasons. It is produced by the sinking of air as it moves equatorward around the

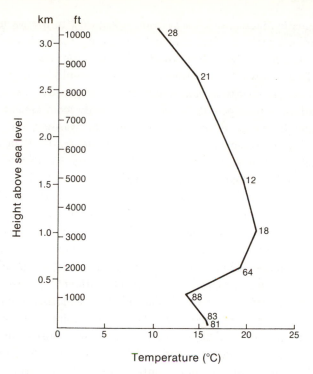

Figure 4.8 Typical sounding at Southern California coast in summer (Santa Monica, California, July 1, 1957). Numbers alongside the sounding curve give relative humidity in percent.

eastern end of the high-pressure areas (anticyclones) that remain over the sub-tropical oceans most of the time (see Chapter 9). A typical sounding showing this type of inversion is given in Figure 4.8. This sounding shows that at Santa Monica, California (near Los Angeles), on July 1, 1957, there was a 400-m layer with an approximately adiabatic lapse rate, above which an inversion extended to a height of about 1 km. The persistent presence of the inversion at about this level is responsible for the high concentrations of pollution for which Los Angeles has become notorious. Vertical motions carry the pollution emitted by the large urban complex up and down readily through the adiabatic layer, to that extent diluting it. But the inversion above acts as a lid that stops vertical motions. The pollution is kept from mixing with the air at higher levels. Confined to this shallow layer, it becomes highly concentrated. Viewed from above, the Los Angeles area looks as though it were covered by a lake of brownish smog, the surface of the lake being as sharp and distinct as a water surface. The clear air of the inversion layer is in marked contrast to the murky layer below.

In the schematic diagram of a warm front in Figure 1.5, the warm air is shown flowing upward over the receding cold air. A sounding made in the cold air ahead of the front would pass upward from the cold air into the warm air. At the frontal boundary surface (really a layer of transition), there would be an increase in temperature with height, that is, a *frontal inversion*. Frequently, the warm air above the frontal surface is only potentially warmer than the cold air below. That is, if it were brought to the same pressure, it would be warmer; but at the higher elevation and lower pressure, it is actually somewhat colder than the "cold" air that is at a higher pressure. When this situation occurs, the frontal boundary is a layer of greater stability than above or below, but not an actual inversion. A more detailed discussion of fronts and frontal inversions is presented in Chapter 11.

While inversions represent the reverse of what is regarded as the normal temperature variation in the troposphere, they occur frequently and in some places and times persistently. Thus, trade-wind inversions are present much of the year, not only in Los Angeles, where the severe smog has called worldwide attention to them, but all along the subtropical west coasts of continents and over the adjoining eastern portions of oceans. Like California and Baja California, the coasts of Chile and Peru, and of North Africa and South Africa, are subject to them. Similarly, it has already been mentioned that ground inversions are almost always present over land during clear nights with light winds. Clear nights with light winds are usually associated with high-pressure centers—anticyclones—around which the air spirals outward near the ground. When anticyclones become stationary or move very slowly over land, there is sinking motion aloft accompanying the outward spiraling of the air near the surface. This reinforces the surface inversions and may cause them to persist through the day as well as the night. If the stagnation of the anticyclones and the accompanying intense, persistent inversions occur over areas where large amounts of contaminants are emitted into the air, high pollution concentrations accumulate. Situations of this sort were responsible for the pollution disasters at Donora, Pennsylvania, in 1948, where 20 people died, and at London, England, in 1952, where there were about 4000 deaths in excess of the normal death rate in one week.

Questions, Problems, and Projects

1. Define the following terms:

   a. Temperature
   b. Pressure
   c. Density
   d. Adiabatic process
   e. Sounding

   f. Inversion
   g. Unstable equilibrium
   h. Lapse rate
   i. Isobar
   j. Potential temperature

   Include these definitions in the glossary you prepare in response to question 8 of Chapter 1.

2. Calculate the Fahrenheit temperatures corresponding to every 10°C from −40°C to 40°C, and make a graph of the relationship on graph paper. Use this graph to make a table of the Celsius temperatures for every 20°F from −40°F to 120°F.

3. If you could make a leakproof balloon of weightless and perfectly elastic material, the pressure inside it would be the same as the pressure of the surroundings, but the density inside it might be different. What could cause the difference? What would happen if the balloon were lifted to a position in the atmosphere where the pressure is lower and allowed to come to thermal equilibrium there?

4. Explain why the air pressure at high levels, for example, mountain tops, is lower than at sea level. Why is it necessary for jet planes to be "pressurized," and why are they equipped with oxygen masks in case of depressurization?

5. When air is pumped into a tire, the tire gets hot. Why? How is this related to the change in temperature of an air parcel when it descends from a higher to a lower level?

6. Consider the variation of temperature with height at various times during a clear day: (a) in the early morning after the air has been cooled by contact with the ground throughout the night; (b) in the midmorning when the ground is being heated rapidly by the sun; (c) at the time of maximum temperature, when it is presumed that there is no longer any heating of the air by the ground. Discuss the kind of stability that is present in the layer of air immediately above the ground at each time.

7. Suppose the maximum temperature on a summer afternoon is 30°C. What is the approximate temperature at 1 km at that time? If the pressure is 1010 mb at the ground, what is the approximate pressure at 1 km?

8. What is the significance of inversions with respect to air pollution?

9. Using the data in Table 4.1, compute the density of air in the standard atmosphere at 1 km, 5 km, 10 km, and 20 km. Make a graph showing the variation of density with height.

10. We saw in Section 4.2 that near sea level in the Standard Atmosphere the pressure decreases 1.2 mb when we go up 10 m. Compute the pressure change in ascending 10 m in the standard atmosphere at 1 km, 5 km, 10 km, and 20 km.

5 Water in the Atmosphere: Water Vapor, Condensation, Clouds, and Precipitation

5.1 Measures of moisture in the air

In the discussion of the composition of the atmosphere in Chapter 2, water vapor was left for later consideration because of its variability. It is this variability, associated with the fact that it is constantly being added to the atmosphere by evaporation and removed by condensation and precipitation, that makes it such an important part of the air. The most conspicuous aspects of the weather—rain, snow, hail, fog, lightning, and so forth—result from the presence of water in the atmosphere.

The amount of water vapor in the air is termed *humidity*, but there are several ways in which humidity is expressed, such as absolute, relative, and specific humidity. The most commonly used of these, relative humidity, is perhaps the hardest to understand. Its significance will become clear after we have discussed some of the other measures of moisture in the atmosphere.

We can begin with the *absolute humidity*. This term is just another name for the water-vapor density, ρ_v; that is, the mass of water vapor per unit volume of air. In other words, if we were able to take all the molecules of water vapor in a unit volume and weigh them separately, we could find ρ_v by dividing by the acceleration of gravity g. (Remember that the weight W is the force acting on a mass M due to the attraction by the earth: $W = Mg$.)

In an analogous fashion, we can think of the pressure that is exerted by the molecules of water vapor as they bounce around. While they obviously collide with nitrogen and oxygen molecules, as well as with other water-vapor molecules, it turns out that the part of the total pressure due to their motions is the same as the pressure they would exert if no other gases were present. The part of the pressure they exert is called the water-vapor pressure, usually represented by the letter e, and the law of partial pressures just stated may be written

$$p = p_d + e \qquad (5.1)$$

where p is the total pressure, and p_d is the pressure due to the molecules composing dry air.

The vapor pressure is also a convenient measure of the humidity. It is related to the absolute humidity through the equation of state [see equation (4.1)]

$$e = \rho_v R_v T \qquad (5.2)$$

where R_v, the gas constant for water vapor, has the value 461.5 J/kg K. (Since vapors near the condensation point do not behave exactly like perfect gases, this equation is not quite correct, but it is a very good approximation.)

The vapor pressure and the absolute humidity of a parcel of air change when the pressure and temperature of the parcel change, even if no water vapor is added to or subtracted from it by evaporation, condensation, or mixing. It is desirable to have a humidity variable that remains constant when the temperature and pressure change. Such a variable is the *specific humidity, q*, or its equivalent, the *water-vapor mixing ratio, w*, defined as follows:

$$\text{specific humidity } q = \frac{\text{mass of water vapor}}{\text{mass of air}} = \frac{\rho_v}{\rho}$$

$$\text{mixing ratio } w = \frac{\text{mass of water vapor}}{\text{mass of dry air}} = \frac{\rho_v}{\rho_d} = \frac{q}{1-q}$$

Since the largest values of q observed are just a few percent, the values of q and w differ at most by only this amount, and the two are used interchangeably in

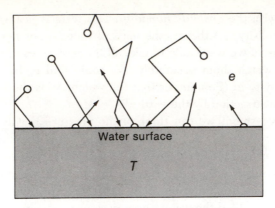

Figure 5.1 Schematic representation of water molecules leaving water surface in enclosed tank and of molecules returning. The saturation vapor pressure, e_s, which the vapor pressure in the space above the surface reaches when the number of molecules returning is the same as that leaving, depends only on the water temperature, T. [Courtesy of NOAA.]

practice. By substituting from equations (5.2) and (4.1) in these definitions, it can be shown that

$$q = 0.622 \, \frac{e}{p} \qquad (5.3)$$

and

$$w = 0.622 \, \frac{e}{p - e}$$

To discuss the relative humidity, we must explain the concept of saturation. This is because the ordinary statement that the relative humidity is the percentage of the water vapor that "the air will hold" is not correct. It has been demonstrated that in the absence of surfaces on which condensation can take place, relative humidities of more than 400 percent can be produced. When we discuss the condensation process, the reason for this will become clear.

Let us consider a sealed tank partly filled with pure water (Figure 5.1). The upper part is initially assumed to contain dry air at the same temperature. Because of the thermal agitation of the water molecules, some of them will escape from the surface (i.e., they will evaporate). As the evaporated molecules bounce around in the space above, some of them will bounce back to the water surface, and eventually the number returning to the surface will exactly equal the number leaving. When this state is reached, the vapor above is said to be in *saturation equilibrium* with the water surface, and the pressure it exerts is called the *saturation vapor*

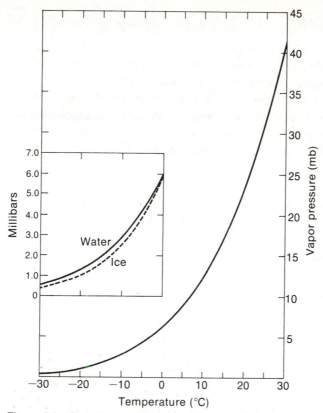

Figure 5.2   Saturation vapor pressure (*right-hand scale*) over plane surfaces of pure water as a function of temperature. *Inset:* Saturation vapor pressure (*left-hand scale*) over water and ice at temperatures below 0°C. [From H. R. Byers, *Elements of Cloud Physics* (Chicago: University of Chicago Press, 1965). Copyright © 1965 by the University of Chicago.]

pressure e_s with respect to a plane water surface at the temperature T that it has attained.

The temperature is important, for the rate at which the molecules leave the surface depends on the temperature. Thus, e_s is a function of the temperature. It depends only on the temperature, as the air pressure has practically no influence on it as long as the temperature is not so high that e_s is equal to the total air pressure above the liquid.[1] In Figure 5.2 the variation of e_s with T in the range $243 \text{ K} \leqslant T \leqslant 303 \text{ K}$ is shown. Its numerical values for various temperatures are

---

1. Actually, there is a small effect of the air pressure because the amount of air dissolved in the liquid water depends on the air pressure, and the dissolved air molecules displace some of the water molecules in the liquid surface. However, this effect amounts to less than one-half percent of the correct value.

tabulated in Appendix D. It has the value 6.1 mb (0.61 kPa) at 0°C (273 K), and approximately doubles for every 10 K up to 20°C (293 K), at which air "can hold" not quite four times as much water vapor as at 0°C. At higher temperatures the doubling takes place at somewhat larger temperature intervals.

When a liquid is heated to the temperature at which its equilibrium vapor pressure is equal to the total air pressure above it, the liquid *boils*, that is, it evaporates without restraint (except for the need to supply the heat required to change the liquid to vapor; see Section 5.4). The normal boiling point of water, 100°C (373 K), is the temperature at which e_s reaches 1013.25 mb. If the air pressure is lower—for instance, at high elevations—the boiling point temperature is correspondingly lower. The rate of decrease of the boiling point temperature with height is approximately one-third degree Celsius (0.33 K) per hundred meters.

The *relative humidity*, U, is defined as the ratio of the observed vapor pressure e to the saturation vapor pressure $e_s(T)$ for the observed temperature. It is usually expressed in percent, so that

$$U = 100\, e/e_s(T) \tag{5.4}$$

For instance, if the observed temperature is 20°C, for which we see in Figure 5.2 or Appendix D that e_s is 23.4 mb, and if the observed vapor pressure e is 12.3 mb, the relative humidity is

$$U = 100 \left(\frac{12.3}{23.4} \right) = 53\%$$

Note that the vapor pressures do not have to be expressed in SI units in this equation, as long as both e and e_s are expressed in the same units. This is not true in general; in most equations it is necessary to express all quantities in SI units (or in cgs units).

The relative humidity can be measured by use of the wet- and dry-bulb psychrometer, discussed in Section 2.3, or with a hair hygrometer (see Figure 5.3). A detailed discussion of the measurement of humidity is contained in Appendix F.

If the relative humidity exceeds 100 percent, the air is said to be *supersaturated*, and the amount of excess, expressed in percent, is called the *supersaturation*, S:

$$S = U - 100$$

Another frequently used measure of the humidity is the *dew point temperature*, T_d. This is defined as the temperature at which the air would become saturated by cooling with no evaporation into or condensation from it, and no change in pres-

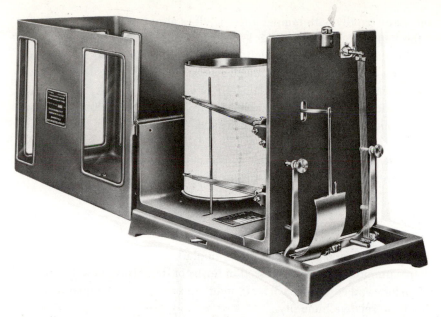

Figure 5.3 Hygrothermograph. This instrument is a combination of a thermograph, for recording temperature, and a hygrograph, for recording the relative humidity. The humidity is sensed by strands of human hair (extending from top to bottom at the right end of the instrument). The hair gets longer when the relative humidity increases. A mechanical linkage transmits the change in length to the lower arm, at the end of which is a pen that draws a line on the chart as the drum is rotated by a clock. Similarly, the upper arm is controlled by a temperature sensor. In the instrument shown, it is a curved tube filled with liquid; when the temperature rises, the liquid expands and straightens the tube. [Courtesy of Science Associates, Inc.]

sure. It is the temperature at which the observed vapor pressure is the saturation vapor pressure,

$$e_s(T_d) = e$$

In the above example, in which the observed vapor pressure is 12.3 mb, from Figure 5.2 we see that the dew point temperature is 10°C.

Corresponding to the saturation vapor pressure, we can define the *saturation mixing ratio*, which is the value of the mixing ratio for which a parcel of air having a given temperature and pressure would be saturated. Note that in contrast to the saturation vapor pressure, which is practically independent of the pressure, the saturation mixing ratio depends on the air pressure as well as the temperature. To see this we apply equation (5.3) and have for the saturation mixing ratio

$$w_s = 0.622 \frac{e_s}{p - e_s} \tag{5.5}$$

Since e_s increases rapidly with temperature, w_s does also; but for any given temperature, the higher the pressure, the lower w_s. Since pressure decreases with height, we see that a kilogram of air aloft can contain more water vapor than one having the same temperature at the earth's surface. However, since air cools adiabatically when it is lifted, a given kilogram of air lifted to greater heights will approach saturation. We shall see that this is the principal way clouds aloft are formed. But clouds at the ground, which we call *fog*, form principally in other ways.

In the example given above, in which $e = 12.3$ mb, the value of the mixing ratio if the total air pressure is 1013.25 mb, turns out to be

$$w = \frac{0.622 \cdot 12.3}{1013.25 - 12.3} = 7.6 \cdot 10^{-3}$$

The saturation mixing ratio can be computed similarly.

The relative humidity can be expressed in terms of the mixing ratio by solving equation (5.3) for e and equation (5.5) for e_s and then substituting in equation (5.4). This gives for the relative humidity

$$U = 100\frac{w}{w_s}\frac{0.622 + w_s}{0.622 + w} \cong 100\frac{w}{w_s} \tag{5.6}$$

Since w and w_s are always much smaller than 0.622, the approximation is very good. When we have the relative humidity and the saturation mixing ratio given, we can get the actual mixing ratio by solving equation (5.6), obtaining

$$w \cong \frac{U}{100}w_s \tag{5.7}$$

5.2   Condensation

When water condenses in the form of drops in the air, it is obvious that it must start with drops of very small radius, which then grow to larger drops. The equilibrium vapor pressure e_r over a small drop of radius r is larger than that over a plane water surface at the same temperature. In fact, it was shown by William Thomson (Lord Kelvin) that e_r is proportional to $1/r$. The curve in Figure 5.4 shows how the equilibrium vapor pressure over drops of radius r varies with r. For very small drops, e_r is many times e_s. C. T. R. Wilson devised the cloud chamber, which subsequently was used so effectively in nuclear physics, in order to study this effect. Wilson found that when all dust and ions were removed from air,

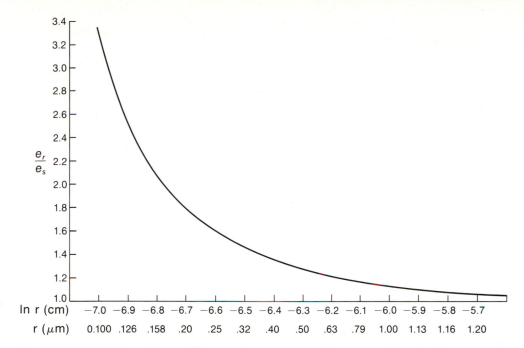

Figure 5.4   Ratio between the equilibrium vapor pressure over drops, e_r, and the saturation vapor pressure over a flat water surface, e_s, at 273 K, as a function of drop radius r. For drops larger than a few micrometers, the ratio is practically unity. For r smaller than 0.1 μm, e_r is many times e_s.

supersaturations up to 700 percent occurred without condensation taking place. Under these circumstances, ionizing radiation left tracks of droplets condensing on the ions, showing the paths of the rays.

In the troposphere large supersaturations are never observed. In fact, the relative humidity rarely exceeds 100 percent by more than a few tenths of 1 percent. The reason for this is that the air always contains a large number of solid or liquid particles that can act as *condensation nuclei*. These particles favor formation of droplets at reasonable humidities in two ways. In the first place, they provide surfaces of larger radius of curvature so that the drops that form on them are in equilibrium at small supersaturations. For instance, from Figure 5.4 we see that if there are particles larger than 0.6 μm, drops can form on them if the supersaturation is 0.2 percent; if there are nuclei larger than 1 μm, condensation on them will occur at 0.1 percent supersaturation. Actually, this is true if the nuclei are "wettable." Some substances are hygrophobic and resist condensation even if they are large enough.

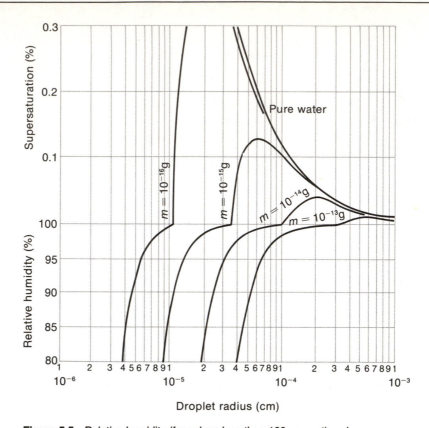

Figure 5.5 Relative humidity (for values less than 100 percent) and supersaturation (R.H. − 100) in equilibrium with drops of salt solution containing various masses, *m*, of common salt (NaCl). [From B. J. Mason, *Clouds, Rain, and Rainmaking* (Cambridge: Cambridge University Press, 1962).]

In the second place, some substances "like" water and tend to absorb it even at humidities less than 100 percent. Such substances are called *hygroscopic*. By dissolving in the water, the hygroscopic nuclei reduce the equilibrium vapor pressure below that over pure water. The combined effects of size and dissolved salt are shown in Figure 5.5 for sodium chloride nuclei of several sizes. In each case the nucleus can start growing at humidities below 100 percent. In order to grow indefinitely, the humidity must exceed 100 percent, but by amounts much smaller than would be required for a nonhygroscopic nucleus of the same size. For instance, a salt crystal of mass 10^{-15} g, with equivalent radius of 0.05 μm, which would require several percent supersaturation if it were not hygroscopic, requires only 0.13 percent supersaturation to grow indefinitely.

Hygroscopic nuclei are introduced into the atmosphere by evaporation of drops of sea spray and by combustion processes, both natural (forest fires) and artificial (domestic heating and industry). They vary in number with time and location, but in general it appears that there are enough large hygroscopic nuclei everywhere (say, more than 100/cm³) so that condensation will occur when the relative humidity exceeds 100 percent by a very small amount—perhaps less than 0.1 percent. For practical purposes it can be assumed that clouds will form whenever the relative humidity reaches 100 percent.

If the temperature is below 0°C, the condensation still takes the form of liquid drops, down to considerably lower temperatures. This is because formation of ice crystals requires special nuclei, called *ice-forming nuclei* (IN), which don't become effective until the temperature is considerably below 0°C. There are two kinds of IN: *freezing nuclei*, which cause liquid droplets to freeze, and *sublimation nuclei*, on which the vapor is deposited directly in the form of ice. (The direct transition from vapor to solid phase is called *sublimation* or *deposition*.) Sublimation nuclei are not as plentiful as condensation nuclei. The effective temperatures of most naturally occurring ice nuclei are lower than $-9°C$, and sublimation nuclei effective at temperatures higher than $-20°C$ are scarce.

As shown in the inset in Figure 5.2, the saturation vapor pressure at temperatures below 0°C is lower over ice than over supercooled water. Air saturated with respect to water is supersaturated with respect to ice. We shall see in Section 5.5 that this has importance in the formation of precipitation.

The relative humidity can increase either by the increase of the vapor content (by evaporation) or by the decrease of the saturation vapor pressure (by cooling). If the relative humidity is raised above 100 percent by either of these processes, condensation will occur. When the humidity is raised above saturation in the air immediately above the ground or sea surface, fog forms; when this occurs aloft, clouds develop. If, however, the ground itself cools below the dew point, but the air remains warmer, dew is deposited on the ground.

In a sense fog or cloud is present whenever condensation has produced water drops in the air, however small or few. However, by convention fog is defined as limiting the visibility to less than 1000 m. When the visibility is greater, the formation is called wet haze or mist, or sometimes light fog. *Dense fog* is defined as fog with visibility less than 400 m. To be perceptible against the sky, clouds must be approximately as dense as fog.

5.3 Formation of fog. Fog types and their distribution

It may seem paradoxical that evaporation can result in supersaturation and the formation of fog. There are two kinds of situations in which this takes place:

(1) When cold air flows over warm water—for instance, polar air passing over an unfrozen lake in winter—there will be rapid evaporation from the water. This

is because the vapor pressure in the cold air is much less than the equilibrium vapor pressure at the temperature of the water. As the vapor mixes with the air, it will be cooled and the vapor pressure of the mixture may exceed the saturation value for its temperature, in which case condensation on nuclei will take place. Since the air is heated from below as well as moistened, the fog that is formed takes the form of rising filaments or streamers. Because of this appearance, over the Arctic Ocean and neighboring waters this type of fog is called *Arctic sea smoke*. The more general name for it is *steam fog*. Examples of steam fog many of us are familiar with occur when cold air moves across a heated swimming pool at night or when a tub is filled with hot water in a cold bathroom.

(2) When warm rain falls through cold air, as takes place sometimes ahead of a warm front, the rain will evaporate into the cold air. This is because the equilibrium vapor pressure for the warm raindrops is much higher than the vapor pressure of the cold air, even if the air is "saturated." If the humidity of the air is high to begin with, the additional vapor will produce supersaturation when the added vapor is cooled to the air temperature. Fog produced in this way ahead of a warm front is called *warm front fog*.

The decrease of the temperature below the dew point value is the more usual way that fog forms. It can take place by radiation, by conduction and turbulent transport of heat by eddies, and by adiabatic cooling due to decrease of pressure. Fog due primarily to radiational cooling is called *radiation fog*. Because it is usually very shallow, it is sometimes called *ground fog*. Radiation fogs frequently occur in valleys, in which case they are called *valley fogs*. Fogs that are formed by cooling of warmer air passing over a cold surface, with transfer of heat from the air to the ground or sea by conduction and turbulent transport, are called *advection fogs*. When fog forms in air that is cooled adiabatically as it flows up sloping terrain, it is called *upslope fog*.

Radiation fogs occur on clear nights when the air near the ground has high moisture content, but the air aloft has very low humidity. Under these circumstances the ground cools rapidly, since the radiation emitted by the ground is only slightly offset by the radiation downward from the air above, because of the low total water vapor content of the air column. Once the ground has become sufficiently cooler than the moist air immediately above it, the moist air layer absorbs less radiation than it emits, and it cools, ultimately becoming supersaturated and producing fog. In light winds—say, 2 or 3 knots ($1–2$ m s^{-1})—the radiative cooling of the air is augmented by some turbulent transfer of heat from the moist layer to the colder ground. If the winds are stronger, however, the turbulence will be sufficient to mix warmer, drier air aloft with the air at the earth's surface, preventing the formation of fog.

The requirement for clear nights, light winds, and low humidities aloft is met principally in stagnating high pressure areas, particularly in the fall and winter. As

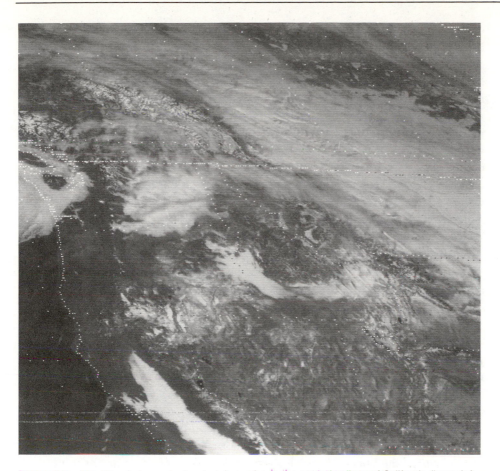

Figure 5.6   Satellite photograph showing dense fog in the central valleys of California (lower-left corner of photo). January 28, 1977. [NOAA-NESS photo. Courtesy of V. J. Oliver.]

the air cools it tends to drain into low areas, such as valleys, so that fog tends to develop there, leaving the hilltops clear. In winter the fog, once formed, may persist throughout the day because the amount of solar radiation, reduced by reflection from the fog top, may not be sufficient during the short daylight period to heat the air enough to dissipate the fog. Thus, in the valleys of the Great Basin and the interior of California, fog may be present for days or even weeks at a time. In the warmer part of the year, the nights are usually too short for cooling to be adequate for radiation fog to form, and the larger amount of insolation dissipates it within an hour or two after sunrise if it does occur.

Advection fogs, like radiation fogs, are formed by the cooling of moist air by loss of heat to a colder surface, but instead of the ground becoming colder than the air *in situ*, in advection fogs the moist air moves, or is "advected," over a cooler surface. When the air is moving rapidly enough toward a colder area to give rise to a sufficiently large temperature difference between the air and the ground, the transfer of heat from the warm moist air to the cold ground lowers the air temperature below its dew point and produces fog. Examples of advection fog take place when air from the Gulf of Mexico moves over the southern United States in winter; when air is carried northward by south winds from the Gulf Stream area of the Atlantic Ocean to the frigid waters of the Grand Banks of Newfoundland; or when air is carried from the warmer portions of the oceans over the cold upwelling waters of the Humboldt Current off South America, the Benguela Current off Africa, and the California Current off the west coast of the United States. Because the temperature contrast between ground and air is greater when the winds are stronger, advection fog can occur in moderately strong winds, in spite of the tendency of turbulent transport to dissipate it. However, if the winds are strong enough, the turbulent mixing lifts the fog off the ground to produce stratus or stratocumulus clouds instead of fog.

Upslope fog forms when moist air flows up a sloping terrain. As the air is lifted, its pressure falls, causing the temperature to decrease adiabatically to the condensation point, and further ascent up the slope produces fog. If the slope is gradual, if the moisture is present only in a shallow layer, and if the upslope component of the wind is light, the fog will not be accompanied by precipitation. For stronger flow of deep, moist air currents up steeper slopes, rain or snow will develop. In some instances when precipitation occurs, instability is released that raises the cloud above the terrain. Thus, the light easterly winds flowing up the sloping plateau east of the Rocky Mountains when a high pressure area moves southward over the Great Plains may produce upslope fog. Stronger westerly flow associated with cyclonic systems moving eastward into Canada from the Gulf of Alaska results in widespread precipitation over the Coast Ranges, the Cascades, and the Sierra Nevada. In this type of situation usually the clouds rest on the mountain slopes, but sometimes they are lifted by convective mixing so that nowhere at the ground is the visibility reduced to the 1-km limit defined as fog.

The frequency of fog throughout the world is shown in Figure 5.7. This figure shows the average number of days per year on which fog occurs. We see in it that coastal regions subject to advection fogs have high frequencies, but that in central Europe, where radiation fogs and upslope fogs in the Alps occur, the frequency is also very high.

Figure 5.8 shows the average number of days per year on which dense fogs (visibility less than 400 m) occur in the United States. Based on data from 256 weather stations, mostly at airports, the details of the distribution may not be

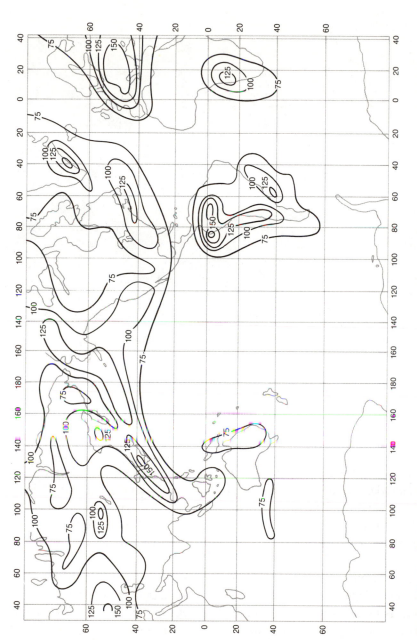

Figure 5.7 Worldwide fog frequency. Average number of days per year on which visibility is less than 1000 m sometime during the day. [From "Fog Modification—A Technical Assessment," by B. A. Silverman and A. I. Weinstein, *Air Force Surveys in Geophysics, No. 261,* March 1973.]

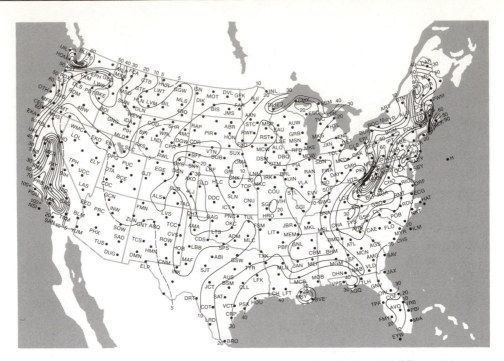

Figure 5.8 Average annual frequency (days per year) with fog that reduced visibility to 400 m or less in the coterminous United States. [From "Heavy-Fog Regions in the Coterminous United States," by R. L. Peace, Jr., *Monthly Weather Review,* Vol. 97, No. 5 (February 1969), p. 118.]

completely correct. For instance, the low values along the coast of central California may be due to the absence of stations right on the shore; the San Francisco airport station, for instance, is on the east side of the peninsula, shielded by hills from the fogs rolling in from the ocean. Again high frequencies are associated with advection fogs forming over cold waters off the Pacific coast, off New England, and in the Great Lakes region, and with upslope fogs on the Appalachian Mountains, the Sierra Nevada, the mountains of New England, and the sloping plateau from western Texas northward to western Nebraska. Another type of advection fog shows in the relatively high values along the coast of the Gulf of Mexico. In this area the fog forms when warm air flows from the Gulf over the colder land in the cold half of the year.

Most fogs are composed of liquid drops, and in most places they are warmer than 0°C. Where supercooled liquid fogs occur, in temperate latitudes in winter they can readily be dissipated by seeding. Warm fogs and ice fogs are much more difficult to disperse artificially. The problem of artificial removal of fog is discussed in Chapter 15.

5.4 Ascent of cloudy air. The saturation adiabatic process

While radiative cooling and heat transfer by conduction and turbulent exchange ordinarily produce condensation in a shallow layer close to the ground, resulting in fog or stratus cloud, when air rises the temperature falls through deep layers because of the decrease in pressure. Condensation resulting from adiabatic cooling in rising air currents is responsible for the formation of almost all clouds and precipitation.

We saw in Chapter 4 that when dry air, or air that is unsaturated, rises without gaining or losing heat, its temperature will decrease 1 K for each 100-m rise. Thus, a parcel of air that contains water vapor will cool as it rises, and if it rises far enough it will become saturated. Any further rise would produce supersaturation or, if nuclei are present, condensation. When condensation takes place the pressure change is no longer adiabatic, because *latent heat of condensation* is released.

Latent heat is the heat energy released or absorbed when a substance changes phase—for instance, from solid to liquid or from gas (vapor) to liquid. A familiar illustration is the course of events when a kettle containing water is placed on the burner of a stove. At first the temperature of the water rises as the heat is added, but once it starts boiling the temperature remains constant (373 K or 100°C at standard sea-level pressure). Heat continues to be added but the temperature does not change, the heat being used entirely to change the liquid water to water vapor (steam). It is because this heat does not show up as a change in temperature that it is called latent heat. The amount of heat required to convert liquid water to vapor at 373 K is $2.26 \cdot 10^6$ J/kg (540 cal/g). When water evaporates at temperatures lower than the boiling point, the amount of latent heat is larger. At 273 K (0°C), it is $2.501 \cdot 10^6$ J/kg. The same amount of heat is given off in condensation as is consumed in evaporation at the same temperature.

When a solid changes to liquid, a similar though smaller amount of latent heat is taken up. At 273 K (0°C), the latent heat of melting of ice (or freezing of water) is $0.334 \cdot 10^6$ J/kg (80 cal/g). If the solid changes directly to vapor without melting first, as when snow evaporates, the latent heat is the sum of the latent heat of melting and the latent heat of vaporization. It is called the *latent heat of sublimation*, totaling $2.837 \cdot 10^6$ J/kg (676 cal/g) at 273 K (0°C).

Let us now consider a rising air parcel that has cooled adiabatically to the point where it is slightly supersaturated and condensation on the nuclei in it begins. As the water condenses on the nuclei, latent heat of condensation is released and tends to offset the cooling due to adiabatic expansion. The rate of decrease of temperature of the rising cloudy air is thus less than the adiabatic process rate in the absence of condensation. To distinguish the two, we refer to the latter as the

Table 5.1 Saturation Adiabatic Process Rate (°C/100 m)

Temperature (°C)	Pressure (mb)				
	1000	850	700	500	300
40	0.30	0.29	0.27		
20	0.43	0.40	0.37	0.32	
0	0.66	0.62	0.59	0.50	0.41
−20	0.86	0.84	0.81	0.76	0.68
−40	0.95	0.95	0.94	0.93	0.90

unsaturated or *"dry" adiabatic rate of cooling,* and to the process rate with condensation as the *saturation adiabatic rate of cooling.* Unlike the unsaturated
process rate Γ, which is constant, the saturation adiabatic rate Γ_s (sometimes called
"wet" adiabatic rate) varies with the temperature and pressure. At high temperatures it is less than 0.54 K/100 m, but at very low temperatures it approaches
0.98 K/100 m, the value of Γ. The values of Γ_s for several pressures and temperatures are given in Table 5.1.

Saturation adiabatic processes can be represented by lines on thermodynamic
diagrams similar to the unsaturation process curves. Since the saturation adiabatic
rate of cooling is smaller, the saturation adiabatic curves, called *saturation adiabats*
or *wet adiabats*, lean less toward lower temperatures. As they go upward toward
lower pressures and temperatures, they become more nearly parallel to the dry
adiabats.

In Figure 5.9 some saturation adiabats are shown, in addition to the other lines
described in Section 4.4, on another type of thermodynamic diagram called the
Skew T-Log P Diagram. This diagram is similar to the one previously described,
except that the isotherms slope upward to the right, so that the angle between the
isotherms and the dry adiabats is about 90°. Since most actual sounding curves and
process curves lie between the isotherms and the dry adiabats, it is advantageous
to use a diagram on which this angle is large. The isobars, as before, are horizontal
lines, and the dry adiabats slope upward to the left. The saturation adiabats are
the curves sloping slightly to the left near the bottom of the diagram and curving
farther to the left as they go up.

In the same way that the potential temperature serves to identify the dry adiabats, the saturation adiabats are labeled by the temperature at which they intersect the 1000-mb isobar. This temperature is called the *wet-bulb potential temperature* of any saturated air parcel whose representative point falls on the
saturation adiabat.

Suppose that a parcel having a temperature of 15°C, pressure of 1013 mb, and
relative humidity of 58 percent moves upward. The representative point for its
initial condition is point *A* in Figure 5.9. As it rises the representative point will

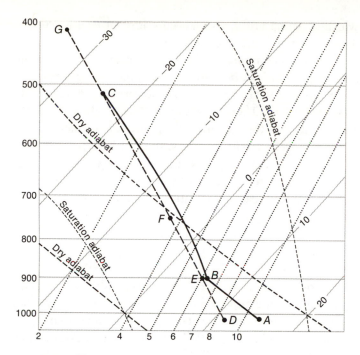

Figure 5.9   Thermodynamic diagram (Skew T–Log P diagram), showing saturation adiabats (*curved dashed lines*) and saturation mixing ratio lines (*dotted lines*). The process that a parcel with an initial temperature of 15°C, a pressure of 1013 mb, and a mixing ratio of $6.1 \cdot 10^{-3}$ would experience if lifted adiabatically is represented by curve *ABC*.

follow the dry adiabatic process curve *AB* until it is saturated, and then the saturation adiabat *BC*.

How do we know it will become saturated at *B*? For this purpose meteorological thermodynamic diagrams have still another set of lines on them, the *vapor lines*, drawn for constant values of saturation mixing ratio w_s. These are shown in Figure 5.9 by dotted lines sloping upward to the right, forming small angles with the isotherms. They are labeled in parts per thousand, frequently expressed as "grams per kilogram." Thus, saturated air at point *A* would have a mixing ratio of $10.5 \cdot 10^{-3}$. Since the relative humidity is 58 percent, by applying equation (5.7) we obtain for the actual mixing ratio

$$w = \frac{58}{100} \cdot 10.5 \cdot 10^{-3} = 6.1 \cdot 10^{-3}$$

When the air has risen adiabatically until its pressure and temperature have the values at which $w_s = 6.1 \cdot 10^{-3}$, which occurs at point B, it is saturated, and from there on it will follow the saturation adiabat.

The pressure and temperature at which the saturation is reached in this fashion are called the *lifting condensation pressure* and the *lifting condensation temperature* of the air represented by point A, and the corresponding height is the *lifting condensation level* (LCL). These values are determined by the initial properties of the air parcel. In turn, they determine the saturation adiabat that the parcel at A would follow if it rises sufficiently. Thus, the wet-bulb potential temperature it would have once it became saturated is determined by the initial properties of the air parcel, and it is considered to have that wet-bulb potential temperature whether or not it actually rises and reaches saturation. We can see that the wet-bulb potential temperature of an air parcel is conserved (does not change) for any adiabatic process, unsaturated or saturated. The potential temperature, on the other hand, changes during saturation adiabatic processes, although it is constant for unsaturated adiabatic processes.

Since the rate of change of temperature when cloudy air rises is different from that of unsaturated air, the criteria for stability of a saturated air column are different from those for unsaturated air. The same type of reasoning as that used in discussing Figure 4.6 leads us to the following criteria for saturated or cloudy air.

If $\gamma < \Gamma_s$, the air column is in stable equilibrium.
If $\gamma = \Gamma_s$, the air column is in neutral equilibrium.
If $\gamma > \Gamma_s$, the air column is in unstable equilibrium.

Note that since $\Gamma_s < \Gamma$, these criteria lead to the possibility that an air column would be in stable equilibrium if it were unsaturated and in unstable equilibrium if it were saturated. This state is called *conditional instability*; it occurs if γ has a value between Γ_s and Γ, that is, if $\Gamma > \gamma > \Gamma_s$.

Because Γ_s varies, it is not easy to apply these criteria directly. The convenient procedure is to plot the sounding curve on a thermodynamic diagram and compare its slope for the saturated portions of the air column represented with the saturation adiabats.

Suppose that the observed sounding is represented by the dashed curve $DEFG$ in Figure 5.9. By comparing the sounding curve with the saturation adiabats, we see that below point F, $\gamma > \Gamma_s$ and above point F, $\gamma < \Gamma_s$. If the air at any place in the column represented by DF is saturated, it is unstable. The air at any position in the column represented by FG is stable, whether or not it is saturated.

Suppose, further, that the air represented by point D has a mixing ratio of $6.1 \cdot 10^{-3}$ and is heated from 11°C to 15°C (by contact with the heated ground). It then will be represented by point A, and a slight upward push would cause it to rise and be continuously accelerated because it would be warmer than its surroundings. Its representative point would follow the dry adiabat AB as before. At the elevation represented by point B, condensation would begin, and the representative point of the air parcel would follow the saturation adiabat BC. When it reached C, it would be in equilibrium with the surroundings, and further ascent would subject it to a downward buoyant force. By this process we would expect a cloud to be formed that would extend from about 900 mb (1 km) to about 500 mb (5.5 km). (We shall see in the next chapter that the process of entrainment influences the growth of clouds produced in this fashion.)

The actual process of heating by insolation is more complicated than postulated above. The air next to the ground does not remain there until it has warmed several degrees, except where the ground is extremely flat and uniform, such as the beds of dry lakes in the desert. In most places the irregularities of the terrain together with the fluctuations of wind tend to cause upward motions as soon as small differences in temperature develop between the surface air over the areas of preferred heating and areas where the heating is not so great. When this happens there will be up-and-down motions that will mix successively deeper layers of air that are affected by the heating. The rising parcels will be replaced by cooler parcels, which in turn are heated by the ground. The process will tend to establish constant potential temperature and constant mixing ratio through the mixed layer.

In Figure 5.10 is shown, in addition to the sounding curve $DEFG$, a curve $D'E'F'$ that represents the observed mixing ratio at each level, plotted with respect to the vapor lines. Since the points in temperature coordinates are the values of the temperature at which air having the observed mixing ratio would be saturated, the curve is also the *dew point curve* for the sounding. The mixing ratio decreases somewhat with height up to 900 mb, and then more rapidly. The heating, which successively establishes constant potential temperature layers HI, JK, LM, and NO, tends to produce a constant value of mixing ratio through each layer, namely, the average of the initial values in the layer. A convenient way of determining this average is to draw the vapor line from the base to the top of the layer in such a way that equal areas are on each side of the line between it and the observed mixing ratio curve. The average mixing ratio line for the layer NO is shown, with the areas on the two sides shaded. Note that after being heated and mixed, the air at O is saturated; a little further heating would cause this air to rise saturation adiabatically. In the example shown, a cloud would form with base at point O 882 mb, about 1150 m, and top extending at least to P, 540 mb, about 5000 m. The result of mixing during heating is a cloud with somewhat higher base and lower top than given by the earlier example, in which it was assumed that the surface air was heated to about the same temperature without being permitted to rise until it reached that temperature. In the more realistic case, the mixing leads to a lower mixing ratio by the time the air becomes saturated; that is why it has a higher cloud base and lower cloud top. Still more realistic consideration would

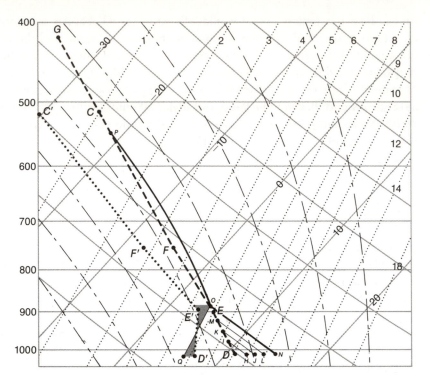

Figure 5.10 Skew T-Log P diagram showing the change in sounding curve *DEFC* due to heating from below. As the heating progresses, first a thin layer is changed from DI to an adiabatic lapse rate, HI; then successively thicker layers become adiabatic. In each adiabatic layer, the mixing ratio tends to become constant due to mixing. When the layer up to *O* becomes adiabatic, the mixing ratio profile changes to *QO*, a constant mixing ratio of $5.8 \cdot 10^{-3}$.

take into account some increase in the mixing ratio of the lower layers due to evaporation at the ground. This might lead to heights of cloud base and cloud top between the two estimates.

The cloud base in the first example is at the lifting condensation level of the air at the surface. This model indicates that clouds will form whenever the heating is adequate to raise the temperature of the surface air sufficiently so that its representative point at the LCL falls to the right (on the warm side) of the sounding curve. In the second example, cloud formation requires that the average mixing ratio for the layer that would be mixed by heating be equal to the saturation mixing ratio at the top of the mixed layer. By averaging the mixing ratio through successively deeper layers it can be determined whether such a layer exists, and if so at what position in the sounding it occurs. That position is called the *convective condensation level* (CCL).

Another type of situation is one in which heating is not present but the wind is strong enough and the terrain uneven enough to produce turbulent mixing of an increasingly deep layer of air next to the ground. As before, the mixed layer attains a constant potential temperature (dry adiabatic lapse rate) and constant mixing ratio; but without heating, the potential temperature becomes the average of the original values through the layer, and so does the mixing ratio. The replacing of a stable lapse rate with an adiabatic lapse rate raises the temperature in the lower part of the layer, but lowers it in the upper part. This decrease in temperature aloft, together with the increase in mixing ratio there when the usual decrease in it with height is replaced with a constant value, may result in saturation and condensation. The pressure, temperature, and height at which saturation is reached by turbulent stirring are called the *mixing condensation pressure, mixing condensation temperature*, and *mixing condensation level* (MCL).

The reason for the lifting of advection fogs when the winds are sufficiently strong, which was mentioned in Section 5.3, can now be seen. Even though heat is lost by the air to the colder ground or sea surface, the turbulent mixing warms the lower part of the mixed layer, raising the condensation level so that stratus or stratocumulus clouds are present instead of fog.

The above discussion provides the basis for explaining the various kinds of clouds. When stable air is lifted uniformly over large areas, as in flow up a warm front, layer clouds form, such as cirrostratus, altostratus, and nimbostratus. If the air is unstable, the layers break up into cumuloform puffs or rolls, cirrocumulus, altocumulus, or stratocumulus. Air that is rendered unstable due to heating as the sun heats the ground or the air moves from colder to warmer regions gives rise to cumulus, cumulus congestus, or cumulonimbus clouds. Mechanical mixing by turbulence without heating or with cooling from below produces low layer clouds—stratus or stratocumulus, the latter if there is some instability. Instability that causes the clouds to take the form of stratocumulus sometimes arises due to radiation from the top of the clouds.

The ways the stability of the air can change, and the relation of the cloud types to flow patterns, will be discussed in Chapter 6.

5.5   Development of precipitation

In the first example in the previous section, we saw that the air parcel having an initial mixing ratio of $6.1 \cdot 10^{-3}$ would undergo condensation as its rose. When it reached point C, its mixing ratio would be the saturation mixing ratio there, which we see from Figure 5.10 to be $1.0 \cdot 10^{-3}$. Thus, a total of 5.1 g of water would have condensed out of each kilogram of rising air. At the pressure and temperature represented by C, the air density is 0.7 kg/m^3, so that the liquid content of the cloud that formed, assuming none fell out of the parcel, would be 3.6 g/m^3 or $3.6 \cdot 10^{-6}$ g/cm^3. By assuming a reasonable value for the number of nuclei on which the condensation occurs, we can estimate the average size of the drops in the cloud. For instance, if 100 nuclei per cubic centimeter were activated, each drop would have a mass of $3.6 \cdot 10^{-6} \div 100 = 3.6 \cdot 10^{-8}$ g. Since the density of water is unity, the average volume of each drop, $4/3 \, \pi r^3$, is equal to $3.6 \cdot 10^{-8}$. From this we find that r, the average radius, is equal to 20.5 μm. As we shall see, a drop of this size is too small to fall as precipitation. Other processes are required to produce precipitation-sized particles.

One might wonder why all clouds do not precipitate. The water drops (or ice crystals) are heavier than the air and should fall. The explanation lies in the fact that the drops in fog and clouds are so small that they fall very slowly through the air. If the air is moving upward (and most clouds form because of upward motion), the air movement will more than compensate for the downward motion of the drops. Only when the drops have grown large enough to have a large velocity of fall will they overcome the upward air motion and reach the ground as rain.

Again, this difference in fall velocity between small and large drops may be puzzling. Galileo demonstrated that gravity acts to produce the same acceleration on all bodies, light or heavy. In a vacuum, small and large drops would behave alike. But the resistance of the air to the motion depends on the size, as well as the speed, of the falling body in such a way that the resistance increases more rapidly than the force of gravity as the size and weight get larger. The result is that there exists a speed, called the *terminal velocity*, at which the force due to air resistance equals the force of gravity. When the speed of the falling body reaches this value, the forces are in balance, and the body is no longer accelerated but falls thereafter at a constant speed. In general, the larger and heavier the body, the greater its terminal velocity.

For water drops, the terminal velocity depends only on the radius (or equivalent radius for large drops that are distorted from spherical shape when falling). It varies directly as the square of the radius for small drops, and approximately as the first power of the radius for larger drops. Table 5.2 gives some representative terminal velocities.

We note that a drop having a radius of 5 μm would fall (in the absence of air motion) about 12 meters in an hour. It would take more than 24 hours for it to fall from a cloud base at 300 m (1000 ft) to the ground. In a much shorter time it would evaporate if the air below the cloud were not saturated. On the other hand, a drop having a radius of 1 mm would fall 300 meters in less than a minute, and it would take large updrafts to offset its falling.

We see that if clouds consist entirely of small drops, with radii of less than 20 μm, very slight upward motion of the air will prevent precipitation; but if they contain drops larger than one- or two-tenths of a millimeter, precipitation is likely.

Table 5.2   Terminal Velocities of Water Drops in Air

Radius (μm)	Velocity (cm s^{-1})	Radius (mm)	Velocity (m s^{-1})
1	0.012	0.25	2.06
5	0.30	0.5	4.03
10	1.2	1	6.49
20	4.8	1.5	8.06
50	27	2	8.83
100	72	2.5	9.09

(Note the change in units between the left and right pairs of columns.)

Observations show that clouds do indeed consist of small drops. For all types of liquid clouds, the most frequent drop radius is between 5 and 10 μm. There are smaller and larger drops but few, if any, are larger than 20 μm. The number of drops ranges between 30 and 300 per cubic centimeter, with 100 per cubic centimeter being a representative value.

It is the large number of drops that leads to their small size by the condensation process. For instance, even if all the water vapor were condensed out of fairly moist air—say, with a mixing ratio of 10^{-2}—the average radius would be 30 μm, if 100 drops per cubic centimeter were formed. In actuality, only a fraction of the water vapor in the air condenses, and consequently the drops are smaller than that. Condensation does not lead to drops large enough to precipitate because condensation nuclei that can be activated at small supersaturations are always present in large numbers.

Another factor is the rate at which condensation can take place. While the initial condensation on nuclei forms a cloud quite quickly once saturation is attained, computations show that the rate of growth decreases as the size increases, and it would take several hours to form precipitation-sized drops by condensation, even if there were few enough condensation nuclei present. Since showers sometimes take place within one-half hour of the first formation of a cloud, it is clear that other processes must be responsible for the growth of cloud droplets to raindrops.

The amount of growth required may be seen by considering how many typical cloud drops are required to make a typical raindrop. We have seen that cloud drops are mostly in the 5–10 μm radius range, while raindrops are of the order of 1 mm (1000 μm). Since the mass or volume varies with the cube of the radius, we see that it would take about *one million* cloud drops to form a single raindrop.

The two processes that have been found to be important in the growth of precipitation particles are the collision–coalescence process (the "warm cloud" process) and the three-phase process (the Bergeron process).

The collision–coalescence process is the easier to understand. Since clouds consist of drops of various sizes, the larger drops will be falling relative to the air

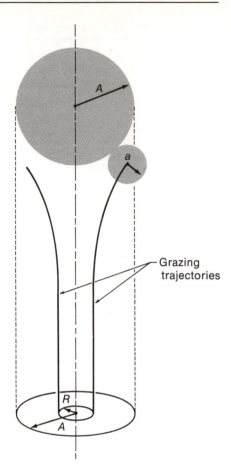

Figure 5.11   Grazing trajectories of a small drop relative to a large drop, illustrating the concept of collision efficiency. The linear collision efficiency, y_c, is defined as the ratio R/A.

faster than the smaller ones. If a smaller drop is directly below a large one, the large one will catch up with the smaller drop and should be expected to collide and combine with it. After that, being still larger, it will fall still faster, collecting more and more small drops as it falls, and eventually (after a million collisions) it will come out of the cloud as rain.

Even this process, however, is more complicated than it seems on the surface. For as the large drop falls through the air, it pushes the air ahead of it out of the way. This air moves aside and tends to drag the small drop aside with it. Thus, even though the small drop is initially in the path of the large one, the small drop may be carried around the large drop by the air rather than collide with it.

The situation is illustrated in Figure 5.11. Here the path of the small drop is pictured approaching the large drop, the way it would appear if one were moving with the large drop. The paths shown are those for which the droplets would just touch. Should the small drop start farther away than R from the axis of fall of the

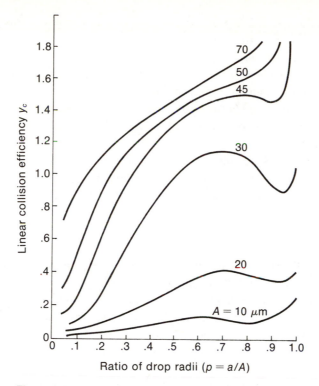

Figure 5.12   Linear collision efficiency y_c for various values A of the radius of a large drop of water falling relative to a small cloud drop of radius a, as a function of the size ratio, $p = a/A$.

large drop, the drops would miss each other. The ratio R/A, where A is the radius of the large drop, is called the linear collision efficiency y_c. It turns out that y_c depends on A and on the ratio $p = a/A$, where a is the radius of the small droplet. Figure 5.12 shows the computed values of y_c for various values of A.

The fraction of small drops collected, the collision efficiency, is proportional to y_c^2. We see from Figure 5.12 that when A is less than 20 μm, the collision efficiency is extremely small. Thus, it would appear that drops larger than 20 μm in radius are needed to initiate precipitation. Since most clouds formed by condensation do not contain drops of this size, it is necessary to call on the second process, the three-phase process, to explain most occurrences of rain. It is called the three-phase process because it requires the presence of ice crystals in addition to liquid drops and water vapor. As such, it can occur only at temperatures below 0°C. In some cases precipitation starts from clouds that are warmer than 0°C throughout their extent. These exceptional cases must have fewer condensation nuclei or some very large condensation nuclei (giant salt nuclei) to lead to the formation of drops larger than 20 μm by condensation.

The three-phase process requires the simultaneous presence of water drops and ice crystals at temperatures below "freezing." The occurrence of supercooled water drops is quite common. In fact, it is unusual for clouds to consist of ice rather than liquid at temperatures between 0°C and − 10°C. Between − 10°C and − 20°C, liquid and ice-crystal clouds occur, and below − 20°C clouds are generally composed of ice particles.

The reason why liquid clouds occur at "sub-freezing" temperatures is similar to the reason why condensation in dust-free air requires great supersaturations. In the absence of appropriate nuclei, ice will not form at temperatures above about − 40°C. Under ordinary conditions in lakes and streams or in the ice-cube trays in a refrigerator, freezing nuclei are present or the freezing is nucleated at the solid shores of the lake or at the walls of the tray. In a drop floating in the air, nuclei that are effective near 0°C are less likely to be present.

As discussed in Section 5.2, ice nuclei are of two types: freezing nuclei, which initiate the change from liquid water to ice, and sublimation nuclei, which facilitate the change from vapor to solid phase. In both cases the nuclei are temperature dependent. The most common natural nucleus, kaolinite (a form of clay), is effective at − 9°C, and this explains why clouds frequently remain liquid at least down to this temperature.

The number of ice nuclei present in the atmosphere that are effective at temperatures higher than − 20°C tends to be much smaller than the number of condensation nuclei effective at small supersaturations. Whereas the latter is of the order of 100 per cubic centimeter, the number of ice nuclei is typically one per liter. Thus, even at temperatures at which ice nuclei are effective there would initially be only one ice crystal amid thousands or hundreds of thousands of liquid drops.

At temperatures below 0°C, the equilibrium vapor pressure over ice is lower than that over water. In Figure 5.2, the inset shows the difference. Curve a is for liquid water and curve b is for ice. If air at − 10°C is saturated with respect to liquid water, it will be supersaturated with respect to ice.

Let us now visualize the situation in which an ice crystal forms in the midst of a myriad of liquid drops. The air initially is saturated with respect to the liquid drops and is thus considerably supersaturated with respect to the ice crystal. Therefore, more vapor molecules will condense on the crystal than will evaporate from it and the crystal will grow. As the vapor condenses, the vapor pressure is lowered below the saturation with respect to the drops, and the drops start evaporating. This process, by which the vapor is evaporated from the drops and condensed on the ice crystal, is called the three-phase process, or, after the Swedish meteorologist who first pointed out its importance, the Bergeron process. As this process continues, the water is rapidly transferred to the ice crystal from the many surrounding drops. The crystal grows much larger than the drops and starts falling relative to them. At this time, in addition to growing by the three-phase process, it will grow by collision and coalescence, once it has grown larger than the minimum size for nonzero collision efficiency.

The fact that most occurrences of precipitation require the three-phase process to initiate them, and that this usually occurs naturally only when the cloud tops are high enough for their temperatures to be below $-10°C$ or $-15°C$, has led to attempts to stimulate precipitation and augment its amount artificially. Vincent Schaefer discovered that ice nucleation could be initiated by dropping solid carbon dioxide (dry ice) pellets through a supercooled cloud, and Bernard Vonnegut found that silver iodide particles were effective ice nuclei at temperatures as high as $-4°C$. Many cloud-seeding experiments have been carried out, both with dry ice dropped from airplanes and with silver iodide smoke generators operated at the ground or on planes. In general, it may be said that while some of these experiments appear to have increased the amount of precipitation, we do not yet have adequate knowledge to discriminate between the conditions under which seeding will augment precipitation and those under which seeding will suppress cloud development and decrease precipitation. (We know that circumstances of both types can exist.) Weather modification is discussed in more detail in Chapter 15.

5.6   Forms of precipitation

Precipitation formed by the three-phase process is initially in the form of ice. When the temperature at all levels right down to the ground is below 0°C, or at the highest only slightly above 0°C, it remains frozen; but if the temperature of the lowest layers is more than a little above 0°C, the particles usually melt and form raindrops. An exception is hail, which consists of pellets of ice that fall so rapidly that they are still solid when they reach the ground, even at fairly high temperatures.

Precipitation includes all forms of water particles that fall to the ground. The following are some of the forms it takes.

Drizzle is a fine mist of tiny liquid drops, smaller than 0.5 mm in diameter, which fall from stratus clouds, usually as a result of the warm cloud process. The amount of precipitation resulting from drizzle is very small, accumulating at most at the rate of 1 mm/hr.

Rain is the name given to all liquid precipitation other than drizzle. By definition the smallest raindrops are 0.5 mm in diameter. There is a natural limit to the largest size. Drops larger than 5 mm in diameter are unstable and break up into

A   B

Figure 5.13 Instruments for measuring precipitation. (A) The standard rain gauge used by the U.S. National Weather Service for periodic observations—for instance, the total daily rainfall. Precipitation that has fallen into the gauge is transferred to a measuring tube with a cross-sectional area one-tenth that of the collector, so that the depth is magnified ten times for measurement. When continuous records are required, a recording rain gauge is used, of which the weighing rain gauge shown in (B) is one example. The precipitation drains from the collector into a container that rests on a scale. The weight of the precipitation raises the pen arm, which records the amount on the chart on the clock-driven drum. [Courtesy of Science Associates, Inc.]

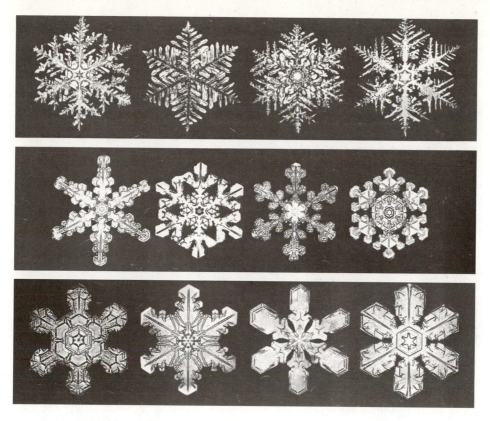

Figure 5.14 Snow crystals in the form of branched hexagonal plates or dendrites. [Courtesy of NOAA.]

smaller ones as they fall. As shown in Table 5.2, the terminal velocity of drops having a diameter of 5 mm is 9 m s^{-1}. In clouds in which the updrafts exceed this speed, no drops can fall toward the ground. Steady, continuous rain falls from layer clouds that are associated with fronts and cyclones. Showers, which consist of rain with rapid changes in intensity and sudden starts and stops, fall from convective clouds. Rain is characterized as light if it falls at a rate less than 0.5 mm/hr, moderate from 0.5 to 4 mm/hr, and heavy if greater than 4 mm/hr.

Snow is the general designation for precipitation of opaque or semiopaque ice particles in the form of individual crystals, small pellets, or flakes formed by aggregation of crystals. While the other forms are not uncommon, in most snowstorms the precipitation is in the form of flakes. Snowflakes range in size from a few millimeters to several centimeters. At low temperatures they tend to be small; at temperatures near 0°C they may be very large.

Ice crystals is the designation for precipitation in the form of single, individual crystals. They may fall from stratus clouds, or they may appear to come from a clear sky, falling as fast as they form in the rapidly cooling air. In the latter case, the designation "diamond dust" is used because the crystals sparkle in the light.

Snow pellets, also called *soft hail* or *graupel*, are white opaque spherical particles having a snowlike character. They are crunchy and easily crushed, in contrast to ice pellets and hail, which are hard ice. They usually occur in snow showers falling from convective clouds, particularly in mountainous areas.

Snowfall is measured both in the actual depth of snow accumulating and in equivalent liquid content. Depending on the type of snow, usually determined by the temperature, the liquid content of snow may vary greatly. Light fluffy snow may be as much as twenty times as deep as the equivalent liquid, while heavy "wet" snow may be as little as six times its depth when melted. A rough average value is that snow is about ten times the depth of the equivalent rain. The amount of precipitation given in weather reports is the equivalent melted depth.

Ice pellets are small transparent or translucent quasi-spherical particles of ice. They are of two types. One, which is also called *sleet,* is formed by the freezing of small raindrops as they fall. In the situation in which sleet occurs, a layer of air having temperatures above 0°C overlies a colder layer near the ground. Precipitation, formed at greater heights in the form of ice crystals, melts in falling through the warm layer, and then the drops formed thereby freeze while falling through the cold air below it. A "sandwich" of warm air like this is present sometimes at warm fronts. More serious from the standpoint of damaging weather are the situations that cause *ice storms*. In these situations, the vertical temperature distribution is similar to those that produce sleet, but the subfreezing air at the ground is not deep enough or the raindrops are too large for the drops to freeze as they fall. In that case, the rain freezes on everything it strikes on the ground. The resulting *glaze* on roads and walks creates a serious hazard to travel, and tree branches and utility wires sag and break because of the weight of the accumulated ice.

The other form of ice pellets is called *small hail*. It is formed when ice crystals or small snow pellets fall through supercooled cloud drops sufficiently rapidly to accumulate a coat of clear ice.

Hail is precipitation in the form of quasi-spherical or irregular lumps of ice. In size hailstones range from 5 mm—smaller particles are by definition termed ice pellets or snow pellets—to 10 cm or more in diameter, the most frequent size being about 1 cm. Hail forms in regions of very strong updrafts and high content of supercooled liquid in cumulonimbus clouds. The hailstones are composed of alternate layers of clear ice and opaque ice, the latter rendered opaque by the presence of numerous small air bubbles. The layered structure is attributed to differences in the rate of accumulation and freezing of the supercooled water.

Figure 5.15 A hailstone measuring more than four inches across. This hailstone fell during a thunderstorm at Weatherford, Oklahoma, July 1, 1940. [Courtesy of NOAA.]

When the hailstone falls through air with high liquid content the water accumulates faster than it can freeze, and a coat of liquid forms, which becomes a layer of clear ice when it freezes. When the hailstone falls through air with smaller and less numerous cloud drops, they freeze immediately on impact, trapping bubbles of air as they freeze.

When storms with moderate to large hailstones occur, they cause damage to crops, to structures, and to vehicles. In addition to attempts at weather modification aiming to increase precipitation where more rain is needed, experiments have been carried out to try to reduce or eliminate crop-damaging hail. These experiments will be discussed in Chapter 15.

Questions, Problems, and Projects

1. What is meant by saturated air? Why doesn't condensation occur when the relative humidity is slightly in excess of 100 percent in the absence of particles suspended in the air?

2. Suppose that the relative humidity is 100 percent and the temperature is 10°C at sunrise, and that during the day the air is heated but its water vapor content does not change. What will the relative humidity be when the temperature reaches 20°C? (*Hint:* Can you get the data you need from Figure 5.2 or Appendix D?)

3. a. From Figure 5.4 find what would be the relative humidity in equilibrium with a pure water drop having a radius of 0.6 μm.
   b. What would happen if the ambient humidity had this value and the drop grew slightly larger, say, to 0.61 μm? What would happen if it became slightly smaller?
   c. From Figure 5.5 find what relative humidity is needed to enable a condensation nucleus containing 10^{-15}g of NaCl to grow indefinitely?

4. Why does the vapor pressure change when an unsaturated air parcel rises adiabatically without mixing with its surroundings? Does its mixing ratio change too? What about its relative humidity?

5. Why is the saturated adiabatic process rate of cooling smaller than the dry (unsaturated) rate? Why is the difference between them smaller at lower temperatures?

6. In Figure 4.8 a typical vertical temperature sounding for Santa Monica is represented. Suppose you heated a parcel of air at the ground to 25°C, and then released it. To what height would it rise if it moved adiabatically without mixing until it came into equilibrium with the surrounding air? Suppose that as you heated it you added moisture until it became saturated and then released it. To what height would it rise in that case? (*Hint*: An approximate value for the saturation adiabatic process rate can be obtained from Table 5.1 by interpolation, assuming the surface pressure to be about 1000 mb.)

7. What is the difference between cloud drops and raindrops? Discuss the ways in which raindrops can form in a cloud.

8. Why doesn't a falling large drop in a cloud collect every smaller drop in its path?

9. A parcel of air having a temperature of 281 K, relative humidity of 50 percent, and pressure of 101.0 kPa at sea level is lifted adiabatically to a height of 1 km, where the pressure is 89.7 kPa. Find its relative humidity at 1 km. (*Hint*: What humidity property remains constant in an [unsaturated] adiabatic process? What is the temperature of the parcel at 1 km? Can you find the value of the humidity property and its saturation value at 1 km?)

Convection. Cumulus Clouds, Thunderstorms, and Tornadoes

6

6.1 The development of convective clouds and thunderstorms

In Chapter 1 a preliminary description of the development of convective clouds was presented. We are now in a position to discuss this phenomenon more completely. We know, for instance, that on a clear morning the heating of the ground by solar radiation leads to destabilization of the layer of air next to the ground, that is, to the development of a lapse rate γ greater than the dry adiabatic lapse rate Γ. Due to unevenness in the amount of heating because of differences in the character of the surface, or due to differences in flow, disturbances occur that release the instability and produce upward and downward motions. These motions are organized in either *cells* or *rolls*, the nature and scale of which depend on the rate of heating and the prior state of flow of the air.

That vertical motions start before clouds develop in them is readily seen at times. If you are hiking on a dusty trail with the sun beating down on you, the discomfort is increased by the clouds of dust that are stirred up by those ahead of you on the trail and carried upward in rising currents of heated air. Similarly, the plowing of fields and trucks traveling on dirt roads raise swirls of dust to considerable heights, and dust devils, small whirlwinds that are rendered visible by the dust rising at their center, occur over heated fields, particularly in the desert. All these provide evidence of upward motions caused by thermal instability.

The nature of the organization of these dry thermals into cells and rolls was pointed out by Alfred Woodcock a number of years ago. He observed that sea gulls seek out the rising air currents to soar in. By remaining in the updrafts they can soar gracefully for long periods without moving a wing. Woodcock noticed that when the wind is light the gulls soar in circles, but when the wind exceeds a certain critical value (about 6 m s^{-1}), the pattern of their soaring changes to one in which they travel in lines along the direction of the wind. He interpreted this change in soaring pattern in terms of the type of pattern of convection—cells at low wind speeds and rolls oriented along the wind when the winds are sufficiently strong.

If the rising air is humid enough it will reach the condensation level and clouds will form in the shape of cells or helical rolls in accord with the structure of the convection that is present. Depending on the stability of the air above the layer that is heated from below, the clouds will be shallow or deep. If the lapse rate there is less than the saturation adiabatic ($\gamma < \Gamma_s$) only *cumulus humilis*, "humble" cumulus clouds of fair weather, will occur. If $\gamma > \Gamma_s$ through a deep layer, the cumulus will grow upward into cumulus congestus and ultimately into cumulonimbus, with lightning, thunder, heavy rain, and sometimes hail. In some instances the instability (with respect to saturated air) extends throughout the troposphere, and the cumulonimbus tops reach the tropopause or penetrate into the stratosphere.

Figure 6.2 shows an instance of the growth of cumulus into cumulus congestus and cumulonimbus. The photographs were taken at intervals of about eight minutes from a distance of about 70 km. In photograph A cumulus clouds in the shape of a roll or a group of rolls are present. Photograph B shows some cumulus congestus towers extending upward at several places. In C, the towers have grown, but the motions are becoming more organized, with the two towers on the right dominating. In photograph D, the right-hand tower has become glaciated (the particles at the top have turned to ice), with characteristic smoothness having replaced the bulbous cauliflower configuration, and the other towers appear to be less energetic. In E and F the anvil shape is evident, and the entire cumulonimbus tower continues to rise. Throughout the development some small clouds remain at about the height of the original cumulus tops.

As Benjamin Franklin demonstrated in his famous kite experiment in 1752, lightning is an intense discharge of electricity. The upper part of a cumulonimbus cloud develops a large net positive charge, and the lower part an equal net negative

Figure 6.1   Lightning discharges in a thunderstorm. [Courtesy of NOAA.]

charge. There are several ways that this accumulation of charge (charge separation) can be explained, most of them associated with the presence of ice crystals in the cloud, and it has not been determined which is the correct one. In fact, it seems likely that several or all play a role in charging thunderstorms. The discharge occurs when the potential difference, or voltage, has become sufficiently large. The discharge can occur within the cloud, from one cloud to another, or from cloud to ground. It is the cloud-ground strokes, of course, that cause most of the damage, injuries, and deaths, although in a few instances lightning has damaged airplanes in flight.

A

B

C

Figure 6.2   Example of growth of cumulus into cumulonimbus. [Courtesy of University of Chicago Cloud Physics Laboratory.]

D

E

F

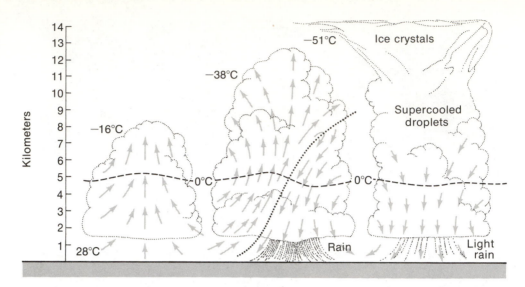

Figure 6.3 Stages in the development of a thunderstorm cell: left, cumulus or growing stage; center, mature stage; right, advanced or dissipating stage.

The thunder is the sound caused by the sudden expansion of the air that is heated by the electric current. The sound waves from the nearest part of the lightning stroke reach the observer first, and those from the more distant parts reach him or her in successively later moments, so that thunder usually takes the form of a sudden clap from the near part of the stroke followed by rolls or rumbles from the parts of the stroke that are farther away. Since sound travels approximately 330 m s^{-1}, each 3 seconds between the time the stroke is seen and the arrival of the first thunderclap corresponds to 1-km distance to the nearest part of the stroke, and each 3 seconds duration of the rumbling of the thunder would correspond to 1-km length of the stroke if the stroke were directed along a straight line from the observer. Usually the stroke moves along other paths, and the actual length is greater than this. For example, thunder that was first heard 6 seconds after the lightning flash and lasted 12 seconds indicates that the nearest part of the stroke was 2 km away and it was at least 4 km long.

Investigations using airplanes that pass through the clouds at several levels at the same time have shown that a well-developed thunderstorm typically consists of several cells at different stages of their development. The stages of individual cells are shown in Figure 6.3. In the growing cumulus stage the motion throughout the cloud at all levels is upward. The cloud drops are still small enough to be carried upward with the air currents. As the cloud gets deeper and the top reaches levels where the temperature is low enough, ice crystals begin to form and the

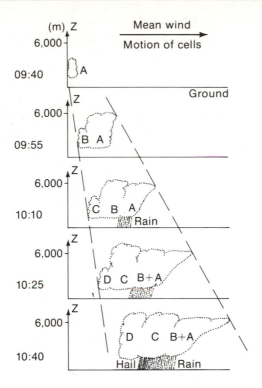

Figure 6.4 Schematic history of first four cells of a thunderstorm at Clermont–Ferrand, France, May 3, 1973, recorded using radar, airborne cameras, and ground-based cameras. Coordinates of each diagram are distance in direction of mean wind and height (Z). The diagrams are labeled at left with the observation times. [After Ramond and Tixeront, *Proceedings of International Conference on Cloud Physics,* July, 1976; published by American Meteorological Society, Boston, Mass.]

three-phase process of drop growth begins. In the mature stage the precipitation is under way, and with it a downdraft has developed within the cloud, induced by the drag of the falling drops. The updrafts and downdrafts reach speeds of 30 m s^{-1} and are turbulent, so that planes flying through thunderstorms at this stage may be subjected to accelerations large enough to produce structural damage. Usually, pilots will try to avoid flying through cumulonimbus clouds both for the sake of the comfort of the passengers and for safety. The inflow of new warm air at the bottom and sides of the cloud is usually inadequate to maintain the large buoyant updrafts, and the cell changes to the dissipating stage, in which the updraft has died and the air is descending throughout the cell.

Figure 6.4, based on measurements made in a thunderstorm near Clermont-Ferrand, France, on May 23, 1973, shows the way successive cells develop. The

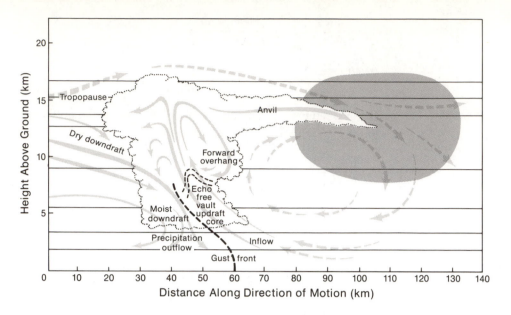

Figure 6.5 Schematic diagram showing structure and air flow in a supercell thunderstorm. The vertical scale is four times the horizontal; the storm cloud is about 15 km high and about 100 km in horizontal extent. [After Hoxit et al., *Monthly Weather Review,* Vol. 104, Nov. 1976, p. 1419.]

stages of the storm are shown at 15-minute intervals. At each stage a new cell has started upwind of the previous ones, and the cells that formed earlier have progressed toward maturity and then dissipation.

While most thunderstorms are of this multicell type, with each cell passing through the life cycle described above, a few of them consist of a single persistent large intense cell that remains in an approximately steady state for long periods, up to 8 or 10 hours. These "supercell" storms produce large hail and cause a large proportion of all damage by hail.

Figure 6.5 illustrates the structure of a supercell storm. The updraft in the forward part of the storm is so strong that all drops that form are carried upward too rapidly to allow them to grow large enough to reflect radar waves, giving rise to what is known as the "echo-free vault" or "weak-echo vault" in the radar view of the storm. The rear part of the storm contains the steady downdraft, with accompanying rain and hail. The cold descending air spreads out at the ground, producing the gust front at its forward edge. The development and persistence of a supercell appears to depend on an appropriate balance between the strength of the buoyant updraft, as controlled by the amount of instability, and the inflow, which is governed by the larger scale convergence and wind shear.

6.2   Influence of the environment: Entrainment

When heated air rises it cannot leave a vacuum, so air that has been surrounding the position it occupied near the ground must flow in to take its place. Similarly, the air above it must move aside to make room for it, and there must be downflow to replace the air that has flowed in at low levels and to make room for the outflow aloft. Thus, a complete circulation is established. The descending air current is usually much more widespread than the rising current, so that its speed is usually much lower. Nevertheless, the descent results in adiabatic heating, so that the temperature of the air around the convective cloud will be higher than the original environmental temperature, and the buoyancy force, which depends on how much warmer the cloud is than its surroundings, will be correspondingly reduced. The result is that the lapse rate must be considerably greater than the saturation adiabatic for numerous large convective clouds to develop. If the lapse rate is only slightly greater than Γ_s, the clouds will be widely scattered.

In addition to the effect of its sinking motion discussed in the preceding paragraph, the environmental air has an effect on the cloud growth because of the turbulent mixing that occurs at the sides and top of a growing cumulus. Unsaturated cooler air from outside is mixed into the cloud by this process, causing the drops either to evaporate or to grow more slowly and producing a decrease in the temperature of the cloud. The addition of environmental air into a cloud by mixing is called *entrainment*. As a consequence of entrainment the rate of decrease in temperature of the rising saturated air parcel is greater than the saturation adiabatic rate, and the buoyancy is correspondingly reduced. Thus, both the sinking of the surroundings and the entrainment of environmental air decrease the difference in temperature between the cloud and the air around it, thereby reducing the rate at which the rising current is accelerated. The criterion for instability, which determines whether or not convection will begin, is unchanged by these influences, but they act as a partial negative feedback mechanism that makes it necessary that γ be considerably larger than Γ_s for widespread active thunderstorms to occur.

6.3   Processes that produce changes in stability

We have seen that the heating of the layers of air near the ground by the sun's radiation during the day makes these lowest layers of the atmosphere unstable (and, in turn, the cooling from below during the night renders them stable again). But this effect extends upward only 2 or 3 km at most, ordinarily. The stability of the layers at higher levels, which determines whether there will be deep convective clouds accompanied by showers and thunderstorms, depends on other factors.

The factors are related to the general flow pattern of the air. The principal ones are *relative advection, horizontal convergence,* and release of *convective instability*.

Relative advection refers to the horizontal movement of air with different temperatures at different levels. Advection means simply the horizontal transport of air and its properties. Suppose that at lower levels the wind is bringing in warmer air, say, from the south, while at upper levels colder air than that which was present is being brought in, say, from the west. The effect of this warm advection below and cold advection aloft will be that as time goes on the temperature decreases more rapidly with height, that is, the lapse rate increases with time. If this process continues long enough, the air over a place could change from stable to unstable. The reverse type of relative advection, with cold air advection below and/or warm air advection aloft, would tend to stabilize the air or increase its stability.

In horizontal convergence the air is moving with different speeds at the same level in such a fashion that air tends to accumulate in an air column as it moves along. This process is illustrated in Figure 6.6A, which shows the horizontal and vertical cross sections of an air column at two different times, together with the sounding curves for the two times on a thermodynamic diagram. The wind along *AC* is assumed to be stronger than the wind along *BD*. After a time, the shape of the horizontal section of the column has changed to *A'B'C'D'*. To contain the same mass (at the same temperature and pressure) the volume must remain constant, and with the reduction in its horizontal area the vertical dimension of the air column must increase. Therefore *EF* moves to *E'F'*. The lifting of the air above *E'F'* leads to an outflow at higher levels so that the pressure at *E'F'* stays at approximately the value it had at the same height before the convergence. In the thermodynamic diagram the sounding curve before the convergence is shown by line *AE*. During the convergence the pressure at the top of the column changes, and the air there goes through the adiabatic process *EE'*. The sounding curve of the air column after horizontal convergence and vertical stretching is *AE'*. This process thus increases the lapse rate. However, it can never render an initially stable unsaturated air column unstable. If the lower part of the air column were initially saturated and the upper part unsaturated, however, the horizontal convergence and the accompanying upward motion of its upper portion could produce instability in the air column.

When an air column has a moisture distribution such that its vertical motion can produce instability, the column is said to be *convectively unstable*. For the convective instability to be released by horizontal convergence alone, it is necessary that the lower portion of the air column be saturated initially. However, if the entire column is lifted (with or without convergence) and the lower part is sufficiently humid to reach saturation before the upper part, instability could arise.

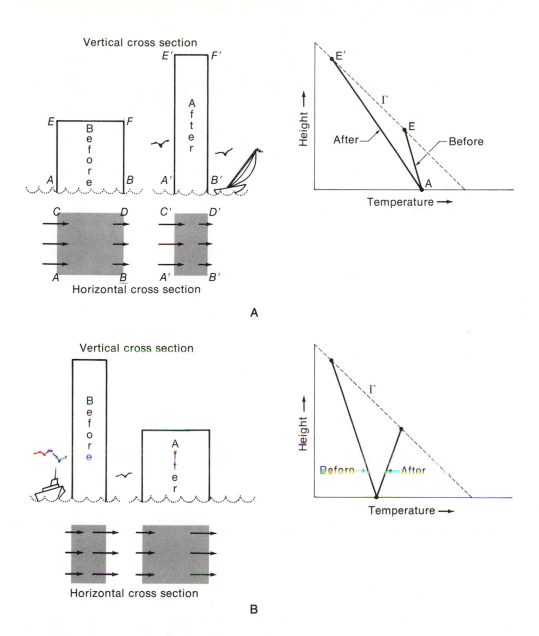

Figure 6.6 (A) Destabilization of a layer by horizontal convergence and vertical stretching. (B) Stabilization by horizontal divergence and subsidence.

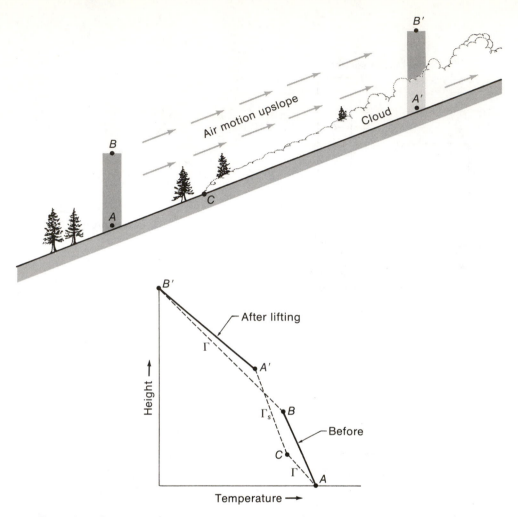

Figure 6.7 Convective instability. The destabilization of a layer occurs when it is lifted bodily and the lower part becomes saturated before the upper part does.

This process is illustrated in Figure 6.7. In this diagram the air column AB is pictured as moving up a mountain slope to $A'B'$. Its sounding is shown on the thermodynamic diagram by the curve AB. It is assumed that, in ascending, the lower portion of the column becomes saturated at C, but the upper portion remains unsaturated. The process curve for the bottom of the column is thus ACA'; for the top it is the dry adiabat BB'. The sounding curve after the ascent is $A'B'$. Since the lower portion is saturated and the lapse rate is greater than Γ_s, the column has become unstable.

Divergence can increase the stability of an air column in the same fashion that convergence can reduce its stability. In fact, inversions are frequently produced or reinforced by the divergence and subsidence associated with stationary or slow-moving anticyclones. The process of inversion formation by horizontal divergence is illustrated in Figure 6.6B. A more detailed example of the formation of an inversion by this process is presented in Section 10.3 and Figure 10.4.

In addition to the changes in stability produced by the influence of the earth's surface because of its warming by solar radiation during the day and cooling by radiative loss at night, air may pass over terrain that has a different temperature than the air and be rendered unstable or stable. For instance, if air moves in winter from snow-covered land onto an unfrozen lake or ocean, it will be heated rapidly from below and rendered unstable. This process leads to cumulus clouds and snow showers at the leeward shores of the Great Lakes in winter, with the amounts of snow frequently totaling several inches. An instance of the opposite effect is the air moving northward from the warm waters of the Gulf Stream to the very cold water off the coast of Newfoundland. The air is strongly cooled from below, producing an intense inversion. Dense fogs are caused by the condensation as the air is cooled.

Radiative exchange between layers of air aloft can also change the lapse rate. Ordinarily, these changes occur slowly and are small. An exceptional case is the radiative exchange at the top of a cloud. The liquid water in the cloud radiates nearly as a black body and loses more energy than the water vapor and CO_2 in the air layers above radiate back. This loss of heat may cool the upper part of the cloud sufficiently to restore or augment the instability and cause the cloud to grow. The occurrence of nocturnal thunderstorms is attributed partially to this process. The clouds develop during the afternoon, started by solar heating, but may not have quite enough energy to penetrate a stable layer aloft. The additional cooling at the cloud top may be sufficient to stimulate updrafts in the cloud that are strong enough to get through the layer that stopped its earlier growth.

To recapitulate, the stability of layers of air can be changed

1. By heating or cooling from below due to solar heating or radiative cooling of the ground;

2. By heating or cooling from below due to passing over a warmer or colder surface;

3. By heating or cooling at upper levels by radiative exchange;

4. By relative advection;

5. By convergence or divergence;

6. By release of convective instability by convergence or lifting.

Frequently, more than one of these processes go on at the same time.

6.4   Relation of convective activity to flow patterns

The factors that contribute to destabilization through deep layers result in a relationship between larger-scale flow patterns and showers and thunderstorms within air masses. Some of the air masses and circulations in which convective activity is favored are

1. Fresh polar air moving cyclonically over warmer land or water;
2. Tropical maritime air moving poleward over land heated by solar radiation in spring and summer;
3. Moist air subjected to convergence effects of sea breezes, valley-mountain winds, or monsoons.

The properties of polar air masses and the changes they undergo after leaving their regions of formation will be described in Chapter 10. For now it is sufficient to say that as they move equatorward they are heated from below, and the portion of them next to the ground is rendered unstable. In general, equatorward motion is divergent, but cyclonic motion tends to be convergent. If the flow is sufficiently cyclonic the equatorward effect will be overcome, and there will be enough convergence to destabilize the upper layers, allowing convective clouds to become thick enough to produce showers. Cyclonic flow of fresh polar air usually occurs immediately behind a cold front. Farther back, the flow straightens and becomes anticyclonic, leading to stability and clearing. Furthermore, the ground becomes cooled as the cold air moves over it, and after a while it no longer warms and destabilizes the air. Thus, the showers in the cold air occur only a short distance behind the cold front.

The properties of tropical maritime air will also be discussed in Chapter 10. Without going into detail at present, therefore, we shall state only that, as the name implies, it is a warm, moist air mass. The moisture is greatest in the lowest layers; aloft the humidity is frequently quite low, so it is convectively unstable. When tropical maritime air comes over land in winter, it is warmer than the ground and becomes stabilized; convection is suppressed and only fog or stratus cloud with drizzle is present. In spring and summer, however, the land heated by the high sun is warmer than the oceans from which the tropical air comes. The air is destabilized and showers and thunderstorms occur. In regions where the flow is poleward (e.g., south winds in the Northern Hemisphere), there is an additional destabilizing effect due to convergence, and the showers and thunderstorms are general. Where the flow is toward the equator or anticyclonically curved, divergence suppresses the convection and the showers are scattered or absent.

The tropical maritime air is usually associated with a large anticyclone. On the east side of the anticyclone, where the flow has an equatorward component, there

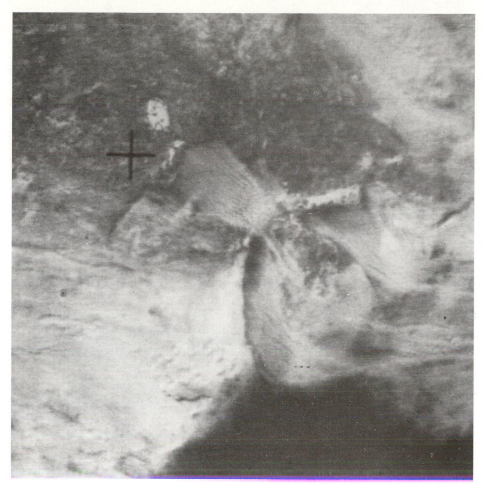

Figure 6.8   Satellite photo (February 1971) showing convective cloud in cyclonic-flowing polar air. Heating over Great Lakes increases convective activity, so clouds take form of lines of cumulus along wind. [NOAA–NESS photo. Courtesy of V. J. Oliver.]

is little or no convective activity; to the west of its center, the poleward flow, with its convergence, produces widespread showers and thunderstorms.

Sometimes the thunderstorms in the tropical air are organized into a line of convective activity called a *squall line*. The line in this case is not simply the roll type of convective motion discussed at the beginning of this chapter. Even a single large thunderstorm may have strong winds associated with the outflow of cool air arising from the downdraft. Detailed analysis has shown that the intense convective system, including the downward motion produced by the precipitation, gives rise

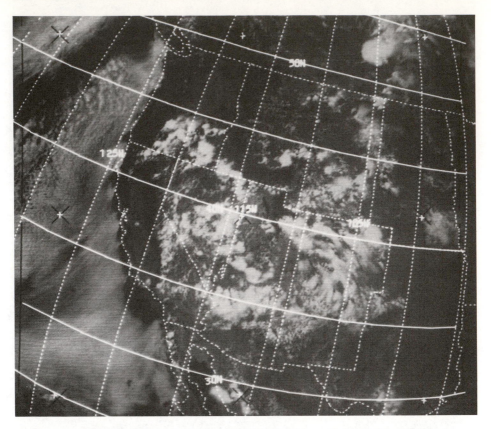

Figure 6.9 Satellite photo (July 5, 1970, 2209 GMT, or 3:09 P.M. Mountain Standard Time) showing thunderstorms in mountains of southwestern United States, associated with convergence induced by heating of high terrain and southerly (poleward) flow to west of the anticyclone (Bermuda or Azores High), which extends over the eastern United States in summer. [NOAA–NESS photo. Courtesy of V. J. Oliver.]

to a small high-pressure center, or "mesohigh," that pushes air that has been cooled by the evaporation of precipitation falling through it outward ahead of the storm in the form of a gust front or pseudofront. This outflow releases the convective instability of the air ahead of it, thus propagating the convection. When several thunderstorms occur side by side, the outflow forms a continuous line of strong gusty winds.

Squall lines may originate entirely within the tropical air mass, or they may be set off by the impetus of the cold fronts to the west. In the latter case, they frequently move faster than the fronts and are soon no longer directly associated with them. With the arrival of a squall line the wind usually shifts direction abruptly, as well as increasing in speed. The temperature drops, and heavy rain

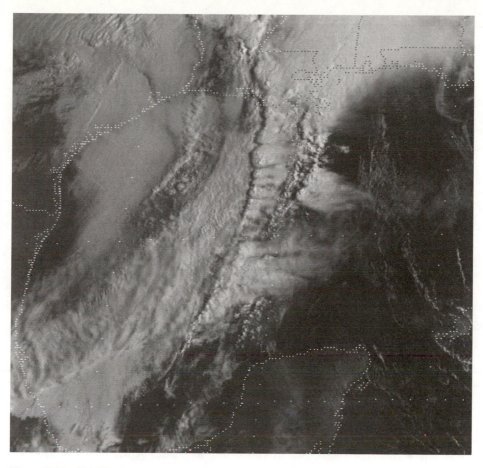

Figure 6.10 Satellite photo showing squall line sweeping across Louisiana and the Gulf of Mexico. Taken at 6:00 A.M. local time (1400 GMT), April 19, 1980, in the visible, the shadow sets off the towering cumulonimbus clouds strikingly, and the cirrus anvils are shown stretching far ahead in the strong winds near the tropopause. [NOAA–NESS photo. Courtesy of V. J. Oliver.]

begins to fall, accompanied by lightning and thunder and sometimes by hail. Hailstones as large as 10 cm across have been observed, although usually they are about 1–2 cm or less in diameter. The winds accompanying severe thunderstorms may reach 30 m s^{-1} or more and cause damage to structures and to trees. The most intense windstorms of all, the tornadoes, are associated with severe thunderstorms arising from squall lines.

Regions where convergence occurs frequently, as a result of winds produced by topographic effects, will have convective clouds and showers if other conditions are favorable. For instance, the Florida peninsula has sea breezes blowing into it from three sides, producing convergence that destabilizes the air and leads to daily showers in summer when moist air is present and the heating by the sun helps set off the convection. Consequently, the region of the United States in which thunderstorms are most frequent is Florida. A secondary maximum in the frequency of thunderstorms is found in the southern Rocky Mountains, where the valley-mountain winds produce upslope motion and convergence over the mountains. The thunderstorms over the mountains of New Mexico and Colorado occur mostly during the afternoon. Over the Great Plains to the east the flow toward the mountains causes divergence during the day, but the outflow from them at night causes convergence. This diurnal pattern of convergence and divergence, together with relative advection effects that are similarly influenced by the mountains, and possibly the radiative exchange at the cloud tops, causes thunderstorms to occur more often at night than during the afternoon over the Great Plains. Monsoon circulations have similar effects on the occurrence of convective showers, with seasonal instead of diurnal periods.

6.5   Tornadoes

Tornadoes represent an extreme culmination of convection, in which the kinetic energy becomes concentrated into an intense vortex or whirlwind. They have the strongest winds occurring at the earth's surface, with speeds estimated to reach up to 130 m s^{-1} (amost 300 mph). Because anemometers are blown away by such strong winds, the highest speed that has been measured in a tornado is much less, 67.5 m s^{-1} (151 mph), recorded in Tecumseh, Michigan, April 11, 1965. They are always associated with severe thunderstorms, and usually with squall lines. They are quite small in extent, ranging from 50 m to about 1 km in diameter, but are locally the most destructive of all storms, with almost all structures and trees in their paths suffering complete devastation.

The characteristic funnel cloud of a tornado (Figure 6.6) appears to be drawn down from the cumulonimbus cloud of the parent thunderstorm, but actually the air rotating around the storm moves inward toward the center and then upward. The dust and debris lifted from the ground, including large objects such as cars, shows the strength of the upward current. The pressure at the center is much lower than at the periphery, with decreases in excess of 50 mb having been measured and values up to 100 mb estimated. The funnel cloud shows the position where the moist air flowing inward toward the center becomes saturated by adiabatic cooling.

The rapid pressure fall when the center of a tornado passes over structures produces an explosive effect which adds to the destruction due to the strong winds.

Figure 6.11 Tornado at Enid, Oklahoma, June 5, 1966. [Photographed by Leo Ainsworth. Courtesy of NOAA.]

Unless buildings have enough open doors and windows to permit rapid equalization of pressure, the force due to the excess pressure inside the building literally raises the roof and pushes out the walls, which the wind then carries away. Injuries and fatalities in tornadoes are mostly due to the victims being crushed by the collapse of structures or struck by objects flying in the wind.

Usually the funnel cloud of the tornado extends downward and reaches the ground for only a few minutes, during which it moves a kilometer or two. However, the small number of tornadoes that cause deaths characteristically remain at the ground for distances ranging between 20 and 50 km, suggesting that the tornadoes that have longer lives and move farther are the more intense ones. A typical tornado moves northeastward with a speed of translation of the vortex of about 20 m s^{-1}, leaving a path of destruction about 400 m wide. (Note the distinction between the speed of the tornado, about 20 m s^{-1}, and the speed of the winds rotating around its center, up to 130 m s^{-1}.) On rare occasions tornadoes have been reported to remain practically stationary, and occasionally they have traveled with speeds up to 30 m s^{-1}. Their path is usually along or parallel to squall lines or cold fronts, but frequently they occur in other situations, for instance, in association with hurricanes. The rotation of tornadoes is usually cyclonic, although instances of anticyclonic rotation have also been reported. In appearance tornadoes frequently have a double structure, the true funnel cloud being surrounded by a separate cylinder of dust and debris raised from the ground.

The circumstances under which tornadoes form are still only partially understood. It is known that they are almost always associated with severe thunderstorms of a type that develops when the release of instability has been delayed by a warm dry layer overlying a deep unstable moist layer. The rapid decrease of humidity results in large convective instability, which is released when sufficient heating and convergence has taken place. The dry air is usually moving rapidly from the west, and the moist air below it is usually moving poleward, so that there is cyclonic turning as one goes upward. This cyclonic turning may be the reason why tornadoes almost always rotate in a cyclonic sense. The swirl may start around a quasihorizontal axis in the region of most rapid turning of the wind with height and get twisted into the vertical by the up-and-down currents of the thunderstorm. Once the axis is vertical the inflow of air will result in increasingly rapid rotation in the same way as the spinning of a skater speeds up when he or she brings in his or her arms. (The principle on which this is based, the conservation of angular momentum, is discussed in Chapter 8.)

While tornadoes have occurred in every state of the United States and in many other parts of the world, they are most frequent in the central and southeastern United States, with maximum frequency in a band extending from the Texas Panhandle northeastward to northwestern Missouri. In a typical year, between 500 and 1000 tornadoes occur in the United States. The total number reported in the 11-year period 1960–1970 was 7428. Of these, 223 tornadoes had fatalities associated with them, producing 1014 deaths.

In the late winter the most probable location for tornadoes is the southeastern states. As spring progresses they are more likely farther west and north; in summer they occur in the Great Plains and the northern states, but with the decrease in air-mass contrasts, they become less frequent. More than half of the tornadoes in the United States have occurred in April, May, and June; of those that have occurred in Kansas and Oklahoma, three-quarters took place in these three months. However, there is no month in the year in which tornadoes have not occurred.

With respect to time of day, tornadoes, like thunderstorms, are most frequent in the afternoon. About one-third of those that have occurred in the United States have taken place between 3:00 P.M. and 6:00 P.M., local time, and 82 percent have taken place between noon and midnight.

Figure 6.12 Devastation left by tornado in Guin, Alabama, in February 1974.
[Courtesy of NOAA.]

The general conditions favorable to severe thunderstorms and tornadoes can be forecast fairly well twelve or more hours in advance. However, specifying the exact time and place of occurrence of individual storms, affecting, as they do, very small areas, defies present methods of prediction in advance of their development. A network of radar stations has been established that covers the area subject to frequent occurrence of tornadoes in the United States. When a tornado is either seen or detected by radar, prompt warnings are issued by radio and television to alert the people in the vicinity.

Over the ocean tornadoes are called *waterspouts*, because of the lifting of water at their base by the wind and updrafts.

Questions, Problems, and Projects

1. Discuss the influence of a convective current on the environment.
2. Why is it necessary that γ be considerably larger than Γ_s for deep convection and widespread active thunderstorms to take place?
3. What kind of weather is characteristic of the eastern portion of a subtropical anticyclone? The western portion? Give the reasons for the differences, if any.
4. Why are the areas of most frequent thunderstorms in the United States the Florida peninsula and the southern Rocky Mountains?
5. Describe two weather situations, in each of which more than one of the processes that are listed at the end of Section 6.3 act to change the stability of the air. Explain how these processes act and what the effect of the changed stability is.
6. Does the fact that the tornado funnel extends downward from the main thunderstorm cloud show that there is descending motion there? If not, explain why it looks that way. Why are tornadic winds so strong?

Motions in the Atmosphere.   7
Small-Scale Circulations

7.1   Air in motion

We know from common experience that air does not stay still but moves about, sometimes gently as a pleasant breeze, sometimes with the gusty violence of a gale or hurricane. It is free to move in all directions, and, indeed, the vertical motions are very important, being responsible largely for the formation of clouds and precipitation. But, ordinarily, horizontal motion predominates. The horizontal motion of air is called *wind*.

The wind direction is the direction *from* which the wind is coming. To remember this sometimes confusing convention the nursery rhyme "The north wind doth blow and we shall have snow" is helpful, at least for those in the Northern Hemisphere. Speeds, in the International System, have the units meters per second (m s^{-1}). In international meteorological usage the knot (kt) is still standard for some purposes, and in the United States miles per hour (mi/hr) is common. The conversion relationship is 1 m s^{-1} = 1.94 kt = 2.24 mi/hr. Wind speeds usually are in the range of a few to a few tens of meters per second.

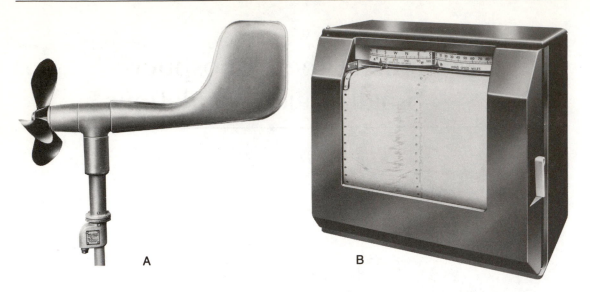

Figure 7.1   Wind measuring instruments. (A) An instrument in which the anemometer, for measuring wind speed, is combined with a wind vane. The large tail keeps the propellor pointed in the wind. The propellor drives a small electric generator. The output of the generator, which is proportional to the wind speed, is recorded on a strip chart (B), along with a record of the direction toward which the vane is pointing. [Courtesy of Science Associates, Inc.]

At the earth's surface wind direction is observed using wind vanes, and wind speed is measured with anemometers, which may be of the windmill or cup type, in which the speed of rotation is a measure of the wind speed, or of the type in which the pressure exerted by the wind is interpreted in terms of its speed. At levels away from the ground the wind speed and direction are evaluated by tracking small balloons as they rise. A more detailed description of wind measurement is given in Appendix F.

In early years, particularly on shipboard, instruments were not available for measuring wind speed, and to facilitate comparability among observations Admiral Sir Francis Beaufort (1774–1857) introduced a scale based on the rigging of a naval sailing vessel. Subsequently, objective criteria based on the effects of Beaufort scale winds on the sea surface and on trees and structures on land were developed, and standard descriptions of these effects have been adopted by the World Meteorological Organization for international use. These descriptions and the corresponding speeds that would be measured by an anemometer at a height of 10 m above the sea or above open flat ground are given in Appendix G. Using the Beaufort Wind Scale one can estimate the wind speed without having an anemometer to measure it.

For simplicity we may use the following approximate terminology:

Calm: < 0.2 m s^{-1}

Light breeze: 1–4 m s^{-1}

Moderate breeze: 5–8 m s^{-1}

Fresh breeze: 9–11 m s^{-1}

Strong wind: 12–15 m s^{-1}

Gale: 16–20 m s^{-1}

Strong gale: 21–24 m s^{-1}

Storm: 25–28 m s^{-1}

Strong storm: 29–32 m s^{-1}

Hurricane: $\geqslant 33$ m s^{-1}

Although winds are gusty and variable they form systematic patterns. Some patterns are very large, such as those formed by trade winds that carried the great sailing vessels of the eighteenth and nineteenth centuries across the Atlantic and Pacific Oceans. An analogous phenomenon is the high-level jet stream, which often adds more than 50 m s^{-1} to the speed of jet airplanes in their west-to-east trips across continents and oceans in middle latitudes. Other large systems include the cyclones and anticyclones of middle latitudes, with diameters of thousands of kilometers. On a smaller scale are the tropical cyclones (hurricanes and typhoons), ordinarily 100–400 km across.

Still smaller are local storms, including the awesome tornado, which, although extremely strong, is but a few hundred meters in diameter. Leaves and dust swirling as eddies form when the wind blows past the corners of buildings and the convolutions of cigarette smoke reveal very small patterns in the motions of the air. Within the large patterns are imbedded smaller ones, so that the total motion of the atmosphere is an incredibly complex superposition of irregular patterns of all sizes.

To put some semblance of order to this complicated situation, we shall examine some of the scales of motion separately, in the hope that the laws that govern them are sufficiently simple and that understanding the parts will help us to comprehend the whole. We start with the factors that lead to air motion.

7.2 Newton's laws of motion

The laws of motion discovered by Isaac Newton (1642–1727) provide the basis for all analyses of the behavior of moving matter. In a fluid such as air, the application of the laws requires some special considerations. Motions in a rotating system such as the earth and its atmosphere likewise introduce differences from motions in an *inertial* system. An inertial system is one that is at rest or moving with constant speed in a straight line.

Newton's laws of motion are essentially contained in the expression of his second law, which says that *the rate of change in the amount of motion is proportional to*

Figure 7.2 Photograph taken from Gemini 12 spacecraft off the west coast of Mexico, November 13, 1966, showing air motions of various types and scales—streaks, waves, and vortices—reflected in the cloud patterns. [Courtesy of NASA.]

the sum (resultant) of the forces acting and is in the direction the (resultant) force is acting.

The amount of motion is defined as the product $M\mathbf{v}$ of the mass and the velocity. If a small change $\Delta(M\mathbf{v})$ occurs in a time Δt, the rate of change is the ratio $\Delta(M\mathbf{v})/\Delta t$. Using this notation we can write Newton's law in the form of an equation:

$$\frac{\Delta(M\mathbf{v})}{\Delta t} = \mathbf{F}_1 + \mathbf{F}_2 + \ldots = \mathbf{F} \qquad (7.1)$$

(Rate of change of motion = Sum of forces = Resultant force)

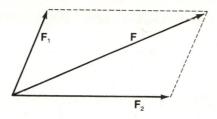

Figure 7.3 Addition of vectors.

The expression looks like an ordinary equation but in fact it is a *vector* equation, for each force may have its own direction as well as magnitude. Quantities that have both direction and magnitude (and combine in a certain way) are called *vectors*.[1] When two vectors F_1 and F_2 are added they combine by the *parallelogram* rule, as shown in Figure 7.3.

In ordinary circumstances the mass M does not change. The rate of change of velocity $\Delta v/\Delta t$ is equal to a, the acceleration. Equation (7.1) can therefore be written simply as

$$Ma = F \qquad (7.2)$$

that is, the mass times the acceleration is equal to the resultant force.

7.3 Forces in a fluid

If we consider a portion of a fluid away from its boundaries, for instance, a parcel of air some distance above the ground, we can see that there are not many forces that can act on it. First, of course, there is the force of gravity, which acts on every mass in the vicinity of the earth. If the parcel carries an electric charge there will be the force of electric attraction (or repulsion), but ordinarily the charges in the air are very small. So, in addition to gravity, the main force acting on a parcel of fluid is the action on it by the fluid around it. This action can be expressed as consisting of two parts, *pressure* and *viscous stress* or internal friction.

If the pressure were the same on all sides of the parcel, its effect would cancel out; that is, the resultant force from the pressure forces on all sides would be zero. We know from Chapter 3 that the atmospheric pressure always varies at least in one direction, the vertical, so that the downward pressure on the top of a parcel of thickness Δh is less than the upward pressure on its bottom. We used the fact

---

1. Vectors are represented in print by boldface letters. Their magnitudes (numerical values) are shown by the same letters in Roman type.

that this difference in pressure, resulting in a net upward force, is usually almost exactly balanced by the force of gravity to derive the hydrostatic equation that gives the rate of decrease of pressure with height. Observations, as represented on weather maps, show that the pressure at the same level also varies from place to place. This variation gives rise to horizontal net pressure forces. At any given place the pressure at each height changes as time goes on.

To summarize the distribution of the pressure (reduced to sea level) on surface weather maps, isobars (lines along which the pressure is constant, with lower values on one side of the line and higher values on the other) are drawn. As we saw in Section 1.3, when this is done relatively simple patterns emerge, with closed isobars enclosing centers of high pressure or low pressure on the surface (sea-level) weather map.

To represent the pressure distribution aloft, isobars could be drawn on maps for various upper levels—say, heights of 1 km, 2 km, 3 km, etc. However, instead of this, the general practice is to plot the heights at which a particular pressure— say, 850 mb, 700 mb, or 500 mb—is reached, and draw contour lines showing the height distribution of that constant pressure surface. It can readily be seen that a height line or contour line of an isobaric surface coincides with an isobar on a constant level. For instance, a 3000-meter contour line on a 700-mb map is exactly the same as the 700-mb isobar on the 3-km map for the same time. The spacing of contour lines of an isobaric surface is similar to the spacing of isobars on a constant level surface: where the isobars are close together the contour lines are close together, and vice versa. Figure 7.4 shows why this is true.

The rate of horizontal variation of pressure represented by isobars on a constant level map or contour lines on an isobaric map is much smaller than the vertical variation. The values shown in Figure 7.4, in which the pressure change of 4 mb takes place in 50 m in the vertical, but over distances of 200–300 km in the horizontal, are typical. The resultant horizontal pressure forces thus may be expected to be much smaller than the vertical, but since they are unopposed by gravity, they lead to motions more readily.

It is obvious that the pressure variation results in a horizontal force pushing the air from high to low pressure, tending to produce flow in this direction in the same way that water tends to flow downhill. To see this quantitatively, let us consider a parcel of air in the form of a rectangular parallelepiped extending from an isobar drawn for pressure p_0 to one drawn for pressure $p_0 + \Delta p$ on a constant level map, with the vertical faces parallel and perpendicular to the isobars, as shown in Figure 7.5.

It is clear that the total force acting on face $KLQP$ of the parallelepiped is exactly equal and opposite to that on $MNSR$, since for every point on $KLQP$ there is a point on $MNSR$ where the pressure is the same. Thus, the net horizontal force will be the resultant of the forces on the two other vertical faces of the parallelepiped.

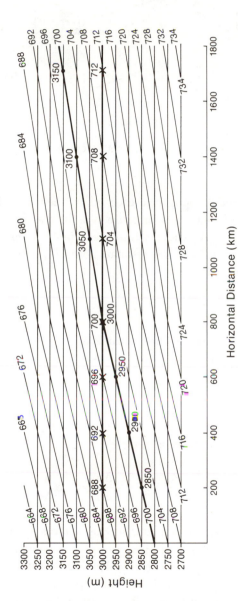

Figure 7.4 Schematic atmospheric cross section showing the relationship between spacing of isobars on a constant level (3 km) chart and height contour lines on a constant pressure surface. The vertical scale is greatly exaggerated (by a factor of 1000). The 3-km level and the 700-mb isobaric surface are shown by heavy lines. The ×'s show the positions of the isobars on the 3-km level; the heavy dots show the positions of the contour lines on the 700-mb isobaric surface. On the left side of the diagram, where the isobars are close together, the contours are close together also. On the right side they are both farther apart.

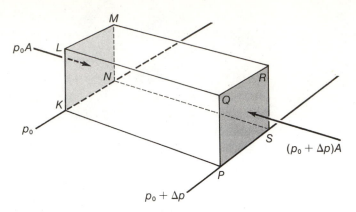

Figure 7.5 Illustration of net horizontal pressure force.

These forces are directed opposite to each other; their magnitudes are given by the pressure times the area, since pressure is defined as force per unit area. The force on face $KLMN$ is $p_0 A$, where A is its area; the force on $PQRS$ is $(p_0 + \Delta p)A$ in the opposite direction, and the resultant force F_M acting on the mass of air contained in the parallelepiped is

$$F_M = p_0 A - (p_0 + \Delta p)A = - \Delta p \cdot A \qquad (7.3)$$

It is desirable to compute the force per unit mass. If the distance $KP = \Delta n$, the volume of the parcel is $A \cdot \Delta n$, and its mass is

$$M = \rho A \cdot \Delta n$$

The magnitude of the force per unit mass [which, by equation (7.2), is equal to the magnitude of the acceleration a] is

$$F = \frac{F_M}{M} = -\frac{1}{\rho} \frac{\Delta p}{\Delta n} = - \alpha \frac{\Delta p}{\Delta n} \qquad (7.4)$$

The minus sign shows that the resultant horizontal force is opposite to the direction of increasing pressure, that is, in the direction of decreasing pressure. Since isobars are drawn for fixed differences of pressure Δp on the weather map, the magnitude of the horizontal pressure force is inversely proportional to Δn, the distance between isobars. The rate of change of pressure in the horizontal $\Delta p/\Delta n$ is called the *pressure gradient*. The force due to horizontal pressure variations is called the *pressure-gradient force*.

The pressure-gradient force acts in the direction perpendicular to the isobars toward lower pressure, with magnitude inversely proportional to the distance between isobars.

In deriving equation (7.3) we should properly have used the average pressure on each of the faces of the parallelepiped, to allow for the variation with height, and in deriving equation (7.4) we should have used the average density. We consider that the vertical dimension of the parallelepiped is taken to be very small, so that the difference between the gradient of the actual pressure and the average pressure can be neglected, and that the density variation is similarly negligible.

To get an idea of how big the forces represented by the isobars on the weather map are, let us compute the acceleration corresponding to a typical situation. Suppose 4-mb isobars are 300 km apart on a sea-level chart and the density has its standard value. In SI units $\Delta p = 400$, $\Delta n = 3 \cdot 10^5$, and $\alpha = 0.775$. Then

$$a = \frac{0.775 \cdot 400}{3 \cdot 10^5} = 1.03 \cdot 10^{-3} \, \text{m s}^{-2}$$

If this force were to act for 1 hour, the speed attained would be $3600 \cdot 1.03 \cdot 10^{-3} \, \text{m s}^{-1} = 3.72 \, \text{m s}^{-1}$, and if it continued for 24 hours the air would attain a velocity of 89 m s^{-1} or 200 miles per hour. Since pressure gradients of this magnitude persist for days, one might expect very large winds to develop unless other forces enter. One such force that is always present is friction, which acts to keep the winds from attaining high velocities much of the time. A more important factor arises from the rotation of the earth and will be discussed in the next chapter.

When the horizontal pressure distribution is represented by the height of a constant pressure surface, the expression corresponding to equation (7.4) for the magnitude of the pressure-gradient force per unit mass is

$$F = -g \frac{\Delta h}{\Delta n} \tag{7.5}$$

where Δn is the distance between height contour lines drawn for a height interval Δh. This expression has the advantage over equation (7.4) that it does not contain α. Since α has different values at different heights, F has quite different values for the same spacing of isobars at two levels, say 3 km and 6 km, whereas the same spacing of height lines drawn for the same height interval on every isobaric surface, be it 1000 mb or 100 mb, represents the same pressure-gradient force. This is the principal reason maps of constant pressure surfaces are used in preference to constant level maps in meteorological practice.

As an example of the application of equation (7.5), let us compute the pressure-gradient force for the height line spacing of the 700-mb surface shown in the right portion of Figure 7.4. We have $\Delta h = 50$ m, $\Delta n = 300$ km $= 3 \cdot 10^5$ m, $g = 9.8$ m s^{-2}. Putting these values in equation (5.5),

$$F = \frac{9.8 \cdot 50}{3 \cdot 10^5} = 1.67 \cdot 10^{-3} \, \text{m s}^{-2}$$

a value about 1⅔ times that we got for 4-mb isobars 300 km apart at sea level. The difference is due mostly to the increase in specific volume, or decrease in density, with height; the spacing of 50-m contour lines at 700 mb corresponds closely to the distance between 4-mb isobars at the 3-km level.

7.4   Causes of horizontal pressure variations

Since the pressure at any point is almost exactly equal to the weight of the air above, per unit area, and since the density depends (inversely) on the temperature, one might expect to have a close correspondence between temperature and pressure, with warm areas having low pressure and cold areas high pressure. In some instances (thermal lows over continental deserts in summer and high-latitude continental highs in winter), this situation dominates, but on the whole conditions are more complicated, with the air column over any given place being cold at some elevations and warm at others. For instance the temperature at 16 km is lower over the equator, where it is in the troposphere, than at 60°N or 60°S, where it is in the stratosphere.

The effect of temperature difference on the pressure-gradient force is most conspicuous in places where differences of terrain produce large differences in the temperature near the ground in short distances over which the conditions at higher levels are practically unchanged. The clearest example is the diurnal variation on clear days at coastal locations. The effect of differential heating and cooling of land and sea produces the *sea breeze* (from sea to land) during the day and the *land breeze* at night.

To see how the temperature variations produce the pressure-gradient forces and accelerations in these places, consider the schematic drawings in Figure 7.6, in which vertical cross sections through the atmosphere over a flat island and its surrounding ocean are represented. In these cross sections the pressure distribution is represented by isobars. It is assumed that initially the pressure at the earth's surface is the same over the land and the sea. As one goes upward, the

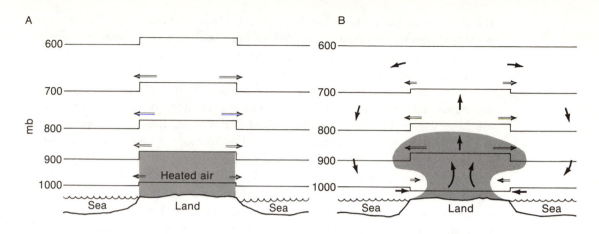

Figure 7.6 Schematic development of a sea breeze: *double arrows*—horizontal pressure gradient forces; *single arrows*—resultant air motions.

pressure drops in accord with the hydrostatic equation, equation (4.4). Where it is cool the pressure drops more rapidly and the distance between isobars is smaller than where it is warm.

Figure 7.6A represents the conditions shortly after sunrise. Over the ocean the temperature at the surface remains low because the heat from the sun penetrates deeply into the water and the air above is relatively cool. Over the island, however, the sun's radiation raises the temperature of the ground, and the air for some distance above it is heated. As a consequence, the isobars, before motion begins, take the form of horizontal lines with vertical steps. Over the sea where the air remains cool, the isobars are close together. In the heated air over the island, they are farther apart. Aloft, the pressure at each level is higher over the island than over the sea.

If the heating were to take place instantaneously, there would be no motions at first, only the expansion of the air and the lifting of the isobaric surfaces over the island. This process would give rise to horizontal pressure-gradient forces aloft that tend to push the air outward from the island, as indicated by the double arrows in Figure 7.6A, but would not give rise at first to forces at sea level. After the pressure-gradient forces had acted a while, the accelerations would result in outward movement of the air aloft. The movement from land to sea aloft would cause an accumulation of air at upper levels over the sea that would raise the sea-level pressure there, and correspondingly a depletion of air and decrease in sea-level pressure over the island. These changes in pressure would produce a pressure-gradient force at low levels from the sea toward the island that would accelerate the air from sea toward land, producing the sea breeze at low levels. The motions resulting from the seaward forces aloft and the landward forces near the ground are shown schematically by the single arrows in Figure 7.6B.

The actual process is a continuous one, rather than the stepwise process discussed above. The heating, expansion of air, development of pressure-gradient forces, accelerations, and air motions occur together, first slowly as the sun's heating begins to overcome the cooling effects of the night, then more rapidly. The sea breeze reaches its maximum strength in the afternoon, when the heating effect has lasted a long time. Then, as the land cools, it dies down.

At night the reverse effect takes place. The temperature of the land surface falls more than that of the sea surface. This is due to the fact that as soon as the sea surface loses some heat, it becomes denser than the water below and sinks, so that the cooling is spread through a deep layer of water. The greater cooling of the land surface results in a corresponding decrease in temperature of the air near the ground and a depression of the isobaric surfaces aloft over land relative to the sea. The resulting acceleration of the air from sea to land aloft causes the pressure over land at sea level to rise and produces a seaward pressure gradient there that results in a land breeze, that is, a wind from the island toward the sea.

At the coast of a continent the situation corresponds to one-half of the picture over an island. Near the shore the motion is from sea to land during the day and from land to sea at night at low levels, with opposite flow aloft. Far enough inland the temperature is unaffected by proximity to the ocean and there is no influence on the pressure gradient or the winds. Thus, at a continental coast the sea breeze is felt only 10–20 km inland. Similar effects occur at the shores of large inland lakes. On islands and peninsulas at low latitudes, the convergence of the sea breeze increases the tendency for daytime cumulus convection and showers.

An effect similar to the sea breeze is produced by differences in land elevation over short distances, producing the valley–mountain wind regime. In this case the air next to the elevated land is heated by the sun more than the air at the same height over the valley. This gives rise to pressure gradients in the way that was described for the sea-breeze situation and produces motion away from the mountain at high levels and toward it near the ground. The flow up the mountains usually is diverted to some extent when there is a U- or V-shaped valley so that the actual flow is a combination of a wind up the valley floor and a *slope wind* directly up the sides of the mountain. At night the winds are reversed, the air cooled by contact with the mountains descending into the valley and flowing downward along its axis. Figure 7.8 shows schematically the flow in a valley (A) in the early afternoon, when the up-valley and up-slope winds are well developed, and (B) in the middle of the night when the down-valley and down-slope winds are near their maximum.

Figure 7.7 Visible effect of sea breeze in Florida. Photograph taken from Gemini V spacecraft at 1231 local time, August 22, 1965, looking south along the east coast of Florida. The air over the ocean is virtually cloudless. As it moves onshore, it is heated, and cumulus clouds form a short distance inland. [Courtesy of NASA.]

At coastal locations in low latitudes the sea- and land-breeze pattern is present on almost all days throughout the year, but at higher latitudes the situation is frequently complicated by the superposition of larger-scale wind systems, which may offset the effect of the sea–land temperature contrast. In middle and high latitudes the tendency for a sea breeze may be completely outweighed by the influence of moving large-scale systems. In winter, when the heating by the sun is weak, sea breezes hardly ever occur at these latitudes, and in summer they

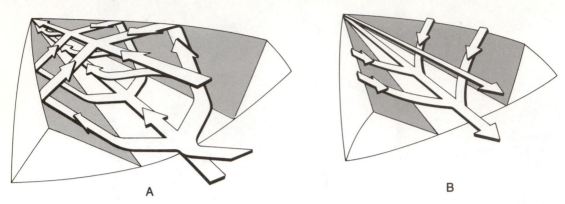

A   B

Figure 7.8 Schematic illustration of flow in a valley: (A) in the afternoon when the valley and up-slope winds are occurring; (B) about midnight when the flow is down the valley and the slope. [After F. Defant, *Compendium of Meteorology*, ed. T. F. Malone (Boston: American Meteorological Society, 1951), p. 665.]

occur only on days when the prevailing wind is sufficiently light. Even then the sea breeze may start only after enough heating has taken place to overcome a wind from the land produced by a larger-scale system. When the latter situation occurs, the inception of the sea breeze may take the form of a *sea-breeze front*, a sudden invasion of cool air from the sea replacing very hot air. Not infrequently, on summer afternoons on the New England coast, temperature drops of 20 degrees or more are experienced accompanying the occurrence of sea-breeze fronts.

In locations where mountains are close to the ocean, the valley–mountain effect combines with the sea–land influence to produce a more vigorous and more extensive diurnal wind oscillation. Los Angeles, California, is an area in which the combination of the two effects occurs. During the summer the land there cools at night only a little below the sea temperature, and the inland desert and mountains become very hot, so that there is practically no land breeze and a stronger-than-normal sea breeze. In winter, on the other hand, the land becomes colder than the sea for several hours, and the land breeze is stronger and of longer duration than the sea breeze.

Figure 7.9 shows the hourly resultant winds (i.e., winds averaged with respect to direction as well as speed) for three different months at downtown Los Angeles. The general response of the winds to the proximity of the ocean to the west and south, and to the mountains to the north and east, is clearly seen. For instance, in September, which is characteristic of the warm months, the winds are from a westerly direction from 10:00 A.M. to midnight, while from midnight to 8:00 A.M. they are light and variable, but mostly from easterly directions. In December,

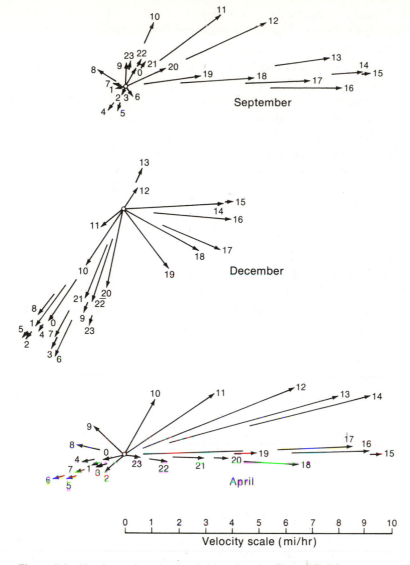

Figure 7.9 Hourly resultant wind at the Los Angeles Federal Building.

characteristic of the cold season, the sea breeze begins later and ends earlier, and its strength is somewhat lighter, while the land breeze is correspondingly stronger and of longer duration. The resultant winds for the month shown, December 1947, were apparently affected by the type of large-scale systems already referred to. It appears that a system that resulted in a period of predominantly northerly flow moved through, adding a net north contribution to each hourly resultant and almost eliminating the southerly component of the sea breeze.

Inspection of Figure 7.9 shows that, in addition to the general reversal of wind direction between day and night, there is a tendency for gradual turning of the wind clockwise (to the right) as the day goes on. This turning is associated with the earth's rotation, which causes moving air to tend to be deflected. This phenomenon will be discussed in the next chapter.

Besides the daily effect of heating and cooling along coasts, in some parts of the world the seasonal variation of the difference in temperature between continents and oceans produces a corresponding seasonal variation in winds. The winds blow from the cooler oceans to the heated continents in summer, and from the centers of cold over the continents toward the oceans in winter. Winds that occur in well-organized seasonal patterns are called *monsoons*, from the Arabic word *mausim*, meaning season. The best-known example is the southwest monsoon of India, which blows from the Arabian Sea across much of the Indian peninsula from June to October, bringing with it general rains that, along the west coast and in the foothills of the Himalayas, are among the heaviest on earth. Similar monsoons are present throughout southern and eastern Asia. For instance, the dominant features of the climate of Vietnam are the southwest monsoon in the summer and the northeast monsoon in the winter. Monsoons are present in greater or lesser degree wherever continents are sufficiently extensive at low latitudes for there to be large areas in which the temperature is markedly different from the adjoining oceans.

There are a number of other local and regional winds. Many of them are due to the local topographic effects on the pressure gradients arising from temperature contrasts produced by the large-scale systems. The *mistral*, a northerly wind blowing from the Alps down the Rhone Valley of France toward the Mediterranean Sea, and the *bora*, a similar northeast wind blowing from the mountains of Yugoslavia onto the Adriatic, are examples of *fall-winds*, in which the cold air flows downward under the direct action of gravity, like the water in a waterfall. Although the air gets warmer because of adiabatic compression as it descends, it is so cold to begin with that it arrives at the coast as a strong chill wind, which temporarily puts an end to bathing at the resorts of the French Riviera.

The *foehn* of the Alps, the *chinook* of the area east of the Rockies, and the *Santa Ana* of southern California are also due to flow downward from high mountains or plateaus, but in these instances the heating by compression is sufficient to result in abnormally high temperatures and very low humidities. The onset of the chinook is frequently accompanied by rises in temperature of 20°C or more in a few minutes, from below freezing to more than 15°C (59°F), and rapid melting or evaporation of the snow on the ground. Similar rises in temperature accompany the foehn in Europe and the Santa Ana-type winds of the west coast of the United States. To the unpleasant, strong, hot winds and low humidities have been attributed physiological responses called "foehn sickness." The reality of a well-defined illness due to the foehn is questionable, but there is no doubt that it causes physical discomfort and stress, which may exacerbate already existing ailments.

Occasionally, the foehn, chinook, or Santa Ana winds attain speeds that cause damage to buildings. If fires break out, the low humidities and strong winds cause them to spread rapidly. Thus, two of the authors of this book had their homes completely destroyed in a fire that burned thousands of acres of brush in the Santa Monica Mountains of Los Angeles and almost 500 homes before a change in wind and the inflow of moist air permitted it to be controlled.

Questions, Problems, and Projects

1. A boat is being propelled across a river at a speed of 4 m s^{-1} perpendicular to the shore. The river is flowing at the rate of 3 m s^{-1}. Make a diagram according to the parallelogram rule showing the resultant velocity of the boat.

2. At a point in a fluid the pressure in every direction is the same. In view of this fact, explain how there can be a resultant pressure force that might cause the fluid to accelerate.

3. Make a schematic graph of the variation with time of the pressure at sea level over a tropical island for a 24-hour period and the pressure at a short distance offshore. Do the same thing for the pressure at a height of about 1 km.

4. If no other forces than the pressure-gradient force were acting, what would be the direction of the winds in relation to the isobars on a weather map? What can you say about the speed of the wind in this case? How would friction with the ground modify this?

5. Suppose that on a map representing the pressure distribution at 3 km, the distance between the 700-mb and 696-mb isobars is 300 km at a place where the temperature is 0°C. Compute the horizontal pressure force on unit mass at that place. Why is it so different from the value we found for 4-mb isobars spaced 300 km apart at sea level?

6. Suppose that, at a coast, the temperature of the air over land 2 or 3 hours after sunrise is 20°C, while over the sea it is 15°C. Assume the temperature is constant with height and that the pressure at sea level is 1000 mb both over land and over the sea. (a) What is the difference in pressure at a height of 100 m between land and sea? (b) If the distance between points having that pressure difference is 5 km, what is the pressure force per unit mass (or the acceleration)? (c) What speed, starting from rest, would it produce if it acts for one hour?
[*Answers:* (a) 20.2 Pa; (b) 0.0034 m s^{-2}; (c) 12.2 m s^{-1}]

8 Large-Scale Motions

8.1 The effect of the earth's rotation

In Section 7.4 it was shown that horizontal pressure differences that arise from the differences in temperature between sea and land at a coast produce horizontal pressure differences that cause winds to blow from cold to warm near the ground and the opposite aloft. Since the regions near the equator are warmer than those at high latitudes we would expect similar pressure-gradient forces and wind patterns on a global scale. The greater warmth at low latitudes would be expected to lead to equatorward flow of air near the ground and poleward movement of it aloft. That the winds do not have these directions most of the time is a matter of common knowledge. Near the ground in low latitudes the easterly trade winds are present, while the surface winds at higher latitudes are characterized by the "prevailing westerlies." In the upper troposphere at almost all latitudes the predominant direction of the winds is westerly. The question arises, What is responsible for this zonal (along parallels of latitude) tendency of the winds when the heating should

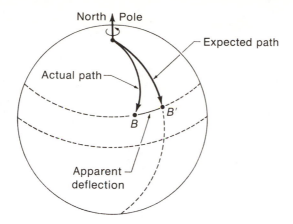

North Pole

Expected path

Actual path

B'

B

Apparent deflection

Figure 8.1   Deflection of an equatorward projectile.

produce meridional flow (that is, flow north–south or south–north along the meridians of longitude)? The answer lies in the effect of the earth's rotation.

The effect of the rotation of the earth on any moving object is to make it appear as though a force is acting on the object. This apparent force is called the *Coriolis force*, after the French mathematician G. G. de Coriolis (1792–1843) who first treated it quantitatively. The effect had been used qualitatively a century earlier by George Hadley (1685–1768) to explain the trade winds. The reason for introducing the Coriolis force is that an object moving over the earth's surface with constant velocity relative to an absolute frame appears to be accelerated relative to a person moving with the rotating earth. Since Newton's law says that if there is an acceleration a force must be acting, we find it convenient to speak of the acceleration as being "caused" by the force that would produce it.

The nature of the apparent acceleration and the corresponding "force" can be seen in the following way. Suppose that a projectile is fired from the North Pole and aimed at a point B at a lower latitude (see Figure 8.1). The earth rotates toward the east, and in the time it takes the projectile to reach B, the target that was at B will have moved to the position B'. An observer on the earth would have expected the projectile to reach the target, since he or she is not conscious of the earth's rotation. The path expected would be PB'. The actual path of the projectile is PB, and thus to the observer the projectile appears to have been deflected to the right. The observer would conclude that there was a force acting at right angles to the motion of the projectile to produce this deflection.

The rotation of the earth toward the east corresponds to counterclockwise rotation when one looks down on the North Pole and to clockwise rotation when one looks down on the South Pole. Consequently, if the projectile were fired from the

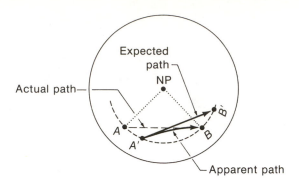

Figure 8.2 Deflection of an eastward projectile.

South Pole instead of the North Pole, the deflection would be in the opposite direction. This leads to the following rule: *The Coriolis force acts to the right in the Northern Hemisphere, and to the left in the Southern Hemisphere.*

A similar deflection or apparent force occurs no matter in which direction the projectile moves. If the projectile is aimed parallel to a parallel of latitude (other than the equator), the deflection will also be to the right in the Northern Hemisphere and to the left in the Southern Hemisphere. It is hard to draw a diagram showing this on a sphere. However, the same effect occurs on a rotating disc (e.g., a phonograph turntable) and Figure 8.2 shows how the deviation arises. The projectile is fired from point A at the target at point B. In the time it takes to get to B the target moves to B' and the cannon to A', so the path expected by the observer moving with the disc (earth) is $A'B'$. In following the actual path AB, the projectile appears to the observer to be following the curve $A'B$, and thus deviating to the right from the expected path.

If the projectile is fired southward (or northward) from a latitude other than the North Pole (or South Pole), the description is similarly complicated by the fact that the projectile will have an eastward velocity due to the rotation of the earth. Thus, in Figure 8.3, a projectile fired northward from O toward a target at C will be moving eastward at the speed of the earth's surface at O. In the time it was expected to reach C, in which O has moved to O' and C to C', it will reach a point C'' to the east of C', where $CC'' = OO' > CC'$. The expected path is $O'C'$; the apparent path $O'C''$ bends to the right of it.

Similar constructions can be made for projectiles fired southward from a position not at the North Pole in the Northern Hemisphere, westward along a parallel of latitude, and from corresponding positions in various directions in the Southern Hemisphere.

Movement in any arbitrary direction can be decomposed into one component along the meridian and one component along the parallel of latitude, for each of which the rotation produces an apparent deflection. Putting the effects of the

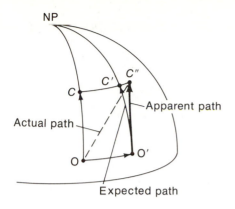

NP

Actual path

Apparent path

Expected path

components back together, it is clear that every such arbitrary motion is accompanied by a tendency to be deflected to the right in the Northern Hemisphere and to the left in the Southern Hemisphere. When one is firing an ordinary rifle in target practice or in hunting, the effect is too small for one to need to take it into account in aiming. But in aiming long-distance projectiles, whether from big cannons or rocket missiles, the Coriolis effect must be included in the ballistic computation.

In large-scale air motion, the effect of the earth's rotation is a dominant factor, and the winds are usually close to the velocity for which the Coriolis force and horizontal pressure-gradient force are in balance. This velocity is called the *geostrophic wind velocity*. In the next sections we shall arrive at quantitative expressions for the Coriolis force and the geostrophic wind velocity.

8.2   Quantitative expression for the Coriolis force

To evaluate quantitatively the magnitude of the Coriolis effect, we must make use of some additional physical relationships. The first of these is *angular velocity*, which is simply the rate of turning of a rotating body. It is the angle through which the body turns in unit time, and it might be expressed in terms of degrees per second, but for convenience in computation it is usually expressed in radians per second, where a radian is $1/(2\pi)$ of a complete circle (2π radians $= 360°$).

Since the earth makes one complete rotation in a day (really a sidereal day), its angular velocity Ω is obtained by dividing 2π radians by the number of seconds in a sidereal day (86,164.1). This gives $\Omega = 7.29 \cdot 10^{-5}$ radians/sec.

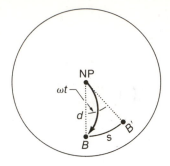

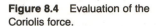

Figure 8.4 Evaluation of the
Coriolis force.

This is the rate at which the earth turns around its axis. At the poles the earth's surface is turning around the vertical at this rate, like a slow-moving phonograph disc. At an arbitrary latitude away from the poles, the earth's surface also rotates around the vertical, but more slowly. You can see that it is rotating around the vertical by watching the stars at night, which make partial circuits around the zenith, except at the equator. Thus, the rate of rotation around the vertical varies with latitude, from zero at the equator to Ω at the poles. If ω is the rate of rotation of the horizontal surface around the vertical at latitude ϕ, it can be shown that $\omega = \Omega \sin \phi$, where $\sin \phi$ is a trigonometric function that is zero when $\phi = 0$, one when $\phi = \pi/2$ (90°), and 0.7 when $\phi = \pi/4$ (45°).

Next, we must consider the linear velocity of a point a distance d from the axis of a system rotating with angular velocity ω. Since there are 2π radians in one complete rotation and ω is the number of radians the system turns in a second, one complete rotation takes $2\pi/\omega$ seconds. The linear distance traveled by a point at a distance d from the axis is $2\pi d$, and thus the linear velocity is $2\pi d \div 2\pi/\omega = \omega d$.

Finally, we need the expression for the distance traveled by a body that is undergoing a constant acceleration, a. If the speed is initially zero, at a time t the speed will be at. The average speed is $at/2$, and the distance s traveled is this average speed multiplied by the time, $s = at^2/2$.

Now we are in a position to derive an expression for the Coriolis force. We return to the treatment in terms of a projectile moving from the axis of rotation. For simplicity we shall consider the motion relative to a rotating disc, which is comparable to the rotation of the horizontal surface at an arbitrary latitude. If (Figure 8.4) it takes a time t for the projectile to move from P to B at a constant speed v, we have

$$d = PB = vt$$

To an observer on the rotating disc, who expected the projectile to reach B', the point to which the target originally at B has moved, it appears that the projectile has been deflected a distance $s = BB'$ at right angles to the direction it was aimed. The linear velocity of this deflection is s/t, and we have seen above that this velocity is also $\omega d = \omega vt$, so that

$$\frac{s}{t} = \omega vt$$

$$s = \omega vt^2$$

Now, the observer on the rotating disc attributes this deflection to the action of a force (the Coriolis force) that produces an acceleration a. For this acceleration

$$s = \frac{at^2}{2}$$

If we set these two expressions for s equal and solve for a, we have

$$a = 2\omega v$$

If we substitute for ω its value for rotation around the vertical at latitude ϕ, and remember that the acceleration is equal to the force per unit mass, we have the expression for the magnitude of the Coriolis force per unit mass

$$F_c = 2(\Omega \sin \pi) v = fv \qquad (8.1)$$

where the single letter f is used to represent $2\Omega \sin \phi$, a quantity that occurs repeatedly in meteorological equations. This quantity is referred to as the Coriolis parameter. Table 8.1 gives the value of f at various latitudes, and the Coriolis force corresponding to three speeds at each latitude.

Table 8.1   Variation of Coriolis Parameter and Coriolis Force with Latitude

Latitude ϕ (deg)	Coriolis parameter f (s^{-1})	Coriolis force (m s^{-2}) for speeds of		
		5 m s^{-1}	10 m s^{-1}	50 m s^{-1}
0	0	0	0	0
15	$3.8 \cdot 10^{-5}$	$1.9 \cdot 10^{-4}$	$3.8 \cdot 10^{-4}$	$18.9 \cdot 10^{-4}$
30	$7.3 \cdot 10^{-5}$	$3.6 \cdot 10^{-4}$	$7.3 \cdot 10^{-4}$	$36.4 \cdot 10^{-4}$
45	$10.3 \cdot 10^{-5}$	$5.2 \cdot 10^{-4}$	$10.3 \cdot 10^{-4}$	$51.6 \cdot 10^{-4}$
60	$12.6 \cdot 10^{-5}$	$6.3 \cdot 10^{-4}$	$12.6 \cdot 10^{-4}$	$63.1 \cdot 10^{-4}$
75	$14.1 \cdot 10^{-5}$	$7.0 \cdot 10^{-4}$	$14.1 \cdot 10^{-4}$	$70.4 \cdot 10^{-4}$
90	$14.6 \cdot 10^{-5}$	$7.3 \cdot 10^{-4}$	$14.6 \cdot 10^{-4}$	$72.9 \cdot 10^{-4}$

It is interesting to compare the values of the Coriolis force given in Table 8.1 with the example of the magnitude of the horizontal pressure-gradient force calculated in Section 7.3. We saw there that if sea-level isobars drawn for every 4 mb are 300 km apart, the pressure-gradient force (per unit mass) is $1.03 \cdot 10^{-3}\,\mathrm{m\,s^{-2}}$. It turns out that at 45°, for instance, the Coriolis force would equal the magnitude of this pressure-gradient force if the wind speed is 10 m s^{-1}. Thus, as suggested at the end of the last section, it is quite possible, for reasonable wind speeds, that the Coriolis force can balance the pressure-gradient force, provided of course that their directions are such that they oppose each other.

Since the Coriolis force always acts at right angles to the direction of motion, it cannot change the speed but only the direction of motion. Acting by itself, it would tend constantly to turn a moving body to the right, and would cause it to go around in circles. The size of the circle would depend on the speed of the moving body (e.g., the projectile), but it turns out that the period of time it takes to go all the way around is independent of the speed, and depends only on the value of f, that is, on the latitude. At 30°, for instance, it takes exactly one day; at 45° it takes 0.7 days. This period is called the *inertial period*.

The inertial period is related to the period of the *Foucault pendulum*, which is frequently exhibited in museums to demonstrate the fact that the earth is rotating on its axis. The Foucault pendulum consists of a heavy weight suspended from a mount that permits it complete freedom of motion. The pendulum, when moved from its equilibrium position and let go, will swing parallel to a fixed plane in absolute space. As time passes the earth rotates beneath it, and to the observer moving with the earth, the plane in which the pendulum is swinging appears to rotate slowly. Usually there are markers that are knocked down as the pendulum "rotates," so that the museum visitor can see how far it moves between the time he first sees it and the time he returns after going to see other exhibits. The period of the Foucault pendulum, called a *pendulum day*, is exactly twice the inertial period.

We can now see why the sea breeze, as illustrated in Figure 7.9, turns with time. When it starts (about 9:00 A.M.), air reaching the observing station has been moving in response to the sea–land pressure-gradient force for less than an hour, and thus the Coriolis force will have deflected its direction very little. But by 3:00 P.M. the air reaching the station may have been moving approximately 6 hours, and the Coriolis force would turn it the portion of a complete rotation equal to the fraction that 6 hours is of the inertial period. At 30° 6 hours is one-fourth of the inertial period, and the sea breeze would be turned one-quarter of the way around (90°), until it parallels the coast. While the presence of other forces, including friction, complicates the situation, the figure shows that the actual turning of the sea breeze at Los Angeles (latitude 34°) is not far from the rate indicated by consideration of the Coriolis force.

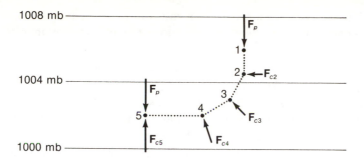

Figure 8.5   Adjustment of wind to geostrophic wind in uniform pressure gradient.

8.3   The geostrophic wind

Let us now consider a situation in which initially there are horizontal variations of pressure, as represented by the isobars in Figure 8.5, but no motion. A parcel at point 1 would be acted on by the pressure-gradient force, represented by the arrow marked $\mathbf{F}_p$, which would accelerate it, displacing it to point 2. At point 2 it would still be acted on by $\mathbf{F}_p$ (which we assume to be constant, for simplicity), but because it is moving it would be acted on also by a Coriolis force $\mathbf{F}_{c2}$, tending to deflect it to the right to point 3. By the time it reaches point 3 it is moving faster, and thus $\mathbf{F}_{c3}$ is larger and bends the path still more. When it reaches point 5 it is moving parallel to the isobars, so that $\mathbf{F}_{c5}$ points toward high pressure, exactly opposite to $\mathbf{F}_p$, which is always perpendicular to the isobars and points toward low pressure. If at this time the speed would have the right value so that $\mathbf{F}_c$ has the same magnitude as $\mathbf{F}_p$ (i.e., $\mathbf{F}_c = -\mathbf{F}_p$), the parcel would no longer be accelerated, but would move with constant speed along the isobars.

When the air is moving in this fashion, so that the Coriolis force is exactly equal and opposite to the pressure-gradient force, it is said to be in *geostrophic balance,* and the wind velocity for which this is true is called the *geostrophic wind*.

It can be shown mathematically that if the initial situation were as postulated above, with no other forces acting, and if the air motion did not produce changes in the pressure distribution, at the time that the motion of the air parcel became parallel to the isobars it would be moving so fast that $\mathbf{F}_c$ would be considerably greater than $\mathbf{F}_p$ and thus it would be accelerated toward higher pressure. It would then be slowed down by the pressure force acting opposite to the direction of motion. The result would be that instead of reaching the equilibrium shown by point 5 in Figure 8.5, it would oscillate around it. It would "try" to attain equilibrium but would always overshoot it.

In actuality the air does not start from rest, but is always in a state much closer to geostrophic balance. In effect, the wind is continually attempting to adjust itself to a balance between the pressure-gradient force and the Coriolis force. Thus, with the exception of locations near the ground, where frictional forces are important, and places where the isobars are strongly curved (which will be discussed in the next section), the geostrophic wind is a good approximation to the real wind. The importance of this lies in the fact that as a consequence the wind flow can be deduced from the isobaric direction and spacing.

The direction of the geostrophic wind is parallel to the isobars, with low pressure to the left in the Northern Hemisphere (to the right in the Southern Hemisphere). The relationship of actual wind direction, which deviates somewhat from the geostrophic direction, and the pressure field was enunciated in the middle of the nineteenth century by a Dutch meteorologist, C. H. D. Buys Ballot (1817–1890) in the form of a rule, referred to as Buys Ballot's law: *If you put your back to the wind, low pressure will be on your left (right) in the Northern (Southern) Hemisphere.*

The relation between the isobaric spacing and the geostrophic wind speed is obtained by setting the magnitude of the Coriolis force, given in equation (8.1), equal to that of the pressure-gradient force, given by equation (5.4). If v_g represents the geostrophic speed,

$$v_g = \left| \frac{\alpha \, \Delta p}{f \, \Delta n} \right| \tag{8.2}$$

The geostrophic wind speed varies

1. With pressure gradient, or inversely with the distance between isobars;
2. With specific volume, or inversely as the density;
3. Inversely with the Coriolis parameter, or inversely with the sine of the latitude.

For 4-mb isobars 300 km apart and specific volume 0.775 m³/kg, for which we computed F_p in Section 7.3, the geostrophic wind speed would be the following:

$$14.2 \text{ m s}^{-1} \text{ at } 30° \text{ latitude}$$
$$10.0 \text{ m s}^{-1} \text{ at } 45° \text{ latitude}$$
$$8.2 \text{ m s}^{-1} \text{ at } 60° \text{ latitude}$$

If the pressure field is represented by the contour lines of a constant pressure surface, as described in Section 7.3, the expression for the geostrophic wind speed is obtained by using the expression in equation (7.5) for the pressure-gradient force. This gives

$$v_g = \frac{g}{f}\frac{\Delta h}{\Delta n} \tag{8.3}$$

where Δn is the distance between contour lines drawn for a height difference Δh. Thus, if the height interval is the same for various constant pressure maps—say, 850 mb, 700 mb, 500 mb, 200 mb, and 100 mb—the same spacing of contour lines will mean the same geostrophic speed (at the same latitude). One does not need to take into account the difference of density, or specific volume, as one goes up to the levels where the pressure is lower.

Near the earth's surface, friction with the ground causes considerable deviations from the geostrophic wind, both in speed and in direction. We shall discuss the effects of friction later. Aloft, friction is negligible and the wind blows almost exactly parallel to the isobars on constant level maps or the contour lines on constant pressure maps with the speed given approximately by equation (8.2) or (8.3).

The effect of friction usually extends only about 500 m to 1 km above the ground. At this height the pressure distribution is not very different from the sea-level distribution, so that drawing isobars on the surface weather map is a good way to summarize the wind distribution there. This is because, above the layer of frictional influence, isobars are approximate *streamlines*, showing the direction in which the air is moving, and their spacing indicates the wind strength, light where they are far apart and strong where they are close together. Similarly, the contour lines on constant pressure charts represent the streamlines, and their spacing indicates the wind speed at upper levels. As we shall see, the direction and speed of the winds near the ground also can be inferred from the isobars.

8.4 Curved flow. The gradient wind

As we have seen in the examples of the weather map in Section 1.3, the isobars are not straight lines but curves that sometimes enclose centers of high or low pressure. In order to remain parallel to the isobars, the wind would have to follow curved paths and thus be accelerated. For curvature of the flow around low pressure, the pressure-gradient force must be greater than the Coriolis force, and the opposite is true for curvature around high pressure. Since the Coriolis force depends on the wind speed, we see that the speed of wind corresponding to the flow following curved isobars is different than for straight isobars. For the same spacing of isobars at the same latitude, the wind speed is less for flow curved around low pressure than for straight flow; for flow curved around high pressure,

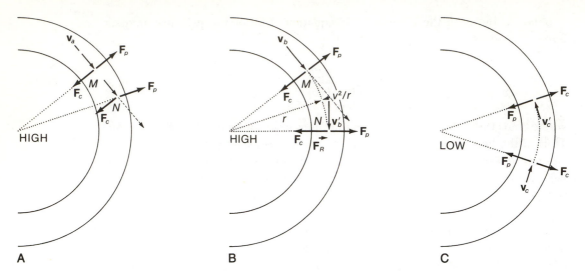

Figure 8.6 Acceleration and balance of forces in flow along curved isobars.

it is greater than for straight flow. To see why this is true, consider a situation with circular isobars curving around a high-pressure center, as illustrated in Figure 8.6A and B.

In diagram A the air at point M is assumed to be moving with the geostrophic velocity v_a corresponding to the direction and spacing of the isobars. Initially the pressure gradient force $\mathbf{F}_p$ is exactly balanced by the Coriolis force $\mathbf{F}_c$ (the magnitude of which is fv_a). The air is unaccelerated and moves in a straight line with constant speed. But a short time later, when it reaches point N, the forces will no longer be in balance, since the direction of the isobars, and thus of $\mathbf{F}_p$, is different there. In the diagram it is seen that the unbalanced resultant force would lead to an increase in the speed of the air parcel, and thus an increase in $\mathbf{F}_c$. In turn, when the Coriolis force increases, there will be an acceleration to the right, toward the high pressure center, and the air will begin to move in a curved path bending around the high.

In diagram B it is assumed that the air at M is moving from the start with a velocity v_b sufficiently larger than the geostrophic so that $\mathbf{F}_c$ is just the right amount larger than $\mathbf{F}_p$ to push the air constantly toward the high center exactly the amount required to keep it moving along the circular isobar.

The Coriolis force, $\mathbf{F}_c$, which is always perpendicular to $\mathbf{v}$, will remain exactly opposite to $\mathbf{F}_p$, and since both $\mathbf{F}_c$ and $\mathbf{F}_p$ are therefore at right angles to the direction of motion at each instant, there is no change in speed. Only the direction changes, but a change in direction of motion is an acceleration, as well as a change in speed. Acceleration toward a center (in this case the center of the HIGH) is called *centripetal* acceleration. Whenever a body moves in a curved path, it is undergoing centripetal acceleration. It can be shown that if the radius of curvature of the path is r, the amount of the acceleration is v^2/r.

The above discussion shows that when the speed is larger than that required for geostrophic balance, so that F_c is larger than F_p, there will be a centripetal acceleration that produces curved flow around high pressure.

Sometimes, rather than speak of the centripetal acceleration associated with curved motion, reference is made to the *centrifugal force*. This is the apparent force outward from the center that the curved motion appears to exert to balance the forces producing the centripetal acceleration.

The situation in which the isobars are curved around a low center is shown in diagram C. In this case F_p is directed toward the center, and in order that the air may move exactly along an isobar around the center, F_c must be smaller than F_p. The speed for flow at constant speed around a low center is thus *less* than the geostrophic speed.

The illustrations in Figure 8.6 are for the Northern Hemisphere, for which the Coriolis force acts to the right of the direction of motion. Inspection of this figure leads to the rule: *In the Northern Hemisphere air normally flows clockwise around high-pressure centers and counterclockwise around low-pressure centers.* This statement is true since normally the winds are as close to geostrophic balance as they can achieve. In the Southern Hemisphere the normal flow is exactly the opposite: clockwise around low-pressure centers and counterclockwise around high centers.

The normal flow around low-pressure areas is called *cyclonic flow*, and the systems consisting of low-pressure centers and the accompanying cyclonic wind systems are called *cyclones*. *Anticyclones* correspondingly refer to the high centers and accompanying anticyclonic flow.

It is important to remember:

1. *Cyclonic flow is counterclockwise in the Northern Hemisphere and clockwise in the Southern Hemisphere.*

2. *Anticyclonic flow is clockwise in the Northern Hemisphere and counterclockwise in the Southern Hemisphere.*

The wind illustrated in diagrams B and C of Figure 8.6 is called the *gradient wind*. The gradient wind is the wind that would flow along curved isobars with no change in speed. From the preceding discussion we see that for anticyclonic flow the speed of the gradient wind is greater than the geostrophic wind speed; for cyclonic flow it is less.

The quantitative expression for the gradient wind speed can be derived by writing an equation for the balance of forces shown in the diagrams. Using Newton's law of motion, equation (6.1), which says that the acceleration is equal to the sum of the forces acting, we write

$$\frac{v_G^2}{r} = F_p - F_c \qquad (8.4)$$

where v_G is the gradient wind speed. If we define the radius of curvature r as positive for cyclonic flow (in which $F_p > F_c$) and negative for anticyclonic flow (in which $F_p < F_c$), this equation is correct for both types of flow. Note that it is also correct for geostrophic flow, straight isobars, for which $r = \infty$ and $v_G^2/r = 0$.

From equation (8.2) we note that $F_p = f v_g$, where v_g is the geostrophic wind speed for straight isobars having the same spacing, and from equation (8.1) the value of the Coriolis force for the velocity v_G is $F_c = f v_G$. Substituting these values for F_p and F_c in equation (8.4), we get

$$\frac{v_G^2}{r} = f v_g - f v_G \qquad (8.5)$$

This is a quadratic equation in v_G, which can be solved directly. For clarity in interpreting the result, it is convenient to divide both sides of the equation by v_G^2, which gives a quadratic equation in $1/v_G$, solve this equation, and take the reciprocal. The result is

$$v_G = \frac{2v_g}{1 + \sqrt{1 + 4\dfrac{v_g}{fr}}} \qquad (8.6)$$

The plus sign is chosen for the radical to select the solution that becomes equal to the geostrophic wind speed when r is infinitely large, corresponding to straight flow.

This equation agrees with our qualitative conclusion that $v_G < v_g$ for cyclonic flow, r positive, since in this case the denominator of the right-hand side of the equation is greater than 2. Similarly, for anticyclonic flow, r negative, the denominator is less than 2 and $v_G > v_g$, in accord with our qualitative discussion. It also shows that for anticyclonic flow there is a limit on the pressure gradient near the high-pressure center if the flow is to follow the isobars. If v_g, which is our convenient shorthand for $(\alpha/f)(\Delta p/\Delta n)$, is greater than $-f\,r/4$ (which is a positive number), the quantity under the square root sign would be negative and v_G would be an imaginary number. To have anticyclonic gradient flow, the geostrophic speed v_g must always be smaller than $-fr/4$, and thus must go toward zero close to the high center, where $-r$ is small. Correspondingly, Δn, the spacing between iso-bars, must be larger and larger as the high center is approached.

There is no such limitation for the pressure gradient for cyclonic flow, and isobars can be very close together near low-pressure centers.

Since for the same isobaric spacing the wind is stronger in anticyclonic than in cyclonic flow, the paradoxical statement is sometimes made that winds are stronger in anticyclones than in cyclones. This is not usually true, because near high centers the isobars have to be far apart, while near low-pressure centers they are frequently very close together, with winds sometimes exceeding hurricane strength.

To get some idea of the amount of difference between the gradient wind and the geostrophic wind, let us consider again the example of 4-mb isobars 300 km apart, for which we found $v_g = 10.0$ m s^{-1} at 45° latitude, where $f = 10.3 \cdot 10^{-5}$ s^{-1}. For $r = 750$ km $= 7.5 \cdot 10^5$ m (cyclonic flow in a circle with radius 750 km) we have

$$v_G = \frac{2 \cdot 10.0}{1 + \sqrt{1 + 4 \cdot \dfrac{10.0}{(10.3 \cdot 10^{-5} \cdot 7.5 \cdot 10^5)}}} = 9.0 \, \text{m s}^{-1}$$

If the flow were anticyclonic ($r = -750$ km)

$$v_G = \frac{2 \cdot 10.0}{1 + \sqrt{1 - 4 \cdot \dfrac{10.0}{(10.3 \cdot 10^{-5} \cdot 7.5 \cdot 10^5)}}} = 11.8 \, \text{m s}^{-1}$$

Sometimes the gradient wind is defined in terms of the curvature of the actual path of the air rather than the isobars, since the path may deviate from the isobars because of changes of the pressure field with time, or for other reasons.

In addition to the deviation of the air trajectories from the isobars, other accelerations are frequently as large as the centripetal acceleration, except in cases of large curvature. Except near the centers of hurricanes and other intense cyclonic storms, these other deviations may render the gradient wind speed no better an approximation of the actual speed than the geostrophic speed.

8.5   Vortices and vorticity

A system in which a fluid is rotating around a center is called a *vortex*. Thus, cyclones and anticyclones are vortices. In general, flow patterns in which there is rotation around centers in the sense cyclones rotate are called *cyclonic vortices*. *Anticyclonic vortices* are defined analogously.

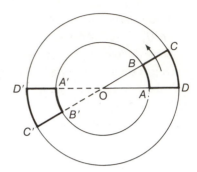

Figure 8.7 Vorticity of air moving uniformly around a cyclonic vortex.

Let us consider a portion of the air that is moving uniformly around a circular cyclonic vortex (Figure 8.7). For convenience we select a sample bounded by two isobars AB and CD and two radii OD and OC. Our sample initially is at $ABCD$. In the time it takes for the air to move halfway around the center, the sample reaches the position $A'B'C'D'$. In addition to the horizontal displacement of the sample, the sample has turned around, so that D, which was to the right of A, is now to its left. The amount of turning of a small portion of fluid is called its *vorticity*. We see that when air moves uniformly around a center in the cyclonic sense it also turns on itself, or locally, in the same sense. It then is said to have *cyclonic vorticity*. Analogously, a parcel of air turning locally in the sense air moves around an anticyclone is said to have *anticyclonic vorticity*.

The parcel does not have to be part of a closed vortex to have vorticity. Vorticity refers to the instantaneous rotation of a parcel, so that any parcel moving in a curved path at a particular time with no variation in speed across it will possess vorticity.

Actually, even a parcel moving in a straight line may also have vorticity if the speed varies across it, so that part of it is moving faster than another part. The variation of the speed of a parcel with distance across it is called *shear*. To see that shear produces vorticity, consider Figure 8.8, in which the lower part of parcel $ABCD$ is moving slower than the upper part. After a time the parcel will be in position $A'B'C'D'$. To illustrate that the motion includes rotation of the parcel, the original direction of the diagonal is drawn at the new position. The diagonal has turned an amount shown by the curved arrow.

If the path of the parcel is curved and the speed varies across it, the vorticity is the result of the two effects. It is for this reason that we specified that the motion around the closed center be uniform when we first illustrated vorticity. If the

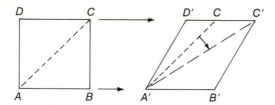

Figure 8.8   Vorticity of air moving in a straight line with shear.

speed varies so that the vorticity due to shear is opposite to that due to curvature, it is possible to have motion around a center that has no vorticity. In this case the motion is called *irrotational*, even though the parcel is going around the center.

For completeness we give the quantitative expression for the vorticity, although it is not needed in our use of the concept. It is usually represented by the Greek letter ζ (zeta), and has the value

$$\zeta = \frac{v}{r} + \frac{\Delta v}{\Delta n} \qquad (8.7)$$

where v is the speed of the parcel, r the radius of curvature of its path (taken positive for cyclonic, negative for anticyclonic flow), and Δv is the change of speed in a distance Δn perpendicular to the wind direction, taken positive if the speed increases toward the right of the wind direction (Northern Hemisphere). It can be shown that this value of the vorticity is equal to twice the angular velocity of the parcel. The first term on the right side of the equation is the vorticity due to curvature, and the second is that due to shear.

Regarded in absolute space, a parcel that is stationary with respect to the earth's surface is rotating and thus has vorticity. As was pointed out in Section 8.2, the earth's surface rotates around the vertical at a rate $\omega = \Omega \sin \phi$. Twice this value, $2\Omega \sin \phi$, which we assigned the name Coriolis parameter f, will thus be the absolute vorticity of a stationary parcel at latitude ϕ. The absolute vorticity of a moving parcel becomes

$$\zeta_o = f + \zeta \qquad (8.8)$$

Under certain circumstances the absolute vorticity of air parcels is conserved. When parcels move equatorward, where f decreases, ζ would have to increase to keep ζ_o constant, and if the increase in cyclonic vorticity takes the form of curved

flow, as it usually does, the parcels would bend eastward. They would gain curvature as long as their motion had an equatorward component. Eventually they
would turn enough to start back toward the pole. Again as they move poleward,
they would lose cyclonic and eventually gain anticyclonic curvature. The result of
this alternation of cyclonic and anticyclonic curved flow is a wave pattern that
characterizes much of the upper atmosphere. These upper-level waves in the
westerlies will be discussed in Section 11.5.

8.6   The effect of friction

In deriving the expressions for the speed of the geostrophic wind and the gradient
wind, we have ignored friction. Sufficiently far from the earth's surface it is small
enough to be ignored, but close to the ground the frictional force cannot be
neglected. A rough value for the thickness of the layer of frictional influence is
1 km. Above this height deviations from the geostrophic wind or the gradient wind
are usually small. The deviation due to friction is large at the ground and decreases
upward.

Near the ground (say, at the anemometer level, which is usually 10 or 15 m
above the ground), the wind is reduced and inclined toward low pressure by an
amount that depends on the nature of the surface. Over the sea the surface wind
has a speed that is about two-thirds of the gradient speed and a direction that
makes an angle of about 15°–25° to the isobars. Over fairly rough ground the
surface wind may be one-half of the gradient wind speed or less and inclined about
30°–40° from the isobars.

A crude interpretation of the effect of friction, which qualitatively explains why
the wind blows toward lower pressure with reduced speed when friction is acting,
is shown in Figure 8.9. The force due to friction with the ground is assumed to act
opposite to the surface wind and be proportional to it. If, as in diagram A, we start
with a situation in which the surface wind is blowing along straight isobars with
geostrophic speed, the pressure-gradient force $\mathbf{F}_p$ and the Coriolis force $\mathbf{F}_c$ will
just balance each other, but friction with the ground will cause a force $\mathbf{F}_F$ to act
on each parcel of air in the opposite direction to the direction of the wind. This
unbalanced force will decelerate the parcel, creating the situation represented in
diagram B, where $\mathbf{F}_c$ is smaller, corresponding to the reduced wind speed. The
excess of $\mathbf{F}_p$ over $\mathbf{F}_c$ pushes the parcel toward lower pressure. Balance is achieved
when, as shown in diagram C, the direction and speed of the wind are adjusted
so that the resultant of $\mathbf{F}_c$ and $\mathbf{F}_F$ is exactly equal and opposite to $\mathbf{F}_p$.

The effect of friction with the ground is transferred through the air by the

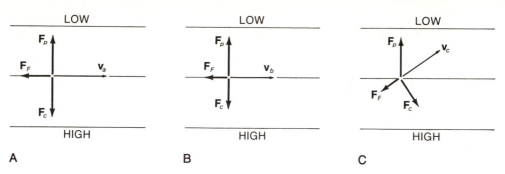

Figure 8.9 Effect of friction on surface winds.

irregular motions of particles of fluid. As mentioned in Section 7.3, the force exerted on a parcel of fluid by the fluid around it consists of two parts, the pressure and the *viscous stress*. The pressure is the force pushing perpendicular to the boundaries of the parcel. The stress is the drag that the fluid around it exerts tangential to the boundaries of the parcel. This viscous stress, or internal friction, exerts a force on the boundaries of the parcel proportional to the variation of velocity with distance across the boundaries. The variation of wind velocity with height is called the *vertical wind shear*, and the resultant horizontal force on a parcel bounded by plane surfaces above and below is the force due to the shear across the upper surface minus the force due to the shear across the lower surface.

If the molecular motions were the only irregular motions causing viscous stress, the effect of friction with the ground would be felt through a very thin layer. The roughness of the ground and an innate instability of the flow cause larger irregularities, or turbulence, which is responsible for the transfer of the effect of the ground up to considerable heights. When heating at the ground causes convective motions the thickness of frictional influence is further augmented, and if convective clouds, showers, and thunderstorms develop the entire troposphere may be affected.

The variation of wind with height in the layer of frictional influence is of the type shown in Figure 8.10, in which the arrows represent the wind velocity at the heights (in meters) indicated by the numbers at their points. The speed of the wind increases with height, and its direction turns to the right (in the Northern Hemisphere) until, at the top of the layer, it approximates the geostrophic velocity.

This change of wind with height is always present in the layer next to the ground. At greater heights there is frequently further change, due to the fact that the pressure systems vary with height. This variation will be explained in the next section.

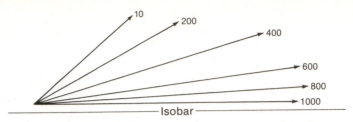

Figure 8.10 Variation of wind with height due to surface friction. Arrows represent wind at heights (in meters) given by numbers at their points.

8.7 Change of pressure gradients and winds at upper levels

The hydrostatic equation [equation (4.4)] shows that where the temperature is high the pressure decreases slowly with height, and where it is low the pressure decreases rapidly. At higher levels the pressure gradient may be regarded as a combination of the pressure gradient at sea level plus a contribution due to the horizontal gradient of the average temperature in the layer up to the particular level.

The change in pressure gradient as one goes upward that is due to the horizontal variation of temperature is demonstrated in Figure 8.11. This figure is a vertical cross section in which the temperature increases from left to right at all levels. For simplicity it is assumed that the pressure has a constant value of 1010 mb at sea level. On the left, where the air is cold, the pressure decreases more rapidly than on the right, where the air is warmer. Thus the distance to the position where the pressure is 1000 mb is smaller on the left than on the right, and, similarly, for the distance from 1000 to 990 mb, and so forth. The slope of the isobars in the cross section increases with height, and so does the pressure difference at constant elevations. Thus, the horizontal temperature gradient leads to a pressure gradient at upper levels. The greater the horizontal temperature gradient (i.e., the closer together the isotherms), the more rapidly the horizontal pressure gradient changes with height.

Corresponding to the change in pressure gradient with height, there will be a change in the geostrophic wind. This change in geostrophic wind with height because of temperature variation is called the *thermal wind*. The magnitude of the thermal wind is proportional to the gradient in average temperature. In fact, the formula for it is very similar to the geostrophic wind formula [equation (8.2)]. It may be written approximately

$$v_T = \left| \frac{g}{fT} \frac{\Delta T}{\Delta n_T} \right| \Delta H$$

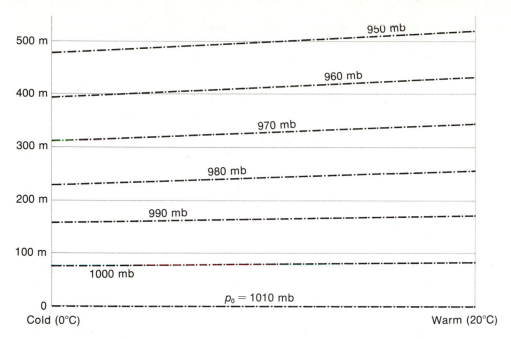

Figure 8.11   Variation of slope of pressure surfaces with height.

where v_T is the magnitude of the change in the geostrophic wind vector in going upward a distance ΔH, and Δn_T is the distance between isotherms (of average temperature through the layer ΔH) drawn for intervals of ΔT. The other terms have been previously defined. The direction of the thermal wind is related to the isotherms in the same way as the geostrophic wind is related to the isobars: *The thermal wind "blows" along the isotherms with low temperature to the left in the Northern Hemisphere (to the right in the Southern Hemisphere).* As an indication of the magnitude of the thermal wind, at 45° latitude a horizontal temperature gradient of 1°C/100 km corresponds to a thermal wind of about 10 m s^{-1} between sea level and a height of 3 km.

Since the isotherms in general are not parallel to the isobars, the direction of the thermal wind will usually be different from that of the geostrophic wind. To estimate the wind at upper levels, the thermal wind must be added vectorially to the geostrophic wind at the lower level. Figure 8.12 shows an example of the thermal wind and the resulting wind at an upper level.

In addition to being proportional to the temperature gradient (and therefore inversely proportional to Δn_T, the distance between isotherms), the thermal wind is proportional to the thickness ΔH of the layer through which the change is being computed. If the isotherms do not change much in direction with height, as is

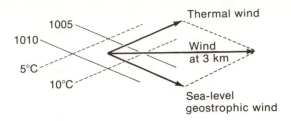

Figure 8.12   Illustration of thermal wind.

often true through most of the troposphere, for large enough ΔH, that is, at high levels, the thermal wind forms the major part of the wind. This means that as you go up, the wind usually becomes more and more nearly parallel to the isotherms, up to the tropopause. Correspondingly, the isobars on upper-level maps (and the isobaric contours on constant-pressure maps) become approximately parallel to the isotherms.

The temperature variation in the horizontal also affects the position and nature of high- and low-pressure centers at higher levels. For instance, if a high center is colder than its surroundings, the pressure will decrease more rapidly at the center than around it, and at a sufficient height the pressure at the position of the high at sea level will no longer be higher than in the surrounding area. In other words *cold highs weaken and disappear with height, while warm highs surrounded by colder air become more intense*. Similarly, *warm lows are shallow and cold lows extend to great heights*. If a center has a temperature gradient across it, it will tilt with height: *low-pressure centers tilt toward low temperatures; high-pressure centers toward high temperatures*. The configuration of the pressure field at upper levels may thus be estimated from the temperature field at sea level. This procedure will be illustrated in Section 11.5, in which maps of the pressure distribution at sea level and at 500 mb are compared.

Questions, Problems, and Projects

1. The moon rotates once every lunar month. Is the Coriolis force at a given latitude on the moon larger or smaller than at the same latitude on earth? By how much?

2. If the wind is 10 m s^{-1} at the equator, what is the magnitude of the Coriolis force there? Can the wind be geostrophic? Why? What does this imply with respect to the relation between wind direction and isobars there?

3. Suppose that you find that at a particular space, say, San Francisco, the isobars at the 3-km level are the same distance apart and in the same direction as the isobars on the sea-level weather map. Would you expect the geostrophic wind speed to be the same at 3 km as at sea level? Greater? Less? Why?

4. Why does the wind near the ground blow toward low pressure while the wind aloft blows almost exactly parallel to the isobars? Does the same rule apply in the Southern Hemisphere?

5. Since the Coriolis force tends to deflect moving air parcels to the right in the Northern Hemisphere, why do the winds around a low-pressure area turn to the left (counterclockwise)?

6. a. Make two drawings showing the surface wind direction around (i) a low-pressure center and (ii) a high-pressure center in the Southern Hemisphere.

   b. Is the gradient wind speed around a low-pressure center in the Southern Hemisphere greater or less than the geostrophic wind speed? Draw a diagram that shows why your answer is right.

7. A hurricane is an intense rotating storm originating at low latitudes, where the Coriolis force is small. If the Coriolis force can be neglected near the center, write the equation for the relation between wind speed and the pressure gradient, neglecting friction. Using this equation, compute what the wind speed would be for concentric 5-mb isobars 10 km apart at a distance of 30 km from the center if $\alpha = 0.800$ m^3/kg. Compute the value of the Coriolis force at 10° latitude, where the value of the Coriolis parameter is $2.53 \cdot 10^{-5}$ and compare it with the pressure gradient force and the centripetal acceleration.

8. If the motion in the hurricane in problem 7 is irrotational, how fast does the wind speed decrease with distance outward from the center? (Express the answer in terms of so many meters per second per kilometer.)
   [*Answer*: 1.15 m s^{-1}/km. At 30 km farther from the center, if the rate of decrease was constant, the cyclonic flow would vanish.]

9. Compute the gradient wind speed at a place at 45°N latitude where 4-mb isobars 250 km apart are curved around low pressure with radius of curvature 500 km and the specific volume is 0.775 m^3/kg.
   [*Answer*: 10.1 m s^{-1}]

9 The Global Circulation of the Atmosphere. Climatic Controls and Global Patterns of Climate

9.1 The nature of the global circulation

At any one time the air moving over the earth as a whole forms a complex array consisting of wave patterns, cyclonic and anticyclonic vortices, and smaller eddies, most of which are constantly forming, developing, moving, and decaying. By averaging with respect to time—say, for a month, a season, a year, or longer—the smaller, transitory patterns are smoothed out. The remaining larger scale and more permanent features are what is known as the *global circulation* or *general circulation* of the atmosphere. This pattern is characterized by permanent or semipermanent cyclones and anticyclones, which are sometimes called *centers of action*, and by persistent wind systems. The most prominent wind systems are the *easterly trade winds* in low latitudes, particularly near the earth's surface, and the *prevailing westerlies* at high latitudes and aloft. If the averages are made by months or seasons rather than years, the Asiatic Monsoon shows up in the summer and winter averages as another conspicuous feature of the atmospheric circulation.

These characteristics of the global circulation are readily recognized in Figures 9.1 and 9.2, in which the average resultant winds[1] at the earth's surface and the upper troposphere are represented by streamlines, lines drawn in the direction of the wind at every point, and isotachs, which show the wind speed. The features of the circulation are also shown in Figures 9.3–9.6, which present the average distribution in the Northern Hemisphere of the pressure at sea level and at the height of the 500-millibar surface in January and July. As pointed out in Chapter 8, the sea-level isobars and the isobaric contour lines are streamlines of the geo-strophic winds, from which the actual winds usually deviate only slightly, except near the ground, where the effect of friction can be allowed for.

In the charts of surface winds (Figure 9.1), the trade winds, northeast in the Northern Hemisphere and southeast in the Southern Hemisphere, are present between 30°S and 30°N. The principal exception is the southwest monsoon of the Northern Hemisphere summer, which blows across the Indian Ocean, the Arabian Sea, the Bay of Bengal, and over India and Southeast Asia. A similar but less extensive monsoon blows from the Atlantic Ocean and the Gulf of Guinea across the portion of Africa south of the Sahara Desert. Anticyclonic swirls of outward flowing air are present, particularly over the oceans at 35°–40° north and south, but also over Siberia on the January chart. Poleward of them the winds are mostly westerly, being northwesterly in some longitudes and southwesterly in others, but predominantly southwesterly in the Northern Hemisphere and northwesterly in the Southern Hemisphere. The data at high latitudes on these maps are inadequate to delineate definitively the cyclonic centers of action, but they are clearly shown in the circumpolar Northern Hemisphere average pressure map for January, Figure 9.3. The low-pressure area between the southern tip of Greenland and Iceland is usually referred to as the *Icelandic low,* and the one in the Gulf of Alaska is called the *Aleutian low.* They are present almost continuously, particularly in the cold season, somewhere near their average position. In the July average-pressure map, Figure 9.4, the entire continent of Asia is covered by low pressure associated with the southwest monsoon, and the Aleutian low, which is displaced northward, shows up only as a trough extending from Kamchatka across central Alaska. Pole-

---

1. Average resultants of vector quantities having direction as well as magnitude are computed by adding vectorially the hourly or daily observations for the period and dividing the magnitude of the vector sum by the number of observations. This procedure, as well as other procedures for averaging vector quantities, is not completely satisfactory, for it may result in a direction that does not occur in the observations and a much smaller "average" magnitude than the individual magnitudes. Thus, if the wind is northwest half the time and southwest the other half, the direction of the resultant wind will be due west, a direction that never occurred in the observations. If, as another instance, the wind is equally frequently due east or due west, with a constant speed of 10 m s^{-1}, the resultant speed will be zero although the average speed is obviously 10 m s^{-1}. However, if the wind direction is distributed in a normal fashion, the average resultant wind gives a good indication of the predominant air motion, and its magnitude is only slightly less than the average speed.

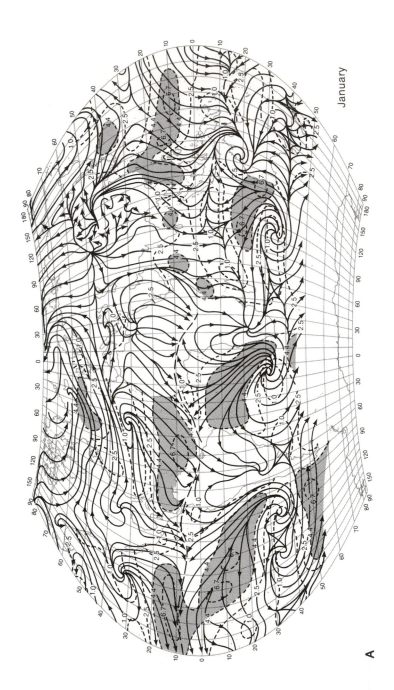

January

A

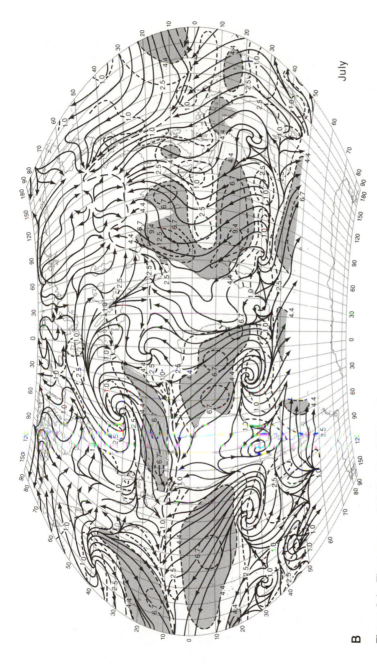

B

Figure 9.1 The average wind direction and speed at anemometer level over the earth in January (A) and July (B). The direction of the resultant wind is shown by streamlines—solid lines with arrows. The resultant wind speed is M S^{-1} is shown by isotachs—in dashed lines—which separate speeds less than the value given in the label on one side from those greater than that value on the other. [Northern hemisphere streamlines after W.M. Wendland and R.A. Bryson, "Northern Hemisphere Airstream Regions," *Monthly Weather Review* vol. 109, no. 2 (February 1981); isotachs and Southern Hemisphere streamlines from Y. Mintz and G. Dean, *The Observed Mean Field of Motion of the Atmosphere*, University of California, Los Angeles, Geophysical Research Papers no. 17, 1952, 65 pp.]

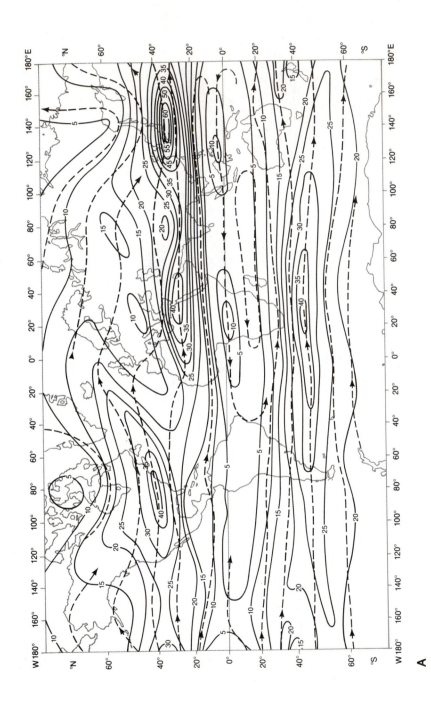

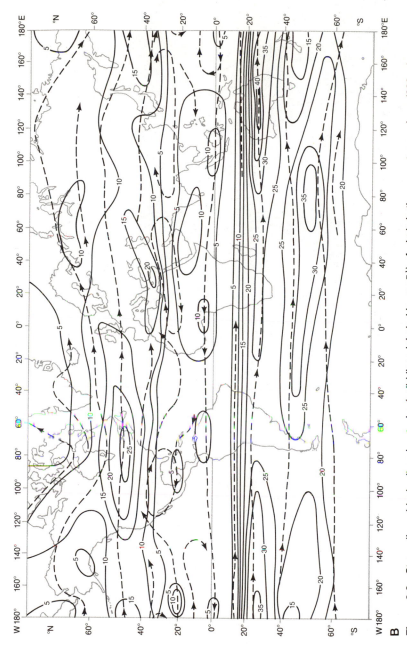

Figure 9.2 Streamlines (*dashed*) and isotachs (*solid lines labeled in m s⁻¹*) of winds in the upper troposphere (300 mb, about 9 km) in January (A) and July (B). [From *Handbook of Aviation Meteorology*, H. M. Stationers Office, London, 1971.]

[213]

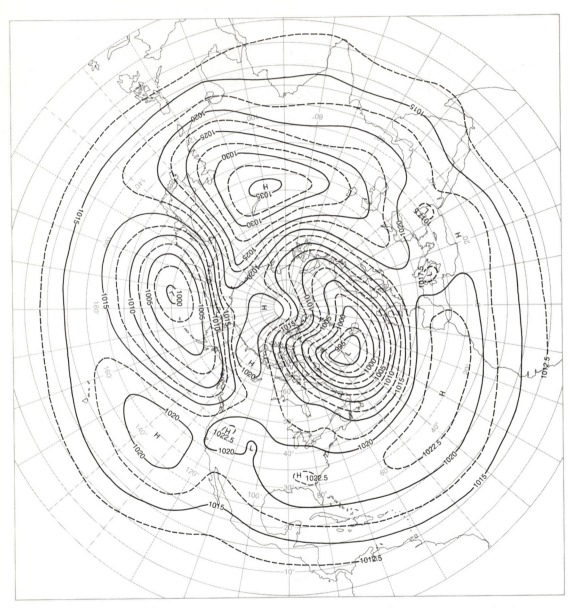

Figure 9.3 Average sea-level pressure in mb, Northern Hemisphere, January. [From U.S. Department of Commerce, Weather Bureau.]

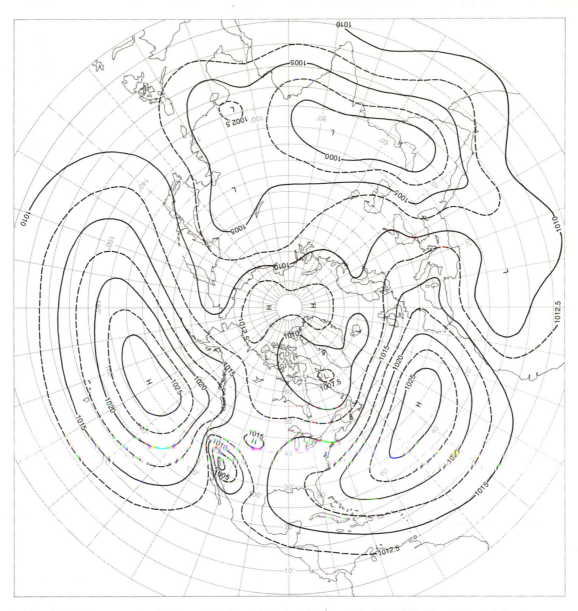

Figure 9.4 Average sea-level pressure in mb, Northern Hemisphere, July. [From U.S. Department of Commerce, Weather Bureau.]

ward of the cyclonic centers in both January and July is a weak high-pressure center, and between this high and the Iceland and Aleutian lows easterly winds, known as the Polar Easterlies, are present.

In the map of winds in the upper troposphere (Figure 9.2), the flow is westerly at almost all latitudes. Only in the immediate vicinity of the equator are the winds easterly. In the Southern Hemisphere the deviations from zonal flow are slight, but in the Northern Hemisphere there are southerly and northerly components in a wave pattern, with troughs over North America, eastern Europe, and off the east coast of Asia. The isotachs show concentrated areas of strong westerly winds in narrow belts, strongest in the winter hemispheres. These are the average positions of the *jet streams*, in the vicinity of 30° and 40° latitude.

In the corresponding upper-air pressure map for January in the Northern Hemisphere (Figure 9.5), the troughs referred to above are shown clearly. The polar high on the surface map is replaced by a low, and the hemispheric flow consists of a circumpolar cyclonic vortex, with three waves in the westerly flow. In the July map (Figure 9.6), the space between contours is greater, corresponding to the light winds; there is a suggestion of six or seven weak troughs, and there is a belt of high pressure most of the way around the earth at about 30° latitude, the residual aloft of the intense anticyclonic centers over the oceans at sea level.

In studying the global circulation, it is convenient to separate the variations with longitude due to differences of terrain and other factors from the part that is symmetric around the earth. This is done by taking the further step of averaging around latitude circles. When this is done a simple pattern emerges, with the zonal flow predominating. The zonal components, being large, are readily evaluated from observations. The meridional components are small, since the northward and southward values tend to cancel each other, and the vertical components, which are not measured, but must be inferred from continuity, are still smaller.

Figure 9.7 shows the observed zonal component of the winds at all levels and latitudes, averaged around the earth separately for the northern winter months (December–February) and the northern summer months (June–August). The vertical scale is height, with the corresponding approximate pressure shown along the right ordinate. The boundary between easterlies and westerlies at the earth's surface is a little farther poleward in the summer in each hemisphere; it slopes equatorward with height and the strength of the easterlies decreases upward in the troposphere. The strength and latitudinal extent of the westerlies increase with height up to the tropopause in both hemispheres, reaching maximum values of more than 30 m s^{-1} at about 30° latitude in the winter hemispheres and about 15 m s^{-1} at 45°N and 25 m s^{-1} at 50°S in the summer. In the stratosphere the winds are westerly in winter and easterly in summer.

The mean meridional circulation for all four seasons is represented by mass-flux streamlines in Figure 9.8. The streamlines are labeled with the values of a quantity

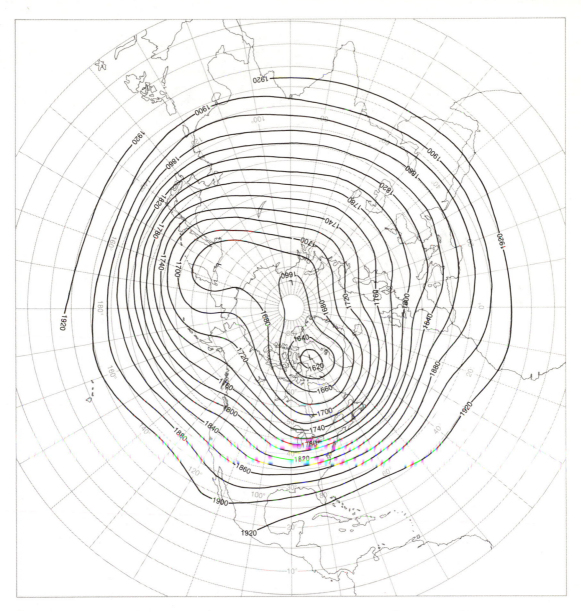

Figure 9.5 Average height, 500-mb surface, in tens of feet, Northern Hemisphere, January. [From U.S. Department of Commerce, Weather Bureau.]

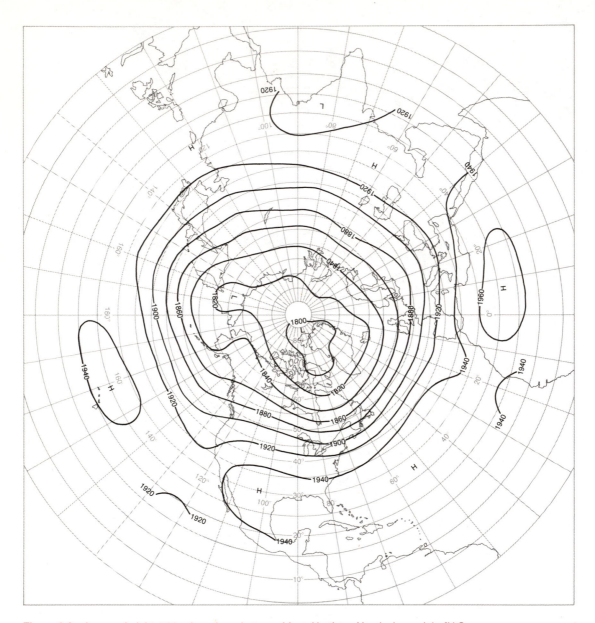

Figure 9.6   Average height, 500-mb surface, in tens of feet, Northern Hemisphere, July. [U.S. Department of Commerce, Weather Bureau.]

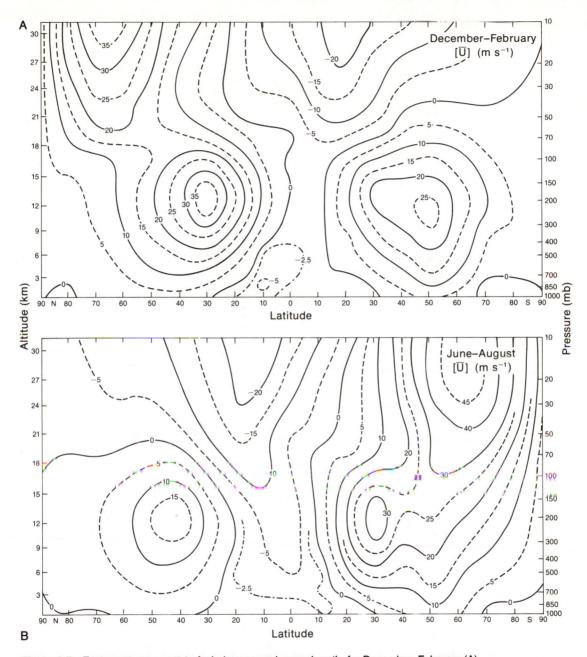

Figure 9.7 East–west component of winds averaged around earth, for December–February (A) and June–August (B). [From R. E. Newell et al., in *The Global Circulation of the Atmosphere*, The Royal Meteorological Society, Bracknell, England, 1978.]

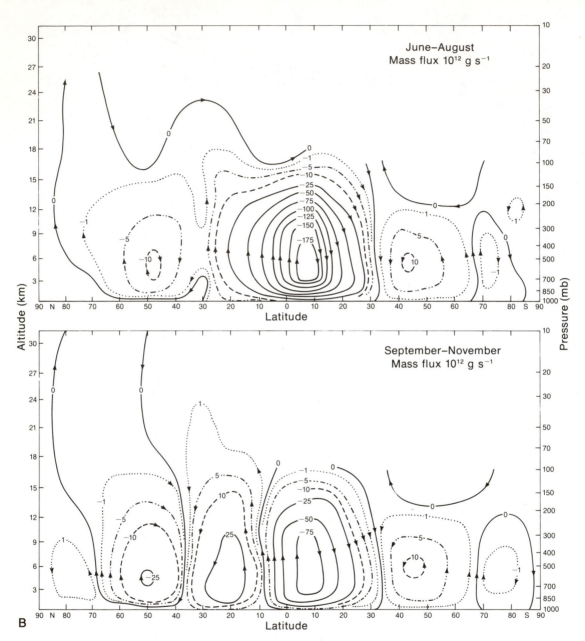

Figure 9.8 Mean meridional circulation for four three-month periods, June–August (A), September–November (B), December–February (C), and March–May (D). [From R. E. Newell et al., in *The Global Circulation of the Atmosphere,* The Royal Meteorological Society, Bracknell, England, 1978.]

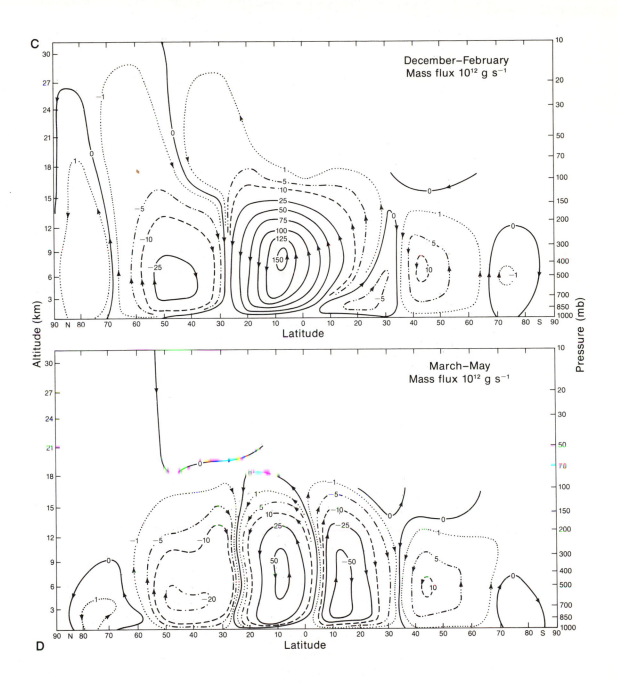

called the stream function, ψ. The south–north average wind speed $\bar{v}$ and the upward component $\bar{w}$ are related to the vertical spacing and the horizontal spacing of these lines by the expressions

$$\bar{v} = \frac{g \, \Delta\psi}{2 \, \pi \, a \cos\varphi \, \Delta p}$$

$$\bar{w} = - \frac{g \, \Delta\psi}{2 \, \pi \, a^2 \cos\varphi \, \Delta\varphi}$$

where a is the radius of the earth, φ is the latitude, g is the acceleration of gravity, and $\Delta\psi$ is the change in value of the stream function in going through a vertical change of pressure Δp in the first equation and through a change in latitude $\Delta\varphi$ in the second equation.

In the spring and fall (March–May and September–November), there are three meridional cells in each hemisphere, with rising air at low latitudes, descending air at 30°–40°, upward currents again in the vicinity of 60°, and descending air near the poles. In the winter this pattern is distorted; the low-latitude descending air in the Northern Hemisphere covers an area extending much farther south; and in summer the rising air reaches all the way from 10°S to 30°N, so that in effect one of the meridional cells is absent. These distortions are due to the effect of the Asiatic Monsoon, which dominates the averages in these latitudes.

The strength of the meridional circulation is much less than the zonal winds. The greatest magnitudes of $\bar{v}$ are about 2 m s^{-1}, and $\bar{w}$ is at most a few tenths of a meter per second in magnitude.

As stated at the beginning of Chapter 8, this observed pattern of wind systems over the earth is far different from that which would be expected if the atmosphere responded to the greater energy received from the sun at the equator than at the poles in the same way as it does to the daytime excess of heating over land at a coast. Instead of the flow being meridional, from poles towards equator near the ground and vice versa aloft, it is predominantly zonal. Instead of a single meridional circulation in each hemisphere, there are three during much of the year, one of which is a reverse cell, with descending air at the lower latitude part (about 30°) and ascending air at high latitudes ($\sim$ 60°). Instead of being uniform around the earth, the pattern varies with longitude.

The variation with longitude is readily understood in terms of the unequal heating of land and sea surfaces, which are distributed unequally around the earth, particularly in the Northern Hemisphere. The occurrence of zonal winds is readily explained by the deflection of meridional motions due to the earth's rotation. The occurrence of three meridional cells per hemisphere, instead of one, and the dominance of zonal flow over meridional are not so readily explained. The circulation must carry the excess energy received at low latitudes to high latitudes where there is a deficit, and how this is achieved in the three-cell circulation is also not readily seen.

The variation with latitude of the net heating due to solar radiation is the driving force for the global circulation and determines the amount of heat transfer it must carry out. Before examining in more detail the reasons the global circulation takes its observed form, how it is maintained, and how it carries out its energy transfer function, we shall look at the variation of net heating with latitude.

9.2   Variation of radiation balance with latitude

Referring back to Chapter 3, and specifically to Figure 3.18, we have seen that, averaged over the entire earth through a whole year, the radiation entering the "top" of the atmosphere is very nearly in balance with that leaving it. That this must be true is seen by considering what would happen if there would be a long-term deviation from such a balance. If more energy were received than was radiated away, the excess would cause the average temperature of the earth as a whole to rise. Since the long-wave radiation emitted by the earth and atmosphere increases with temperature, the outgoing radiation would increase until the temperature had risen to the point at which outgoing and incoming radiation were equal. Similarly, if the incoming radiation were less than the outgoing, the temperature would fall and the outgoing radiation would decrease. Thus, there is a tendency to restore the radiative balance whenever it is disturbed. While seasonal changes in solar radiation keep this balance from existing at any instant, the temperature changes constantly strive to maintain it, and in the average for the entire year the seasonal imbalances offset each other. Except for very slight departures due to variations of the earth's orbit around the sun, and perhaps due to variation of the solar emissions and to the increase in the amount of carbon dioxide and particulate matter in the atmosphere in recent years, it is to be expected that conditions in the earth and its atmosphere have been sufficiently close to constant for radiative balance to be attained.

Satellite measurements have recently given the first opportunity to test this balance. It was found that the average planetary albedo is approximately 0.30, and the average infrared radiation leaving the earth is 0.34 cal/cm^2 min (237 W/m^2). The average incoming radiation is $S_o (1 - A) /4$, where S_o is the solar constant, A is the albedo, and the factor 4 is due to the fact that the energy is intercepted by the cross section of the earth and spread over its entire surface. Inserting the values of S_o and A, we find the incoming radiation is equal to the outgoing within the accuracy of the measurements.

While the radiation is nearly in balance for the earth as a whole, at any particular

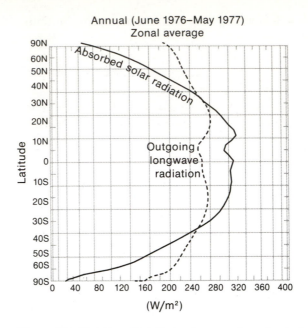

Figure 9.9 Variation with latitude of average incoming and outgoing radiation for year June 1, 1976 to May 31, 1977. [From *Earth–Atmosphere Radiation Budget Analyses Derived from NOAA Satellite Data, June 1974–February 1978*, Washington, D.C., NOAA–NESS, 1979.]

latitude the incoming and outgoing radiation are not equal, even when averaged for the entire year. In addition to differences in solar inclination and length of day, differences in cloudiness and in type of surface produce differences in albedo that cause the net incoming solar radiation to vary with latitude. As we saw in Figures 3.11 and 3.12, the average planetary albedo is lowest in the tropics, where most of the surface is ocean, and highest at high latitudes, where much of the time the earth's surface is covered with snow or ice and clouds are frequently present. The outgoing long-wave radiation depends mainly on the temperature at the earth's surface and throughout the atmosphere. In Figure 3.17, which is a map of the average outgoing radiation for a year, the lowest values, less than 200 W/m^2, are mostly at high latitudes and the highest values, greater than 275 W/m^2, are at low latitudes, but the differences are not great, only about 30 percent, compared with the albedo, which varies by a factor of more than 3.

Figure 9.9 shows the variation with latitude of the average net incoming and outgoing radiation for the year June 1, 1976–May 31, 1977, computed from satellite measurements and averaged around the earth at each latitude. The net incoming solar energy exceeds the radiation emitted to space by the earth and atmosphere between 34°S and 31°N. Poleward of these latitudes there is a net loss of energy. The departures from balance are considerable. For instance, between 10°S and 10°N the excess of incoming radiation is more than 40 W/m². This amounts to more than 3.4 MJ/m² per day, an amount of heat that would raise the temperature from the ground to the top of the atmosphere more than 1 K every 3 days. To keep the atmosphere from heating up at low latitudes and, correspondingly, from cooling continuously at high latitudes, the excess heat must be carried poleward. In the next sections we shall examine how it is accomplished.

9.3   Hadley cells and angular momentum

We begin by considering what might be expected due to the combined effects of the differential heating described in Section 9.2 and the rotation of the earth.

The action of the Coriolis force in controlling the global circulation is similar to the adjustment of the wind to a pressure field described in Section 8.3. Analogous to the sea-breeze situation, the heating at low latitudes and cooling near the poles would produce high pressure aloft at the equator and low pressure aloft at the poles. At the earth's surface the pressure would be low at the equator and increase toward the poles. Initially this would give rise to a meridional circulation, with the heated air rising near the equator and flowing poleward aloft, and the cooled air descending at high latitudes and flowing equatorward at the ground. The stream of air moving poleward aloft would be deflected eastward by the Coriolis force, producing west winds, and the air flowing equatorward at the ground would be deflected into easterlies. In the absence of friction with the earth's surface, a balance between the pressure force and the Coriolis force would be attained after the circulation proceeded awhile, with easterlies at all latitudes at the ground and westerlies at all latitudes aloft.

Because of surface friction, complete balance between Coriolis force and pressure force would not be established. Friction would maintain cross-isobaric flow toward the equator near the ground. The air tending to accumulate there would be heated and tend to rise, and the accumulation would increase the pressure aloft at low latitudes, making the pressure force aloft slightly greater than the Coriolis force, so that the air aloft would be pushed poleward. At high latitudes the air would be cooled and tend to descend, completing the meridional circuit. Thus, a meridional circulation would be maintained in spite of the tendency of the Coriolis force to compel the air to move zonally. The components of this circulation pattern, one in each hemisphere, are called *Hadley cells*, after George Hadley (1685–1758),

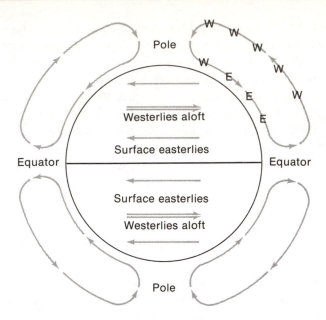

Figure 9.10 Hadley model of general circulation consisting of two Hadley cells: *single arrows*—winds near ground; *double arrows*—winds aloft.

who in 1735 showed that the easterly trade winds were due to the effect of the earth's rotation on the meridional circulation produced by the unequal heating of the earth's surface by the sun. Figure 9.10 shows schematically this circulation model.

Hadley cells of this type would perform the function that is required, namely transferring the excess received by the sun at low latitudes to high latitudes. As we saw in Section 9.1, this type of circulation is present at low latitudes, where the radii of latitude circles change gradually with latitude, but at middle and moderately high latitudes there are meridional circulations in the opposite direction. These reverse cells are called Ferrel cells, after William Ferrel (1817–1891), an American meteorologist who included them in a general circulation model he promulgated in 1852. They would tend to pump heat equatorward instead of poleward. Before discussing how the heat is transferred poleward across the latitudes where the reverse cells exist, we shall consider why Hadley cells cannot extend over regions where the radii of parallels of latitude change rapidly.

There are two reasons why the general circulation cannot consist simply of Hadley cells (one in each hemisphere). One has to do with the conservation of angular momentum of rings of air as they move southward and northward, and the other with the conservation of angular momentum of the earth-atmosphere system.

Applied to rotating bodies, Newton's laws of motion say that the rate of change of angular momentum (amount of rotational motion) of a body is equal to the sum of the torques (twisting forces) acting. A torque is expressed as a force acting at right angles to a radius from a center of rotation. An example of a torque is the twisting action used to tighten a bolt with a wrench. The force is applied perpendicular to the wrench handle. In the absence of torques the angular momentum is constant. Since no external torque is exerted on the earth-atmosphere system, its angular momentum is conserved.

Angular momentum is measured by the product Mvr, where r is the distance of the mass M from the center of rotation and v is the linear velocity due to the rotational motion. If no torques are acting on an isolated mass we have the result

$$Mvr = \text{constant}$$

so that if r decreases, v increases, since M is constant. If a body consists of masses at different distances, we must add up the angular momentum of all its component masses.

A familiar example of conservation of angular momentum is that of a figure skater twirling first with her arms extended, then with her arms pulled in toward her body. As her arms move in, more of the mass becomes concentrated close to her center of rotation and she consequently rotates faster.

If mass remains the same and distance decreases, then to conserve angular momentum, the speed must increase.

Another example is that of an object being swung around on the end of a string. Let it rotate at a constant rate and then pull the string inward so that the distance from the axis of rotation decreases. Since the angular momentum remains the same, the speed of the object increases.

We shall now apply the principle of conservation of angular momentum to the rings of air in the Hadley circulation shown in Figure 9.2. If we assume that the air rising near the equator is stationary relative to the earth, in absolute motion it has a speed v of about 464 m s^{-1} or 1038 mi/hr from west to east. As it moves poleward at upper levels, its distance r from the earth's axis decreases, and in order that vr may be constant v must increase correspondingly. When it reaches 45°, the distance is reduced to 0.707 of its original value. To maintain the constancy of the product vr, the absolute velocity v must increase to 656 m s^{-1}. The speed of the earth's surface at 45° latitude, on the other hand, is smaller by the same proportion as r, 328 m s^{-1}. The air would thus be moving relative to the earth with a speed of 328 m s^{-1} (734 mi/hr). While such extremely high wind speeds (and the corresponding pressure gradients) could conceivably exist, the theoretical analysis shows that they would result in great instability, in the sense that any small disturbance from the zonal flow would grow and cause a breakdown in the zonal circulation. Only in the low latitudes, where the distance reduction is small enough, does conservation of angular momentum result in speeds of zonal flow that are relatively stable, so that the zonal motion so produced can persist. For this reason the Hadley circulation is limited to low latitudes.

The other reason why the simple pressure pattern that the Hadley circulation requires cannot exist over the entire earth is concerned with the effect of friction between the atmosphere and the earth's surface. The pattern at the surface, with low pressure at the equator and high pressure at the poles, would require easterly winds at all latitudes for approximate geostrophic balance. Friction between the easterly winds and the earth's surface would produce a torque, with the atmosphere tending to slow down the earth and the earth tending to speed up the atmosphere. This condition could not persist. Since there is nothing for the air to "hold onto" while pushing on the earth to slow it down, it would quickly reach equilibrium in which it would move with the earth, with no wind at all. Thus, surface winds in one direction over the entire earth cannot exist; westerlies at some latitudes are necessary, in addition to easterlies at others, to maintain a constant average flow pattern.

Actually, Hadley recognized the need for westerlies at the earth's surface to compensate for the frictional drag of the easterlies, and provided for them by having the westerlies aloft brought to the surface near the poles and then losing westerly momentum and becoming easterly as the air proceeded equatorward. Not knowing about the Coriolis force as such, he did not realize that geostrophic adjustment to surface westerlies at high latitudes would require higher pressure at their equatorward boundary and lower pressure at their poleward limit, so that friction would cause poleward flow in the regions with westerlies near the surface.

The actual global circulation, as described in Section 9.1, conforms to this pattern. A simplified schematic representation of it is shown in Figure 9.11. In tropical regions, from about 30°S to 30°N latitude, the trade winds are present—northeasterly in the Northern Hemisphere and southeasterly in the Southern Hemisphere. Where the two systems come together, there is a narrow zone of weak winds, called the doldrums by early mariners but now referred to as the *intertropical convergence zone (ITCZ)*. Located in the rising branches of the Hadley cells, this region is characterized by cloudiness and rain. Some of the photographs from satellites have shown two bands of cloudiness in this area, suggesting that there may be a double structure, or two intertropical convergence zones, at times. The averaging of all the available cloud data from satellites shows that a single well-developed ITCZ is present most of the time.

In middle latitudes are the belts of westerlies, with waves and cyclonic vortices in them. Associated with the waves and vortices in the westerlies is highly variable weather, with rapidly changing temperatures and periods of clear skies followed

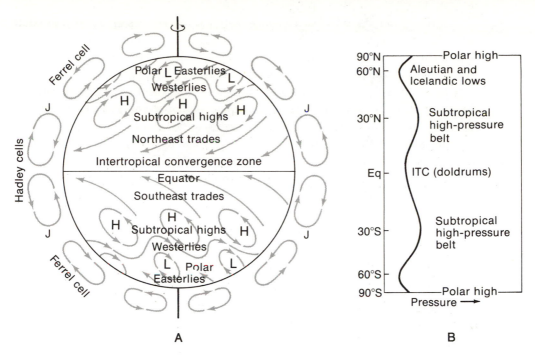

Figure 9.11   (A) Schematic representation of features of general circulation. (B) Variation of average sea-level pressure with latitude.

by periods of precipitation. In winter the temperature may change as much as 20°C in a single day.

Between the westerlies and the trade winds are the subtropical high-pressure belts, called "Horse Latitudes" in the old days. These are regions of calms or light winds. Representing the poleward limits of the Hadley cells, where their descending branches occur, they are generally regions of fair weather and little precipitation. The deserts of the world are mostly at these latitudes.

In polar regions the winds on the average are easterly. The polar easterlies are not as persistent or steady as the trade winds, however. They are frequently interrupted by the passage of storms.

As shown in Figures 9.2 and 9.7, the winds in the upper troposphere are westerly in all latitudes except for a narrow belt near the equator. The west-wind maxima, the average positions of the jet streams, indicated by the letter *J* in Figure 9.11, are at the poleward limits of the Hadley cells.

Since friction would tend to slow down both the easterlies and the westerlies at the earth's surface, there must be a process by which westerly momentum is transferred from the easterlies of low latitudes to the westerlies of temperate latitudes in order to keep both of these currents going. Similarly, it is necessary that westerly momentum be transferred upward from low levels to the upper troposphere in the tropics and downward in higher latitudes.

The net energy furnished to an air column by solar heating is initially divided between sensible heat (higher temperatures), potential energy, and latent heat due to evaporation at the sea (and other water) surface and transpiration by vegetation. Some of this energy is transformed to kinetic energy, which is divided among the mean zonal circulation, the meridional cells, and the stationary and transient waves and vortices. Whatever form the energy takes, its transfer across the polar limits of the Hadley cells, where meridional motion is absent, and across middle latitudes, where the reverse cells would transport energy in the wrong direction, requires another process to carry its excess amounts from tropical to polar regions.

9.4   Transport of heat and momentum by waves and vortices

It was pointed out in Section 9.3 that the high zonal velocities that would be produced by conservation of angular momentum if Hadley cells extended to high latitudes would result in great instability. In fact, even the limited extent of the observed Hadley cells would produce much higher speeds than are observed, and the actual speeds are lower due to vertical and lateral mixing because the zonal flow is unstable and breaks down. This breakdown takes the form of waves in the zonal flow and the development of complete swirls—cyclonic and anticyclonic vortices. These disturbances have such character that they transfer heat and momentum north and south even though the net meridional movement of air across parallels of latitude averages out to be approximately zero at each level. That is, if at some place and time there is a northward flow, at another place and time there is a southward flow of the same amount of air at the same level.

To see how there can be a transport of heat and momentum by north–south movements even though the flow of air cancels out, suppose that the northward-moving air in the Northern Hemisphere is warmer than the air moving southward at the same latitude at every level. Less heat would be carried southward across the parallel of latitude than would be carried northward, and there would be a net transport of heat from south to north. If, on the average, the absolute zonal speed of the northward-moving air were greater (less strong easterly or stronger westerly components of wind) than that of the southward-moving air, there would similarly be a transfer northward of westerly momentum. Deviations of this type from the average for the latitude are to be expected. In the Southern Hemisphere, of course, it is the southward flow that is on the average warmer and more westerly than the northward. Thus, the transfers necessary to carry poleward the excess heat received at low latitudes and to maintain the easterlies and westerlies near the ground (and the westerlies aloft in higher latitudes) can be achieved.

While already in 1926 the English geophysicist Harold Jeffreys (1891–   ) pointed out theoretically that cyclones, anticyclones, and waves in zonal currents, which are grouped under the term *large-scale eddies*, can perform the transfer of energy and momentum required, it was not until after World War II that observational data became plentiful enough to attempt to evaluate the transports quantitatively. Beginning in 1948 extensive computational projects were carried out at UCLA (University of California, Los Angeles) under J. Bjerknes (1897–1975) and at MIT (Massachusetts Institute of Technology) under Victor P. Starr (1909–1976). Even at present there are large areas of the world from which data are not available, but sufficiently accurate values have been obtained to demonstrate that the large-scale eddies do most of the transport of energy and momentum except at low latitudes, where meridional cells play an equal role. A considerable amount of energy and momentum is carried poleward by ocean currents.

Thus, the eddy motions that result from instability of the zonal flow produced by deflection of the meridional circulations that solar radiation tends to establish carry out the energy transfer that the meridional cells are prevented from accomplishing. In turn, the transfer of momentum by the eddy motions gives rise to the westerlies in middle latitudes and aloft, and surface friction in the westerlies produces the weak reverse meridional circulations.

The details of how the transfer of heat and momentum occurs and why the disturbances from the mean flow have the character that is observed have been studied in model experiments in the laboratory and by the use of high-speed electronic digital computers. These studies have helped us understand the large-scale behavior of the atmosphere.

9.5   Experiments on the general circulation

While the full complexity of the atmosphere cannot be simulated in the laboratory, certain essential features of the general circulation can be reproduced by simple experiments called "dishpan" experiments, in which a pan of water is heated on a rotating table. The general configuration of the experiment is shown in Figure 9.12. The pan is heated at the rim (equator) and cooled at the center (pole).

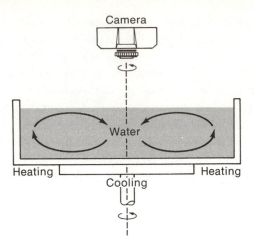

Figure 9.12 Diagram illustrating apparatus used in "rotating dishpan" experiment.

Particles of dust suspended in the water enable its motions to be seen, and a motion-picture camera rotating at the same rate as the pan records how the motions would look to an observer on the pan (earth). If the pan is not rotating, the motion of the water in the pan is just the meridional motion of the Hadley circulation shown in the figure. When the pan is rotating, zonal motions develop, and the character of the flow changes with the rate of rotation.

For very slow rotation the flow consists of a simple zonal motion on which a slow Hadley circulation is superimposed. As the rate of rotation increases, the simple flow breaks down into a series of waves and vortices. The size and number of the eddies thus produced depend on the relation of the heating rate to the rate of rotation. For small heating rates, which produce small temperature differences, and for large rotation rates the eddies are numerous; for large temperature differences and small rotation rates, fewer large eddies or waves occur. Figure 9.13 is a photograph of a rotating dishpan experiment in which the flow has broken down into a large number of waves and vortices. It is readily seen that this type of flow pattern would produce a rapid exchange of heat and momentum between the "pole" and the "equator."

In addition to laboratory model experiments, studies of the general circulation can be carried out using high-speed electronic digital computers to solve the equations of atmospheric motion. In a typical computation it was assumed that initially the atmosphere was everywhere at rest relative to the earth and had a constant temperature. The radiation from the sun was then allowed to heat the atmosphere

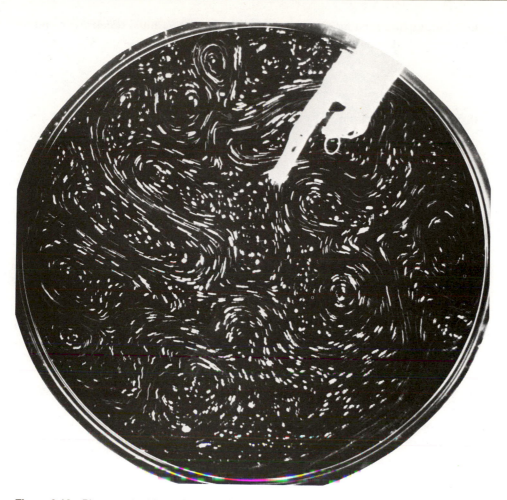

Figure 9.13 Photograph of flow pattern in "dishpan" laboratory model of the general circulation. [Courtesy of D. Fultz, University of Chicago Hydrodynamics Laboratory.]

and the ground, and the effect of this heating on the motion of the air was computed. About 30 days after "the sun was turned on," the motions had settled down to a pattern similar to that observed on any day in the real atmosphere, with a Hadley circulation and easterlies at low latitudes, westerlies at higher latitudes and aloft, anticyclonic and cyclonic eddies in middle latitudes, and waves in the upper westerlies.

The theoretical computations and the laboratory model experiments both reproduce the principal features of the general circulation, showing that to transfer the heat energy in a rotating system the Hadley circulation must break down and be replaced at higher latitudes by a system of waves and vortices of the type described in Section 9.4.

9.6   Global patterns of climate

In Section 9.3, in the discussion of Figure 9.11, the effect of the global wind pattern on the distribution of cloudiness and precipitation was outlined from a schematic standpoint. The observed variation with latitude of the annual precipitation averaged around the earth is shown in Figure 9.14A. The heaviest precipitation is in the latitude belt where clouds and showers are produced by the rising branches of the Hadley cells, that is, in the intertropical convergence zone. Secondary maxima are present at latitudes where stormy weather takes place in the westerlies, where the cyclonic eddies produced by the instability of the westerly winds occur. In the subtropical latitudes between them, the average precipitation is lighter because clear skies and dry weather predominate where the descending currents of the meridional cells are present.

The distribution of the yearly temperature averaged around the earth, shown in Figure 9.14B, is even more simply arranged; the temperatures are highest in equatorial and subtropical latitudes and decrease rapidly in middle latitudes to lowest values near the poles.

The continents and oceans cause large variations with longitude from the average zonal pattern, as does the presence of mountain ranges. We have already seen these variations in the summer and winter temperature pattern, in Figure 2.7, in which the isotherms bend poleward over the oceans and equatorward over the continents in the cold season (January, N.H., and July, S.H.) and vice versa in the warm season (July, N.H., and January, S.H.). The effects on the precipitation pattern are more complicated, for the amounts depend both on the ways the circulation controls the vertical motion of the air and the ways it determines the nature of the air, that is, its moisture content and its stability. For instance, the air moving poleward around the western ends of the subtropical anticyclones comes from warm portions of the oceans where it has become humid by evaporation, and it is rendered unstable by convergence. Pulled onto the heated eastern portions of continents in the warm season, it produces heavy rains. In contrast, the air moving equatorward around the east ends of the subtropical anticyclones comes from cold ocean waters, where it picks up relatively little moisture, and is subjected to horizontal divergence and subsidence. Consequently, there is little or no precipitation on the west coasts of continents in the warm season.

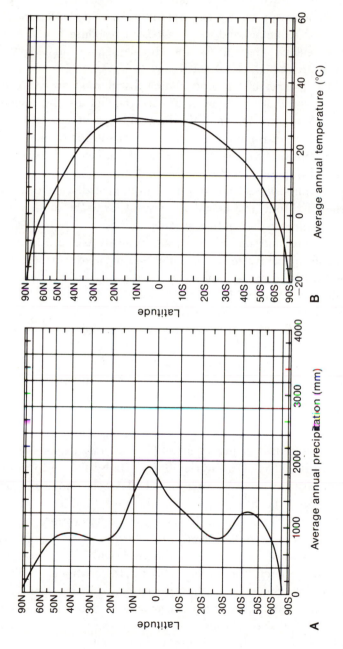

Figure 9.14   Variation with latitude of annual precipitation (A) and temperature (B) averaged around the earth.

The effect of mountain ranges on precipitation is twofold. They force the air to rise and produce precipitation on their windward sides, and they cause descending motion of the dried-out air and produce a rain-shadow on their leeward sides. They also influence the wind patterns, inducing wave patterns in deep currents that flow over them and producing seasonal upslope and downslope winds analogous to the diurnal mountain and valley winds in response to the heating and cooling of those regions, in particular the Himalayas and the Tibetan plateau, where large areas are at high elevations.

Looking at the specifics of the precipitation distribution over the continents (Figure 9.15), we see that the areas of heaviest rainfall are the regions of the Americas, Africa, southern Asia, and the East Indies centered on the equator, which are along the intertropical convergence zone; subtropical and middle latitude eastern portions of Asia, the Americas, Australia, and Africa, which are to the west of the subtropical anticyclones; Europe, the higher-latitude west coasts of the Americas, and New Zealand, which are in the paths of the cyclonic storms in the westerlies; and India and neighboring countries to the south of the Himalayas, which are subject to the Southwest Monsoon in summer. The driest regions are the deserts of north and southwest Africa, southwest Asia, central and western Australia, and subtropical western North and South America, where the air circulating around the subtropical anticyclones descends; the middle-latitude portions of the interior of Asia and North America and southeastern South America, which lie in the rain shadow of high mountain ranges; and the extremely high-latitude continents, where the air is so cold it cannot contain much moisture to be precipitated.

The world pattern of cloudiness is similar to that of precipitation, except that there are persistent areas of nonprecipitating low cloud and fog, particularly over the eastern portions of the oceans in the subtropics, and at high latitudes. The average cloud cover over the Northern Hemisphere at 1400 (2 P.M.) local time, in summer and winter, as determined by polar-orbiting satellites during the four years 1967–1970, is shown in Figure 9.16. The general character of the global circulation is clearly reflected in the cloudiness. High values are present in the intertropical convergence zone between the Hadley cells and at the latitudes of the westerlies, where the flow poleward in the Ferrel cell converges with the equatorward flow from the polar cell. Relatively cloud-free conditions are present in the area of descending flow associated with the subtropical high pressure belt. The seasonal shift of these features is clearly shown, with all of them occurring at higher latitudes in summer. The seasonal effect of the continents is also distinct, with high values of cloudiness in summer, particularly over mountains, due to the increased convective activity, and almost clear weather over continents in winter, except in regions affected by storms in the westerlies. The seasonal contrast in India and southeast Asia due to the monsoon is especially striking.

General pattern of annual world precipitation (inches)

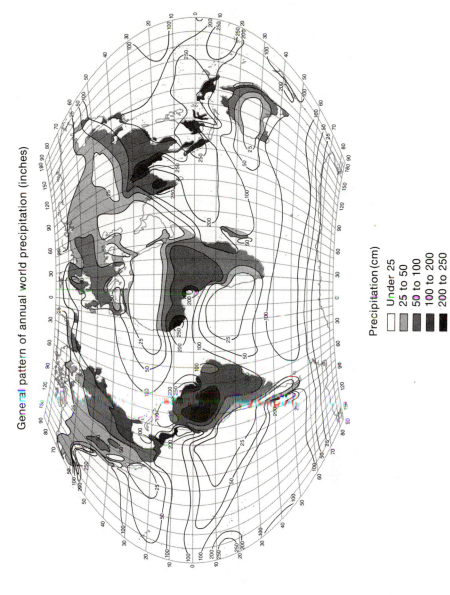

Precipitation (cm)

Under 25
25 to 50
50 to 100
100 to 200
200 to 250
Over 250

Figure 9.15 World distribution of average annual precipitation. [From *Climates of the World*, ESSA Environmental Data Service, Asheville, N.C., 1969.]

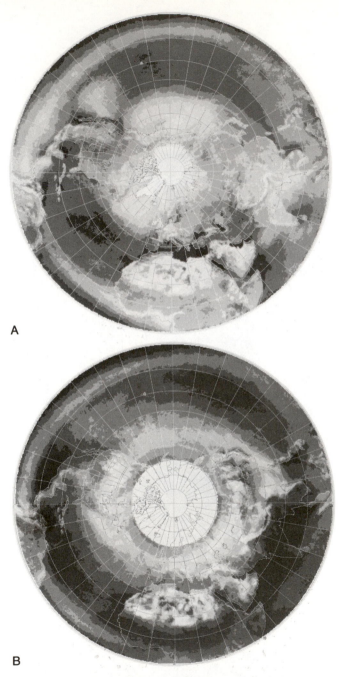

A

B

Figure 9.16 Average cloudiness in Northern Hemisphere in summer (A) and winter (B), as observed by satellite during 1967–1970. [NOAA–NESS photo.]

We have seen that the global circulation and its consequences in controlling the distribution of climate are not represented solely by the average flow patterns, but require consideration of the transient features of the circulation. In the next two chapters these features will be considered.

Questions, Problems, and Projects

1. Discuss why there cannot be a simple Hadley circulation with easterlies at the ground at all latitudes from equator to pole.

2. Why is westerly momentum transported poleward by waves and vortices, even though the winds at low latitudes are easterlies? (*Hint:* Consider the conservation of the angular momentum of rings of air displaced poleward, as discussed in Section 9.3.)

3. a. What would the the best route to sail from San Francisco to Hawaii? What is the best route to sail back?

   b. Why does it usually take jet aircraft longer to fly from New York to San Francisco than from San Francisco to New York?

4. a. Why are the low-pressure centers at about 60°N in Figure 9.3 over oceans and not over land?

   b. Why are the subtropical high-pressure centers in Figure 9.4 over the oceans and not over continents?

5. a. By looking at Figure 9.7, determine at which latitude the average temperature changes with latitude most rapidly in the Northern Hemisphere (i) in winter and (ii) in summer.

   b. Why does the boundary between the easterly trade winds and the westerlies tilt equatorward as one goes upward through the troposphere?

6. Why are the low-pressure troughs in Figure 9.5 mostly over the continents and not over the oceans?

7. Why is the atmospheric circulation in middle latitudes so changeable? Why can't it be steady except for gradual seasonal changes, approximating the average conditions for the time of year at all times?

8. Why does the weather in middle latitudes vary from year to year? For instance, in some years winters are much colder than others; in some years there are floods, in other years droughts.

10 Air Masses

10.1 The nature and classification of air masses

If air remains for a long enough time over a portion of the earth's surface, its properties tend to become typical of that surface. If the character of the surface is approximately the same over a large area—for example, a broad expanse of warm ocean—the properties of a large body of air will become nearly uniform in the horizontal. Such a body of air, with properties (in particular, temperature and moisture content) nearly uniform over horizontal distances of the order of thousands of kilometers, is called an *air mass*. The areas where they form are called *source regions*.

For an air mass to form, the air must stagnate or move for a long time over a large area with uniform properties. In general, this means areas dominated by stationary or slow-moving anticyclones, since anticyclones usually have extensive areas of calms or light winds. For the formation of air masses, light winds and uniform surface properties must be present together. The light winds ensure that the air will stay over the source region long enough to come approximately to equilibrium with it. The principal characteristics of source regions that determine the air-mass properties are their temperature and whether they are land or sea. As an indication of of their temperature, source regions are classified by latitude belts—equatorial, tropical, polar, and arctic (or antarctic)—and the surface character is designated as continental or maritime. Based on these categories of the source regions, air masses are classified as follows:

Arctic	A
Maritime Polar	mP
Maritime Tropical	mT
Continental Polar	cP
Continental Tropical	cT
Equatorial	E

The abbreviations are sometimes used on weather maps to show the air mass over an area, particularly when the air masses have moved from their source regions. Arctic air is typically continental in its properties, particularly in winter, and equatorial air is typically maritime.

   Within the source regions at low and high latitudes, the weather is continuously that characteristic of the air masses, with infrequent interruptions. In middle latitudes the weather is constantly changing, as one air mass after another passes over the place, caused by the prevailing westerlies. Polar and arctic air masses move predominantly equatorward and eastward; tropical and equatorial air masses move mostly poleward and eastward. The weather is alternately that characteristic of one or the other air mass, or of the effect of their interactions.

10.2   Properties of air masses

The properties of the air masses, being derived from their source regions, need almost no elucidation. The continental air masses are generally dry, at least in absolute moisture content, while the maritime air masses have high humidities. The equatorial and tropical air masses are warm; the arctic and polar (by which is really meant subpolar) are cold.

   Arctic air masses are present principally in winter. They form in the polar anticyclone and are characterized by very low temperatures. Their absolute humidity is very low, because of the low temperature, but their relative humidity may be high. They are very stable near the ground, with strong inversions extending through the lowest kilometer or two.

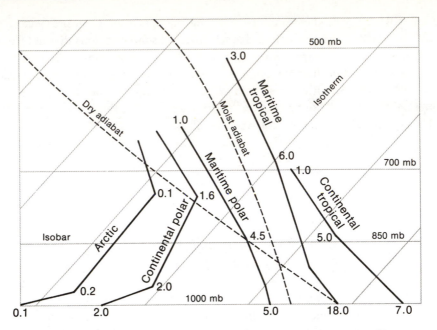

Figure 10.1 Schematic sounding curves for various air masses, plotted on Skew T–log P diagram. Numbers next to sounding curves are mixing ratios in parts per thousand.

Continental polar air masses in winter are similar in properties to arctic air, but not quite as cold. They form, in the Northern Hemisphere, when anticyclones stagnate over Alaska, Canada, Russia, and Siberia. In summer they are moderately cold, with more variable humidity and much less stability.

Maritime polar air forms over oceans at high latitudes. While air rarely stagnates in these regions, the source regions are sufficiently extensive so that the air can attain their characteristic properties even though it is moving fairly rapidly. In winter mP air is relatively mild, compared with cP or A; in summer it is cool; in both seasons it is moist and conditionally unstable.

Continental tropical air forms over subtropical land areas: North Africa, south-western United States and Mexico, and the desert areas in Asia (particularly in summer). It is hot and dry. It is unstable, but clouds are rarely present in it because of the very low humidity.

Maritime tropical air forms over the oceans at low latitudes on the equatorial side of the subtropical anticyclones. It is warm, moist, and unstable in the lower layers, but at a short distance above the sea there is an inversion, above which the air is hot and dry because of subsidence. As the mT moves around the anticyclone and starts poleward over the western portions of the oceans, the moist and (conditionally) unstable layer deepens.

Equatorial air forms near the intertropical convergence zone. It is warm and moist near the ground and unstable to high levels except over the eastern parts of the oceans where the sea surface is relatively cold because of the upwelling of deep water.

In Figure 10.1 are schematic sounding curves representing the typical variation of temperature and humidity of the various air masses with height in their source regions.

When air masses move from their source regions they are modified by interacting with the surfaces over which they pass and by the various processes described in Section 6.3 that alter their stability. They also may interact with each other. Their interactions with each other are described in Chapter 11. Some of the transformations they undergo are discussed in the next section.

10.3   Transformation of air masses

When arctic and continental polar air masses move equatorward, they are usually heated from below and thereby rendered unstable. Frequently, evaporation from warm water surfaces or from moist soil and vegetation increases their moisture content. Consequently, convective clouds develop in them. If incorporated in a cyclonic flow, there may be convergence that contributes to the development of instability through deeper layers, enabling the clouds to become large enough for showers to occur. More often, however, southward flow (in the Northern Hemisphere) is anticyclonic and accompanied by divergence that gives rise to a stable layer that limits the height to which the clouds grow even though the air continues to be warmed from below.

When A or cP moves out over the Atlantic or Pacific Ocean, the rate of heating is so great that the instability and moisture are quickly transported to considerable heights and the air is rapidly transformed to mP. Sometimes the air crossing the ocean is caught up in the circulation around the subtropical high and, passing over the warmer portion of the ocean, reaches land again as mT.

If mP moves from the Pacific over North America, it must ascend the mountain ranges of the west. Beginning with the coastal ranges, this ascent causes cloudiness and precipitation. Since the mP air usually has convective instability that is released by the ascent, the precipitation is heavy. On descending the mountains, the air is heated by adiabatic compression and reaches the plains to the east much warmer and drier than it had been. When mP replaces cP over the Great Plains in winter it brings mild temperatures, frequently with clear weather, after the bitter cold and snow showers of the cP. Sometimes the mild mP air arrives suddenly, with strong winds, and produces increases of temperature of as much as

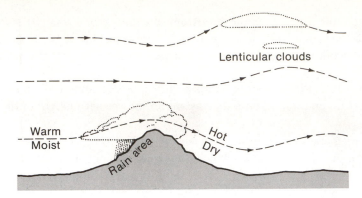

Figure 10.2   Transformation of mP air flowing over mountain range.

20°C in 15 minutes. As discussed in Chapter 7, the warm, dry wind descending the east slope of the Rockies is called the *chinook*. A similar warm wind descending the Alps is called the *foehn*. In Southern California the warm, dry wind descending from the elevated terrain inland is called the *Santa Ana*, but the Santa Ana consists of air that has crossed the Coast Ranges and the Sierra Nevada some days earlier and participated in subsidence over the Great Basin before turning southwestward, descending from the plateau and flowing down through the passes in the mountains to reach the coastal plains.

The transformation of moist air as it ascends over a mountain range and is heated dry adiabatically as it descends is illustrated in Figure 10.2. The precipitation occurs principally on the windward side. Sometimes on the lee side a standing wave develops in the flow, in which clouds form as the air ascends and dissipate as it descends. These clouds, like those over the windward side of the mountains, remain in place as the air passes through them. Because they frequently are lens shaped, they are called *lenticular clouds* (Figure 10.3). Occasionally, several lenticular clouds occur above each other, like a stack of pancakes, or other odd and spectacular arrays of them are present.

Tropical maritime air moving poleward over the ocean or over continents in winter tends to be rendered more stable, and fog or stratus clouds develop in it. However, when mT air moves over land in summer (and throughout the year at low latitudes), it may be rendered unstable, since the land surfaces are warmer than the oceans. Cumulus clouds with showers and thunderstorms occur in mT moving poleward over land in summer.

The structure of air masses is also changed by the convergence or divergence associated with large-scale flow patterns. The effect of convergence and divergence

Figure 10.3   Lenticular clouds formed in wave in lee of mountains. [Courtesy of NOAA.]

on the stability of polar continental air masses has already been mentioned. Divergence is associated with equatorward motion and anticyclonic flow, which therefore tend to stabilize the air and produce inversions, while poleward and cyclonic flow are convergent and thus tend to reduce the stability. The effects of these motions are felt principally at higher levels, since the amount of vertical motion (vertical stretching with convergence and shrinking with divergence) is proportional to the thickness of the column in which the convergence or divergence occurs.

An outstanding example of the effect of divergence on air-mass structure is the flow in the eastern portion of the semipermanent HIGHS over subtropical oceans. Here equatorward motion combines with anticyclonic flow to produce an extensive area of persistent divergence. The result is that the air descends from the middle troposphere and is heated by compression, producing the trade-wind inversion discussed near the end of Chapter 4. The air from high levels cannot descend all the way to the sea surface, and, in fact, the air near the sea gains heat and moisture as it moves equatorward over warmer water, so that a shallow unstable "marine" layer is formed, topped by the intense inversion. The depth of the marine layer depends on the relative intensity of the heating from below and the subsidence of air from above. Near the continents to the east, where it is shallow, the marine

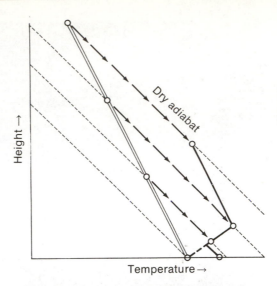

Figure 10.4 Schematic illustration of formation of inversion by horizontal divergence and vertical shrinking of an air column: double line shows the initial sounding; heavy line, the sounding after divergence, with the dashed portion of the inversion replaced by adiabatic layer because of heating at sea surface.

layer contains stratus clouds or fog. Farther west, where it is deeper, characteristic "trade-wind cumulus" are present. Figure 10.4 illustrates the process of modification of mP air to form the trade-wind inversion.

As a final example of the modification of air masses, the process by which mP air is transformed to cP or A air when it moves from the open oceans to a snow-covered continent or the ice-covered Arctic Ocean will be discussed. When the relatively warm maritime air begins to stagnate over the frigid land, the lowest levels are cooled by conduction and radiational exchange, establishing a shallow but intense inversion (Figure 10.5). The increased stability keeps the cooling from being "carried upward" by convective or turbulent transport of heat from above. Further adjustment of the temperature occurs only by radiative exchange.

Even though snow is highly reflective in the visible, as shown by its intense white color, at the infrared wavelengths at which most of the energy is radiated at its temperature it radiates nearly as a black body. Since air emits only in the wavelengths at which water vapor and CO_2 absorb radiation, the return radiation from the layers of air up to the top of the inversion and above it, even though they are at a considerably higher temperature, will be much less than that emitted by the snow. Thus, the snow will continue cooling until it reaches a temperature T_s

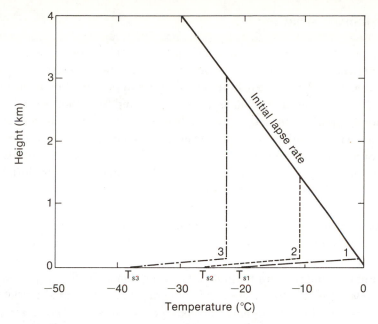

Figure 10.5   Successive stages during transformation of mP air into cP air over snow in winter.

at which the emission from the snow is equal to the return radiation. If the air temperature were to remain constant, the snow surface would approach the equilibrium temperature T_s and stay at that temperature. However, when the snow reached that equilibrium, the air temperature would not be in equilibrium.

Consider the layer of air in the vicinity of the inversion top in sounding curve 1 of Figure 10.5. It would be emitting radiation in those wavelengths at which it radiates at a temperature of about 270 K; the radiation reaching it from above and below at the same wavelengths, which it could absorb, comes from air or snow at lower temperatures. Thus, it would be losing more energy than it gains from radiative exchange, and it would be cooled. Once it cooled, the snow surface would no longer be in equilibrium and would start cooling some more. The cooling would affect progressively thicker layers of air, as shown by curves 2 and 3 in the figure. In this fashion the mP air is transformed to A or cP air through depths of several kilometers by radiative cooling.

We have not considered short-wave radiation from the sun in this discussion. Since the albedo (reflectivity) of snow for solar radiation is more than 80 percent, and the intensity of solar radiation at these latitudes is small except at the time of the summer solstice, the role of solar radiation is negligible.

10.4   Air-mass weather

The weather within an air mass depends on the air-mass properties, particularly its humidity and stability, and on the flow characteristics. In the previous section we have seen several examples of this: cloudiness and precipitation resulting from humidification and destabilization as air masses moved from their source regions.

In general, maritime air masses tend to have cloudiness and showers, and continental air masses tend to be clear. When mT air moves poleward over colder water or onto colder continents in winter, it is stabilized and fog or stratus clouds occur in it, and when cP air moves equatorward it is moistened and destabilized, and convective clouds or showers may occur, particularly if the flow becomes cyclonic.

Air masses frequently coincide in extent with stationary or moving anticyclones. In the eastern portion of the anticyclones the flow is equatorward and is usually divergent. This combination ordinarily results in stabilization of the air masses and clear weather. In the western portion of the anticyclones the flow is poleward and may be convergent. On the western side of anticyclones we would expect the air masses to become less stable and convective cloudiness and showers to occur. Thus, within a single air mass there may be enough variation of horizontal properties for clear weather to be present in one part and showers to occur in another, particularly when the air mass has moved from its source region.

While much of the time the weather at a place is determined by the properties of the air mass that overlies it, the most severe weather usually is associated not with a single air mass but with the interaction of two air masses in the vicinity of the boundary between them, that is, at *fronts*. In the next chapter we shall examine some aspects of this interaction.

Questions, Problems, and Projects

1. a. List the source regions from which you would expect air masses to reach your locality.
   b. For each such air mass, and separately for summer and for winter, describe (i) its properties at the source, (ii) the modifications it would experience in traveling to your locality, and (iii) the kind of weather your locality would have when it is present.
2. a. In what parts of the world would you expect continental polar air to form in winter?
   b. Discuss the changes you would expect continental polar air to experience in moving over the ocean.
3. Discuss the kind of weather you would expect on the western side of the coastal mountains of North America as maritime polar air crosses them in winter, and the kind of weather you would expect east of the Rocky Mountains when maritime polar air reaches there.

Fronts and Cyclones $\quad$ 11

11.1 Introduction

The association of decreasing pressure with the approach of bad weather and rising pressure with clearing skies has long been recognized. In fact, following the custom started in the eighteenth century, many barometers still have legends on them with the word "Fair" at high values of pressure and "Storm" at low values. In general, this association is valid: the approach of centers of high pressure (anticyclones) brings fair weather, while cloudiness and precipitation come with cyclones.

In the early years of weather forecasting this relationship was the dominant consideration, with pressure changes being concentrated upon rather than the motion and interaction of air masses. However, even then some meteorologists attempted to understand the physical causes of the weather by constructing cyclone models that contained some of the features of modern theories of storm development. For instance, Heinrich W. Dove (1803–1879), in 1837, concluded that temperate-latitude weather systems were associated with the interaction of warm and cold air currents, and Robert FitzRoy (1805–1865) published a diagram in 1863 showing cyclones forming at the zone of interaction between warm, moist air coming from subtropical latitudes and cold, dry air coming from polar regions. These early studies were largely ignored by practicing forecasters until the formulation of the Polar Front Theory of Cyclones by J. Bjerknes in 1918. This theory provided a systematic basis for understanding and forecasting the development of cyclonic storms in middle and high latitudes. The theory has been extended and modified through the years, but it remains one of the important guides to understanding weather patterns and the dynamic processes associated with them.

11.2   Fronts

The starting point of the polar front theory is that the boundary between two air masses approximates an abrupt discontinuity, rather than a wide zone of gradual transition. This boundary, of which the intersection with the ground shows up as a line on the weather map, is a surface, not vertical but sloping in such a way that the warm air extends above the cold air. On this frontal surface waves form that increase in amplitude and become cyclones, with low pressure at the center and winds circulating counterclockwise (Northern Hemisphere) around them.

In actuality, rather than a mathematical surface (of zero thickness), fronts are narrow zones in which the rate of change of air-mass properties with distance is large. Typically, the frontal zones of rapid transition from one air mass to the other are of the order of 100 km in width. Nevertheless, these zones permit wave behavior similar to that occurring at the interface between two fluids with differing densities and velocities. The nature of waves in "baroclinic" zones (layers in which the temperature structure varies rapidly in the horizontal) will be discussed briefly in the next section, along with the more readily visualized waves at a mathematical interface.

Figure 11.1 shows a portion of a weather map on which a typical wave cyclone appears.[1] The fronts are shown by dark lines with solid half-circles or triangular points. The contrast in temperature across the front shows distinctly. For instance, in Illinois, just east of the LOW the temperature (in °F) is in the 70's south of the front and in the 40's north of it. Similarly, in Missouri, southwest of the LOW, the temperature east of the front is 78°F; west of it, it drops rapidly to about 38°F.

---

1. The interpretation of the symbols on the weather map was explained in Section 1.3 and is given in more detail in Appendix E.

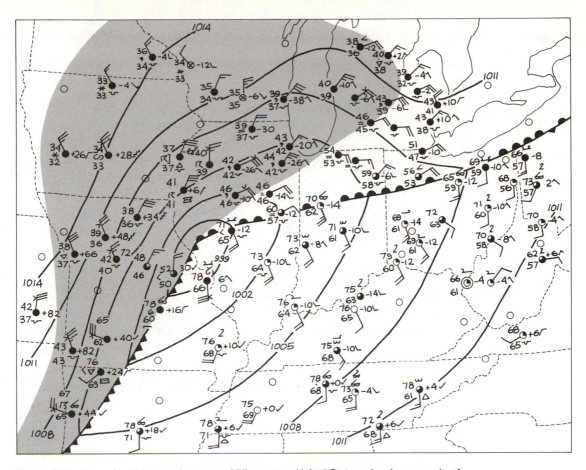

Figure 11.1 Map of surface weather over middle western United States, showing example of frontal wave. Shading indicates area of current precipitation.

Temperature is the principal property used to identify air masses and to locate the fronts between them. Other properties that change at fronts and help the weather-map analyst locate them are the following:

1. The humidity, represented by the dew-point temperature.
2. The pressure gradient. As a consequence of the sense in which the pressure gradient normally changes across a front, isobars crossing the front bend in such a way that the vertices of the angles they form point toward high pressure. Frequently, the front lies in a trough of low pressure.

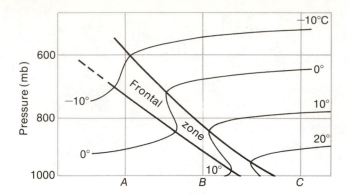

Figure 11.2 Schematic vertical cross section showing temperature distribution across a frontal zone.

3. The wind direction and/or speed. The change is such that *cyclonic shear* is present at the front. By cyclonic shear is meant horizontal variation of the wind in such a sense that a small volume of air at the front would be rotated cyclonically by the wind.

4. Cloudiness and precipitation. The nature of the cloud and precipitation patterns associated with fronts and frontal wave cyclones will be discussed later in this chapter.

In general, fronts can occur between any two air masses that form side by side or are brought together by their motions. However, the air circulations in low latitudes are such that fronts do not ordinarily develop between equatorial and tropical air. The principal front from the geographical standpoint is the *polar front*, which is the boundary between polar and tropical air. It is not continuous all the way around the earth, but is interrupted by regions where the transition between tropical and polar air is gradual. The front between arctic and polar air masses is called the *arctic front*. It, too, is not continuous around the earth. The fronts move with the air masses, and they undergo changes in sharpness and intensity as the air masses are transformed and the flow fields change. The average positions of the polar and arctic fronts undergo seasonal variations. In North America the average position of the polar front in winter is slightly south of Florida (extending northeastward toward the British Isles), while in summer it is north of the Great Lakes. Since cyclonic activity is principally associated with the polar front, there is a similar seasonal shift in the zones of maximum cyclone formation.

Fronts frequently occur between mP and cP air, or between an mP air mass that has undergone considerable modification and a fresh mP air mass.

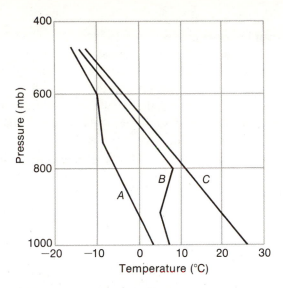

Figure 11.3 Temperature soundings at points *A, B,* and *C* in Figure 11.2.

When a front is moving it is called a *cold front* if cold air is replacing warm air at the ground (or sea surface), and a *warm front* if warm air is replacing cold air. If the air masses are moving parallel to the front so that the front does not move, it is called a *stationary front*. On manuscript weather maps, cold fronts are represented by blue lines, warm fronts by red lines, and stationary fronts by lines composed of alternate red and blue segments. On printed maps, cold fronts are represented by lines from which solid triangular points extend into the warm air; warm fronts are represented by lines with solid semicircles extending into the cold air, and stationary fronts by lines with alternate triangles on the warm air side and semicircles on the cold air side. In Figure 11.1 the front extending eastward from the LOW is represented as a warm front and the one extending southwestward from it is a cold front.

Figure 11.2 shows a schematic vertical cross section through a front with isotherms drawn to represent the temperature distribution. The front is shown as a zone of rapid change in temperature. The temperature decreases about 10°C through the frontal zone, and continues to decrease in the cold air. The horizontal temperature contrast is somewhat smaller at higher levels than near the ground, and at high levels the temperature decreases in the vertical through the front, whereas at low levels the frontal zone is an inversion. This difference is shown in Figure 11.3, in which sounding curves corresponding to positions *A, B,* and *C* in Figure 11.2 are shown. At position *B*, where the front is low, there is a strong inversion, but at position A the frontal zone is represented only by a slightly more stable lapse rate than below or above.

The steepness of the front is greatly exaggerated in Figure 11.2, in which the vertical scale is much larger than the horizontal. In actuality, fronts are inclined upward toward the cold air side with slopes in the range 1:50 to 1:300. By this is meant that if one went 50 to 300 km into the cold air from the surface position of the front, one would find the frontal surface at a height of one kilometer. Because friction slows the air near the ground more than the air higher up, cold fronts tend to be steeper and warm fronts less steep at low levels than aloft. The average slope of cold fronts is about 1:100; of warm fronts about 1:200. Fronts thus make very small angles with the horizontal. Nevertheless, this small angle is enough to provide the lift to the warm air flowing over the cold air to produce clouds and precipitation. Flow of the warm air up over the cold air develops typically in connection with frontal waves. In the next section we shall examine the nature of the frontal wave cyclone.

11.3   Frontal waves. The wave cyclone

At the surface between two fluids of different densities moving with different velocities, waves are likely to be present. The most familiar example is the air-water interface at the top of lakes or oceans. When the wind blows over a water surface, waves develop. These waves have various lengths and amplitudes, and the speed of their movement depends on these factors (and the water depth if it is shallow) and is different from that of the wind or the water.

By analogy, it is natural to expect waves to occur at the surface between two air masses of different densities. As visible evidence of one instance of such waves, billow clouds, the wave-shaped altocumulus or cirrocumulus clouds, are seen frequently. It has been shown that these clouds are due to wave motions at an inversion between warmer air above moving relative to the cooler, moist air below. The clouds form with the upward motion of the wave, and the clear spaces are in the places where the air has descended enough to be heated adiabatically sufficiently to cause the cloud to evaporate. Billow-cloud waves have wavelengths of the order of 1 km. The waves that are of significance in the formation of cyclones have lengths of the order of 1000 km. If waves of this length are unstable, they develop into cyclonic storms.

The concept of stability of a wave is similar to the stability of a parcel in an air column. If the frontal surface is perturbed (say, by displacing part of it slightly toward the cold air), the perturbation will tend to be propagated along the front as a wave. As it moves it may also tend to be damped out, or else to grow in amplitude. If it tends to be damped it is stable, if it tends to grow it is unstable, and if it moves along the front without change in amplitude it is neutral. Stable or neutral waves are of little interest, for their amplitudes ordinarily are too small for them to be noticed on the weather map. If waves of considerable length are unstable they develop into cyclonic storms, with significant weather phenomena associated with them, and contribute to the exchange of energy and momentum in the general circulation.

The existence of frontal waves was discovered in 1918 by J. Bjerknes, by careful examination of surface weather maps. Figure 11.4 shows the detailed model of the wave cyclone and the accompanying weather pattern that was published soon afterward by J. Bjerknes and H. Solberg. In the plan view, or weather map, the wave-shaped front is shown by the dashed line, with a *warm sector* protruding into the cold air that sweeps around it. There is a broad band of cloudiness and precipitation ahead of the warm front, caused by the warm air moving faster than the cold air and climbing up the frontal surface (compare Figure 1.9). At the cold front the precipitation is confined to a narrow band where the warm air is being pushed upward by the cold air. Farther back, the model shows the warm air above the cold frontal surface moving faster than the front, and thus descending and producing rapid clearing.

The model shown in Figure 11.4 applies to the Northern Hemisphere. For the Southern Hemisphere a mirror image would apply, with the reflection taking place across an east–west axis. The discussion in the remainder of this chapter will also be carried out from the Northern Hemisphere viewpoint, with the understanding that to apply it to the Southern Hemisphere requires similar reflection of directions.

The similarity between the idealized model and the actual situation shown in Figure 11.1 is readily seen, as well as some differences between them. Differences are to be expected, since the form of the wave and the extent and character of the clouds and precipitation depend on the particular flow pattern and air-mass properties that precede the formation of the wave. If, as was the case in the example shown in Figure 11.1, the warm air ahead of the cold front is convectively unstable, showers and thunderstorms occur at the cold front. This situation frequently occurs in the central and eastern United States. At the warm front the upward motion is usually more gradual and the warm air is stable, so that the precipitation takes the form of light, continuous rain or snow over a wide area ahead of the front. Occasionally, the warm air ascending the warm front is convectively unstable and gives rise to thunderstorms there, too, and conversely sometimes the precipitation at cold fronts takes the form of continuous rain or snow in a wide band behind the front.

Consider the weather changes that would be experienced during the passage of a wave cyclone. Imagine yourself at the location represented by the middle of the extreme right side of Figure 11.4. The wave, with its accompanying weather, is moving from left to right. Initially, the wind would be from the south or southeast.

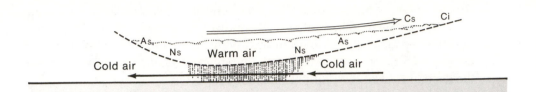

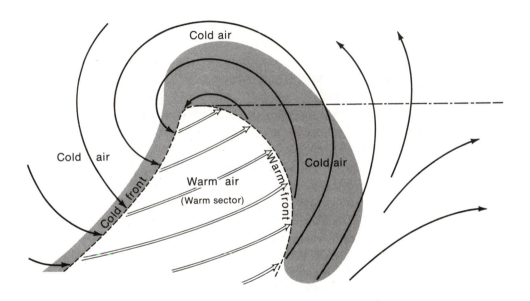

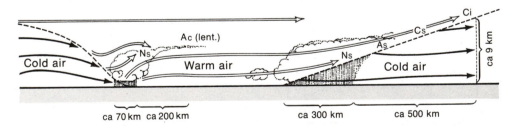

Figure 11.4   Idealized wave cyclone model: *upper*—vertical cross section north of wave cyclone; *middle*—representation of frontal wave and streamlines on surface weather map; *lower*—vertical cross section through warm sector. Shading shows area of current precipitation. [From J. Bjerknes and H. Solberg, *Geofysiske Publikasjoner* vol. 3, no. 1 (1923).]

The approach of the storm would show up first as a thickening cirrostratus cloud sheet approaching from the west. The pressure would fall—slowly at first, but steadily—and the clouds would become lower and thicker changing to altostratus. The cloud sequence as the warm front approached would be that described in Chapter 1. As the warm front moved to within perhaps 300 km, rain would begin to fall, at first light and intermittent, from clouds with bases near 3 km. Then the rain would become heavier and continuous, with the cloud bases lowering to 500 m or less. Fog would be likely as some of the rain evaporated into the shallow layer of cold air beneath the front.

With the passage of the warm front, the wind would shift to southwesterly, the temperature would rise, and the fog, precipitation, and clouds would disappear. This whole sequence would take place, on the average, in about 24 hours.

As the cold front approached, the pressure would begin to fall more rapidly. Some middle or high clouds might arrive as precursors of the line of cumulus or cumulonimbus that was approaching from the west. The arrival of the cold front would bring an abrupt shift of the wind to the northwest, a rapid decrease in the temperature, rising pressure, and showers or thunderstorms of greater intensity but shorter duration than the precipitation at the warm front.

Wave cyclones usually move in the direction of the motion of the air in the warm sector, which corresponds to the motion that would be indicated by the wave action of the warm and cold fronts. Ordinarily the direction of movement is eastward, with a northward or southward component in individual cases. The speed is quite variable, but is of the order of 10 m s^{-1}, corresponding to movement of about 1000 km per day. The daily displacements are larger in winter and smaller in summer.

If the pattern always conformed to the ideal model, it would be much easier than it is to predict the weather. As has been stated, differences occur because of differences in the conditions when the frontal wave forms. Topographic influences also cause differences in the characteristics of the waves. A large factor affecting the predictability of the weather is the fact that wave cyclones are not static entities that move horizontally without change, but dynamic organisms that undergo a "life cycle."

11.4   The life cycle of a cyclone. The occlusion process

Typically, the initial stage of a wave cyclone is a slight deformation of the polar front, frequently induced by a topographic feature. For this reason certain geographical regions are preferred areas for the formation of cyclonic waves. Among the favored areas for cyclogenesis (formation of cyclones) in the United States are the areas east of the Rockies, in the vicinity of the Ozarks, and off the east coast of the Carolinas. In the Pacific Ocean cyclogenesis is frequent in the region east and south of Japan. A favorite location for the formation of cyclones affecting Europe is the Mediterranean Sea west of Italy. In all of these locations the flow of air is disturbed by the change in elevation of terrain, or the flow from land to sea, and part of the front moves slower than other parts, thereby producing a wavelike shape in the front. If the wave is unstable the frontal deformation becomes larger, with a bulge of warm air intruding into the cold air mass.

As the amplitude of the perturbation increases, more and more cold air is replaced by warm air, contributing to a decrease in pressure and the development of cyclonic flow around the region of low pressure at the tip of the warm sector. Within 24 hours of the initial disturbance of the front, a well-defined wave cyclone will have developed, with warm front, cold front, and pattern of cloud and precipitation corresponding to the ideal model in Figure 11.4. In an unstable wave the cold air behind the cold front moves faster than the air receding ahead of the warm front. The air in the warm sector is constantly being squeezed upward thereby. Eventually the cold front catches up with the warm front; at this stage the warm air is completely lifted from the surface and the cyclone is said to be *occluded*. The front that is formed by the merger of the cold front and the warm front is called an *occluded front*. The process of the cold air closing off the warm air from the ground is called *occlusion*.

Various stages of the life cycle of a wave cyclone and the occlusion process are shown in Figure 11.5. Diagram A represents the newly formed wave, with the pressure at the crest reduced just a little, and the precipitation bands quite narrow. In diagram B the wave is fully developed, with the characteristics described in the previous section. At this stage the air throughout the troposphere is participating in the wave motion, so that the tropopause is deformed, as shown in the vertical cross section at the top of the diagram. Diagram C shows the system after the wave has become partially occluded. The lifting of the warm-sector air lowers the center of gravity of the system, thereby converting potential energy into kinetic energy of the increased winds around the deeper low-pressure center. At a later stage (diagram D), the cold front has overtaken the warm front out to a larger distance from the center, practically eliminating the warm sector at the ground. The pressure at the center is still lower, with more closed isobars, forming a large vortex completely surrounded by cold air. After this, no further intensification can take place because the available supply of potential energy has been exhausted, and the cyclone gradually weakens and fills because of the action of friction.

As the occlusion process goes on, the movement of the wave cyclone and its associated low pressure center slows down. When it is fully occluded, so that the low center is completely surrounded by cold air, it may become practically stationary. The lifetime of an occluding cyclone ranges from 48 to 96 hours for the

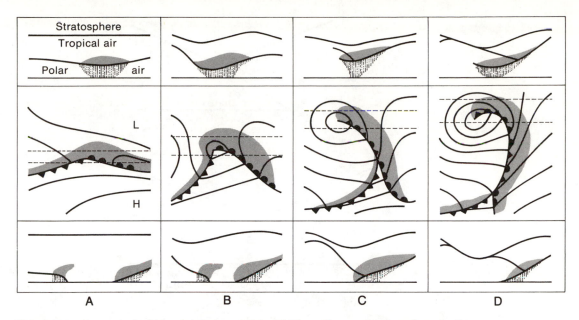

Figure 11.5 Stages in the life cycle of wave cyclone: Middle section represents surface weather maps at various times; upper and lower sections, vertical cross sections at positions of dashed lines in middle section. (A) Growing wave. (B) Mature wave. (C) Partially occluded wave. (D) Wave near completion of occlusion process. [After C. L. Godske, T. Bergeron, J. Bjerknes, and R. C. Bundgaard, *Dynamic Meteorology and Weather Forecasting* (Boston: American Meteorological Society, 1957).]

occlusion process, and it takes an equal or longer time for the dissipation of the cold low.

In diagram C the vertical cross section through the occluded front (lower part of the diagram) has been drawn on the assumption that the cold air behind the cold front is warmer than that ahead of the warm front and climbs up over it, leaving the front at the ground in the form of a warm front. In this situation the front is called a *warm-front-type occluded front*. It may occur when the air behind the cold front comes over the ocean while the air ahead of the warm front has had a trajectory completely over land, or when the cold air immediately behind the front has subsided from higher levels. The opposite situation frequently occurs, in which the air behind the cold front, coming more directly from the north (Northern Hemisphere), is colder than the air ahead of the warm front, which has spent a longer time since leaving its source region. In this case, the air behind the cold front pushes under the air ahead of the warm front, and the front is called a *cold-front-type occluded front*. Figure 11.6 shows vertical sections through the two types of occluded fronts, plus a vertical section through an occluded front with the

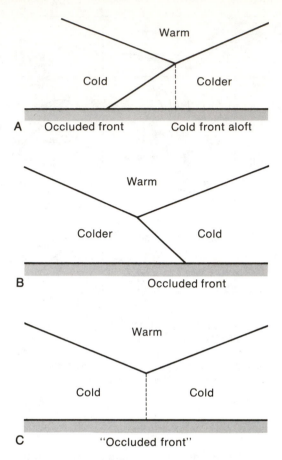

Figure 11.6 Representation of the types of occluded fronts: (A) warm-front type; (B) cold-front type; (C) occluded front with no temperature contrast at earth's surface.

same temperature on both sides. It should be noted that the precipitation comes principally from the warm air that is being lifted in a V-shaped trough in all three types, so that there is no difference in precipitation except with respect to the position of the precipitation in relation to the surface front.

In the warm-front-type occlusion the trough of warm air aloft is ahead of the front at the ground, at the position where the original cold front has climbed up the warm frontal surface. Since this position may be considerably ahead of the surface front and has significant weather associated with it, it is often shown on the surface weather map as a *cold front aloft*, represented by a dashed blue line or a broken printed cold front symbol. Occluded fronts are represented by purple

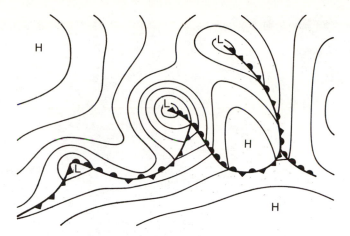

Figure 11.7   Schematic representation of family of wave cyclones at various stages on front.

lines on manuscript maps and lines with alternate semicircles and triangular points on the side toward which the front is moving on printed maps.

Frequently, a succession of waves forms on a front, so that a "family" of cyclones in various stages is present along the front. Figure 11.7 is a schematic illustration of such a wave-cyclone family. Families of this type occur frequently, particularly over the Atlantic and Pacific Oceans. The wave cyclones reaching the American west coast from the Pacific are usually completely occluded members of such families. Many of the cyclones remain over the ocean, passing northeastward along the front as they occlude and merging with the semipermanent Aleutian or Icelandic lows.

11.5   Upper-level flow in relation to wave cyclones. Waves in the westerlies

When the models of the wave cyclone were first developed, observations of the temperature structure and flow patterns aloft were practically nonexistent. Since then, upper-air observations have clarified the relationship between the frontal waves and the flow at upper levels. At the same time, the observations have made it clear that other processes frequently lead to cyclogenesis without the prior existence of fronts. Once the cyclone-forming process is under way, it may give rise to the formation of fronts within the cyclone. Thus, there are two general ways

Figure 11.8   GOES West satellite photo (taken 2345 GMT May 18, 1977) showing a family of wave cyclones over the Pacific Ocean. A well-occluded cyclone is close to the North American coast; some distance southwest is a wave in the early stages of occlusion; and farther southwest there are two small bulges on the front suggestive of waves in the early stage of development. At the western edge of the picture there is an occluded cyclone on another frontal system, and along the southern edge the suggestion of another front with a wave on it. [Courtesy of NOAA.]

in which cyclones may form: frontal waves at low levels that induce wave-shaped flow patterns aloft, and flow patterns aloft that produce low-pressure centers at lower levels and subsequently may draw air masses together, producing fronts between them.

In accord with the discussion in Section 8.7, the horizontal variation of the temperature in a frontal wave cyclone leads to variation with height of the pressure and wind fields. At upper levels the low-pressure center tilts toward the cold air, and since the isotherms on the cold-air side of the front are wave-shaped, some of

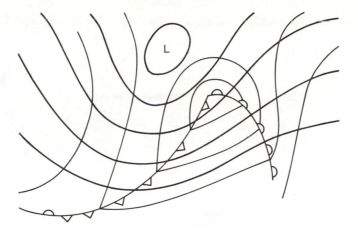

Figure 11.9 Schematic representation of relation between sea-level and upper-level flow in a wave cyclone. Heavy lines represent 500-mb contours; light lines are sea-level isobars and open front symbols represent the frontal wave there.

the closed isobars tend to be replaced by wave-shaped isobars (or wave-shaped isobaric contours on the surfaces of constant pressure). Figure 11.9 shows schematically the upper-level pressure field (say, 500-mb contours) corresponding to the surface isobars in a wave cyclone. The low center at the upper level is displaced northwestward from the surface position, toward the cold-air side of the front, with only one closed isobaric contour around it. The rest of the flow, as shown by the contours, forms a wave, with the ridge ahead of the sea-level position of the warm front and the trough on the cold-air side of the cold front.

Maps showing the observed conditions at sea level and 500 mb on one particular day are shown in Figure 11.10. The general features just discussed are present in relation to each of the cyclonic systems: the low centers and troughs displaced to the cold-air side, and wave-shaped contours aloft replacing the closed sea-level isobars. The consequence is that the flow aloft (regarded as approximated by the geostrophic wind corresponding to the 500-mb contours) forms a wavy band across the entire map. In this instance there are three upper waves represented on the portion of the Northern Hemisphere covered by the maps, with troughs, each associated with an occluding wave cyclone, 40–50 degrees of longitude apart. There would be seven or eight such waves going all the way around the earth.

Waves of this sort are characteristic of the westerly winds that prevail at upper levels in the troposphere. They vary in length. The shorter ones, associated with frontal waves at sea level, move rapidly, about 10–15 degrees of longitude per day. Longer waves, such that there would be four to six around the earth, move more

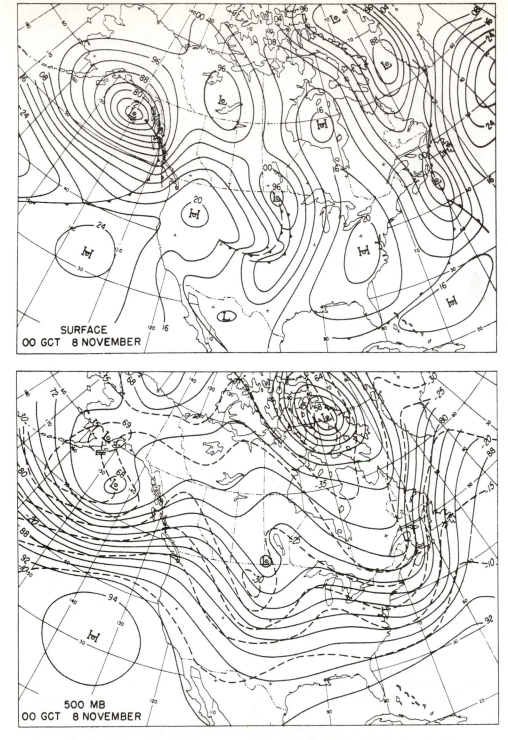

Figure 11.10 Surface and 500-mb charts at the same observation time. Dashed lines on the 500-mb chart are isotherms, labeled in °C.

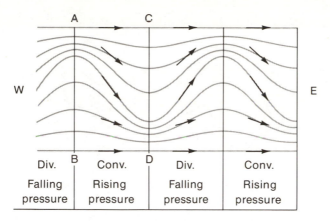

Figure 11.11   Schematic pressure field in a wave in the westerlies at a level in the upper troposphere. [After J. Bjerknes, *Meteorogische Zeitschrift*, vol. 54, 1937, p. 463.]

slowly and are associated with the larger circulation features, such as the Aleutian and Icelandic lows and the subtropical highs. The flow in the long waves, known as *Rossby waves,* after C.-G. Rossby (1898–1957), a Swedish–American meteorologist who studied the dynamics of the upper waves, tends to "steer" the shorter waves, which move southward when behind the trough of a long wave and northward when ahead of it. In addition to changing direction, the troughs of the short waves intensify as they overtake the trough of a long wave and weaken as they catch up with the ridge.

The movement of these waves in the upper westerlies and their relationship to the cyclones at the earth's surface were first discussed by J. Bjerknes in an article published in 1937. Subsequently, C.-G. Rossby studied the dynamics of these waves in detail and derived formulas for their speed. Bjerknes' treatment was based on the effects of the convergence and divergence of the flow in producing pressure changes, while Rossby's analysis was based on the principle that absolute vorticity is conserved under conditions that are approximated in the midtroposphere.

Figure 11.11, taken from Bjerknes' 1937 paper, shows schematically the pattern of convergence and divergence, and the corresponding pressure changes, in a wave at a level in the upper troposphere where the air is moving faster than the wave. Where the wind blows toward closer-together isobars, it is accelerated and deflected toward low pressure; where it overtakes isobars that are farther apart, it is decelerated and deflected toward high pressure. In the pattern postulated, this results in convergence of the flow and rising pressure ahead of the ridge and divergence and falling pressure ahead of the trough. At low levels, where the wind

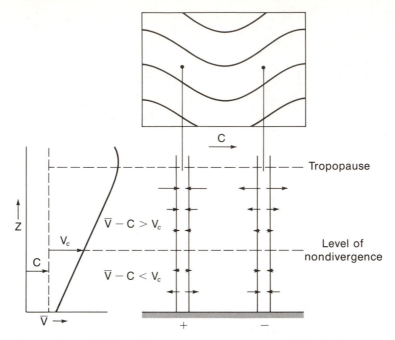

Figure 11.12 Variation of divergence and convergence with height in a wave in the westerlies in which the wind speed increases with height. [After J. Bjerknes and J. Holmboe, *Journal of Meteorology*, vol. 1, 1944, p. 12.]

speed is less than the wave speed, the pattern of convergence and divergence is reversed.

The net pressure fall at sea level ahead of the trough is the result of convergence in low levels somewhat offsetting the divergence aloft, and similarly in reverse for the pressure rise ahead of the ridge. The combined effects at various levels is illustrated in Figure 11.12. At levels where the wind speed exceeds the wave speed by more than a critical amount v_c, the pattern is that shown in Figure 11.11. Where it is less, the pattern is reversed. At the level where the wind speed minus the wave speed is just equal to the critical speed, there is neither convergence nor divergence; this level is called the *level of nondivergence* (LND).

Rossby's approach was to make use of the fact that for the atmosphere as a whole the convergence and divergence almost offset each other, so that in effect the behavior of the entire atmosphere is closely approximated by the behavior at the level of nondivergence, which is found usually in the midtroposphere, between 600 mb and 500 mb. If there is no convergence or divergence, the absolute vorticity of air parcels remains constant as they move. The paths of such parcels are called constant absolute vorticity (CAV) trajectories. By mathematical analysis of the behavior of CAV trajectories, Rossby derived a quantitative relationship for the speed at which waves in the upper westerlies move. In the following we shall examine qualitatively how this relationship arises.

In Section 8.5 the relative vorticity ζ, the rate of turning of a small piece of air due to its motion relative to the earth, was shown to consist of a curvature term v/r and a shear term $\Delta v/\Delta n$. The absolute vorticity, ζ_o, was given as $\zeta_o = \zeta + f$, where f, the Coriolis parameter, is the vorticity due to the earth's rotation around the vertical at the latitude where the air is located. Rossby pointed out that in the center of the westerly current, where the wind speed is a maximum, the shear is zero, and the vorticity is just the curvature term. If we substitute the curvature term for ζ in the equation for ζ_o and use the fact that ζ_o is constant along the path of an air parcel at the LND we have

$$v/r + f = \text{constant}$$

Since f increases with latitude (see Table 8.1), v/r must decrease as parcels move poleward and increase as parcels move equatorward. (Remember that v is always positive and changes relatively slowly along the trajectory; r is by definition positive for cyclonic and negative for anticyclonic curvature.) Thus, the curvature of the path of poleward-moving parcels becomes less cyclonic or more anticyclonic, and the curvature of the path of equatorward-moving parcels becomes more cyclonic or less anticyclonic. Let us see how this relationship influences the movement of westerly waves.

Figure 11.13 is a schematic representation of the CAV trajectory for an air parcel that starts out with negative (anticyclonic) curvature at 50°N. Bending around the ridge of high pressure, it moves southward, and the decreasing value of f results in an increase in ζ from its negative value to zero (no curvature) at 45°N, and to positive values (cyclonic curvature) as it continues southward. The increased cyclonic curvature eventually (at 40°N in this example) bends the path enough so that the air parcel heads northward, leading to decreasing vorticity, which straightens the path and ultimately bends it anticyclonically again. Thus, the conservation of absolute vorticity results in wave-shaped paths of the moving air parcels.

In wave motion streamlines have shapes similar to the trajectories, but with different wavelengths and amplitudes unless the waves are stationary. Thus, in Figure 11.10 the contour lines in the 500-mb map at corresponding latitudes have the same general wave shape as the schematic CAV trajectory in Figure 11.13. (Remember that since the wind away from the earth's surface usually deviates only slightly from the geostrophic, the contour lines of isobaric surfaces are approximate streamlines.) In Figure 11.13, in addition to the CAV trajectory, the streamlines for the initial time t_A when the parcel was at the anticyclonic bend A and the time

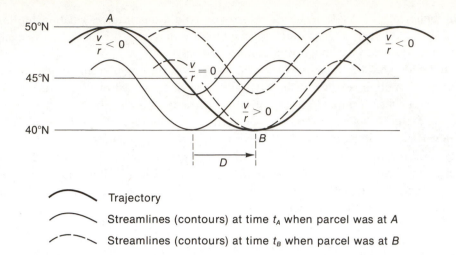

Trajectory

Streamlines (contours) at time t_A when parcel was at A

Streamlines (contours) at time t_B when parcel was at B

Figure 11.13 Schematic representation of the relation between a constant absolute vorticity trajectory (CAV) and streamlines (SL's). *Heavy line*: CAV; *light solid line*: SL at time t_A; *light dashed line*: SL at time t_B.

t_B when it reached the cyclonic bend B are shown. It should be remembered that the trajectory is a picture of the location of the air parcel at successive times, while the streamlines show the direction of motion of many parcels at one particular time. At any instant the streamline at the position of the parcel must point in the same direction as the trajectory. Thus, the streamline at A at time t_A is tangent to the trajectory, and A is the position of a ridge in the streamlines; the streamline at B at time t_B is tangent to the trajectory there, and a trough in the streamline is at B at that time. The movement of the wave is indicated by the position at time t_A of the trough that reaches B at time t_B, as shown by the arrow in the figure.

The distance D that the wave in the streamlines moves in the time $t_B - t_A$ depends on the amount the wavelength of the streamlines is shorter than the wavelength of the CAV trajectory. In other words, the shorter the wavelength of the streamlines the faster the wave moves. If the streamlines were to have the same wavelength as the trajectory, at time t_A the trough would already be at position B, its position at time t_B. Thus, the trough would not be moving, and the wave would be stationary.

The wavelengths of CAV trajectories depend on the wind speed. The stronger the westerlies, the longer their wavelengths. Consequently, for a given wavelength of the streamlines, the stronger the westerlies the greater the distance D and the faster the waves move. Rossby's formula thus involves the average speed of the wind U and the average length L_s of the waves in the streamlines. The formula may be written

$$C = U - bL_s^2 \qquad (11.1)$$

where C is the wave speed and b is a factor that varies with the latitude (taken to be the latitude at the middle of the wave). At 45°, b has the value $4.1 \cdot 10^{-13}$ for L_s in meters and U and C in m s^{-1}. For example, the average westerly component of the geostrophic wind in the wave over western North America and the eastern Pacific Ocean in Figure 11.10 is 30 m s^{-1}, and its wavelength is about 4000 km. Entering these values of U and L in equation (11.1), we get for the wave speed $C = 30 - 4.1 \cdot 10^{-13} (4 \cdot 10^6)^2 = 23.4$ m s^{-1}. At this speed, in 24 hours the ridge over the western provinces of Canada and Washington and Oregon would move 2000 km, to Saskatchewan and the Dakotas. Movement of waves by one-half their length, while rapid, is not unusual in high latitudes in winter.

The length of waves that are stationary is obtained by setting $C = 0$ in equation (11.1), giving

$$L_s = \sqrt{U/b} \qquad (11.2)$$

In Figure 9.7 we see that the average west winds at 45°N and 500 mb are about 14 m s^{-1} in winter and 8 m s^{-1} in summer. The values of L_s for these values of U are about 5850 m and 4400 m respectively, corresponding to 4.8 and 6.4 waves around the earth. In the average 500-mb maps (Figures 9.5 and 9.6), somewhat fewer waves are indicated, roughly four in January and five or six in July, although the contour lines do not form simple wave shapes that make the number easy to decide.

11.6   The jet stream

Since the horizontal gradient of temperature is large along the polar front, the thermal wind in its vicinity is large, leading to a narrow band of very strong winds at high levels. This region of strong winds has been likened to a jet, for instance, the water spurting from a garden hose, and was named the *jet stream*. While its existence had been recognized earlier, it did not receive general attention until World War II, when bombing planes on their way to Japan encountered headwinds so strong that they could hardly make any progress relative to the ground. Winds

Figure 11.14   Infrared GOES satellite photo (1900 GMT November 28, 1978) showing high clouds associated with subtropical jet stream extending across Mexico and eastern United States.

as strong as 135 m s^{-1} (300 mi/hr) have been observed in the jet maximum. The average speed in the center of the jet stream ranges from 35 to 55 m s^{-1} at various longitudes in winter, and is somewhat lower in summer. The maximum speeds in the jet occur just below the tropopause. Jet planes traveling eastward save much time and fuel by flying in the jet stream; traveling westward, they avoid it either by choice of route or by flying below the level of maximum speed.

Because other regions of concentrated high-wind speeds exist, the one associated with the polar front is called the *polar front jet stream* (PFJ). It has been likened to a meandering river winding its way from west to east around the earth. Unlike a river, it is constantly shifting from north to south as the polar front moves, and its meanders correspondingly change position and shape. Like the polar front, it

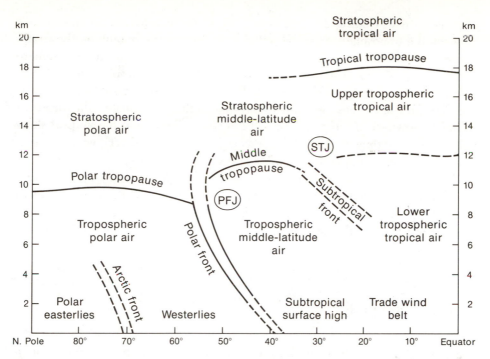

Figure 11.15 Schematic representation of relationship between fronts and jet streams. [After E. Palmén and C. Newton, *Atmospheric Circulation Systems* (New York: Academic Press, 1969).]

is not continuous around the earth, and frequently it splits into two streams, which subsequently reunite. Being a large concentration of momentum and kinetic energy, it plays an important role in the dynamic processes of the troposphere. Its position frequently becomes stabilized and seems to govern the displacement of the polar front and the movement of cyclones along it, rather than the reverse.

In places in the upper troposphere where the poleward branch of the Hadley cell is well-marked, there is a rapid transition in temperature between the warm air carried by it and the air of middle latitudes. This transition has been called the *subtropical front*. The subtropical front is present only at upper levels (8–12 km). The packing of isotherms in it produces a large thermal wind and consequently a concentrated band of strong winds, known as the *subtropical jet stream* (STJ), above it. The STJ is steadier and less variable in latitude than the PFJ. Because of this, the wind maximum in average maps, such as Figure 9.2, tends to be at relatively low latitudes, even though the PFJ at high latitudes is usually stronger.

The relationship between the jet streams and other circulation features is illustrated in Figure 11.15. The tropopause is shown with breaks at the polar front and the subtropical front, and the PFJ and STJ are above these fronts but below the tropopause in the tropical air. The breaks in the tropopause permit lateral exchange between the troposphere and stratosphere. Because of the stability in the stratosphere, vertical mixing can go on only extremely slowly. Within the troposphere vertical motions occur readily, except in the region of the subtropical anticyclones, where the descending branch of the Hadley cell produces the subtropical inversion. The horizontal shear on the poleward side of the jets promotes the turbulent exchange between the troposphere and the stratosphere, and the vertical shear below the jet maxima is responsible for clear air turbulence (CAT), which is experienced as severe bumpiness in plane flights at these levels.

There is also a phenomenon known as the *low level jet stream*. Occurring principally at night in the western Great Plains, it is characterized by wind speeds from 12 to 25 m s^{-1} at levels below 1.5 km, with much lower speeds above that. It constitutes a serious hazard for airplanes during landing and takeoff because of the rapid variation of wind speed with height.

11.7   Cyclogenesis

The mechanism by which cyclones are formed has been studied from several different approaches. In early theories, the development of low-pressure centers was thought to be due to heating and a convective circulation. Except for tropical cyclones, these theories were disproven when measurements of upper-air temperatures became available, which showed that cyclones in temperate latitudes had low temperatures aloft, and not the high temperatures required from hydrostatic considerations in order that low pressure may be present. The instability of frontal waves, as discussed in Section 11.4, provides one way in which cyclones form when the potential energy of position of the cold and warm air masses is transformed into kinetic energy of cyclonic circulation. Potential energy is likewise available whenever there is a variation of temperature in the horizontal, or more accurately, whenever the density varies in surfaces of constant pressure. Regions in which such a variation is present are said to be *baroclinic*, whereas regions in which the density does not vary along the surfaces of constant pressure are *barotropic*. Waves occurring in a baroclinic region may release potential energy and set up cyclonic circulations even when the baroclinicity is not concentrated in a narrow frontal zone. These waves may be set off in the westerlies of the upper troposphere, and the cyclonic circulation may work its way down to the ground. Sometimes when this occurs warm and cold air masses may be brought together by the induced circulation, that is, the baroclinicity becomes concentrated and fronts are formed. After the fronts have formed, the further development may resemble the life cycle of a frontal wave.

A large part of the problem of weather forecasting has been the determination of when and where cyclones will form and deepen, as well as in what direction and how fast they will move. Previous to the use of electronic digital computers, the forecaster depended a great deal on empirical rules that guided the early detection of incipient frontal waves and the estimation of their rate of development and their movement. By applying numerical methods of solving the equations of atmospheric motion, using computers, much of this empiricism has been replaced in weather forecasting. However, because of the complexity of the equations and the inadequacy of the observations of the initial conditions, the numerical solutions are only approximate, and, particularly with respect to forecasting cyclogenesis, they are frequently inadequate. It is expected that improved observations, better mathematical models, and larger computers will lead to better forecasts in this regard.

Questions, Problems, and Projects

1. Make graphs showing the variation of temperature with distance in Figure 11.1, (a) along an east–west line approximately dividing the map in two, and (b) along a north–south line approximately following the Mississippi River. In the graphs mark the positions of any fronts that the lines intersect. Describe how the temperature changes in crossing the fronts and how it varies within the air masses on each side. Describe how the cloudiness and precipitation change in crossing the fronts and how they vary within the air masses on each side.

2. a. Why does the pressure at the low center decrease and the winds around the low center get stronger as a wave cyclone occludes?
   b. Why do wave cyclones in middle latitudes usually move with a component toward the east?

3. a. Why does the westerly component of the wind usually increase with height in the lower troposphere?
   b. Why are the strongest westerlies frequently concentrated in a narrow band or jet?

4. Why are thunderstorms more likely to be associated with cold fronts than with warm fronts?

5. Using the estimate made on page 269 for the average westerly wind speed in Figure 11.10, find how long the waves centered on 45°N would have to be in order that the waves be stationary. How fast would they move if they were one-third of this length? Two-thirds?

12 Weather Systems in the Tropics

12.1 The atmospheric circulation at low latitudes. The Intertropical Convergence Zone

We have seen that at middle and high latitudes the temperature and state of weather undergo continuous variation, determined by the properties of one or another moving air mass and the passage of cyclones with their associated systems of cloud and precipitation. At low latitudes, between the two subtropical high-pressure belts, the variability is much less and except on infrequent occasions the circulation pattern on any day departs very little from the average for the season. Nevertheless, the small departures in flow patterns that take place produce significant variations in daily amounts of cloudiness and precipitation, and in many places there are seasonal changes, with rainy weather part of the year and dryness at other times. There are also variations from year to year; the drought in the Sahel region of tropical West Africa in the early 1970s is an example of a marked deviation from normal that lasted several years.

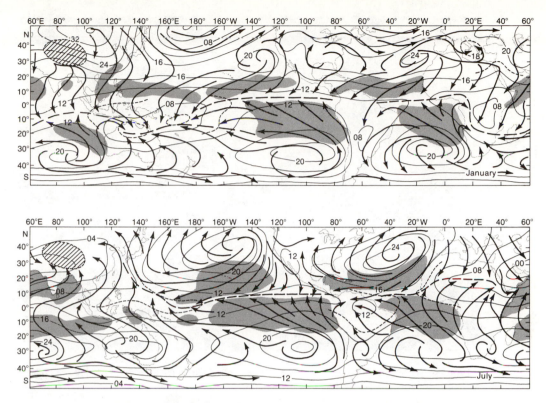

Figure 12.1   Average sea-level isobars and streamlines at surface over tropics, January and July. Regions of steady winds, with wind direction within 45° of the average streamlines more than 80 percent of the time, are shaded. The ITCZ is shown as a dashed line between the northeast trade winds and the southeast trade winds. [From E. Palmén and C. W. Newton, *Atmospheric Circulation Systems* (copyright 1969 by Academic Press).]

Near the equator the influence of the earth's rotation on horizontal motions is small. The horizontal component of the Coriolis force decreases with latitude and vanishes at the equator. The closer to the equator, the more the winds deviate from the geostrophic direction and speed, and the less useful the isobars are to summarize the wind field. Instead *streamlines* are used, which are lines having the property that at every point along them the tangent is in the direction of the wind.

Figure 12.1 shows the average isobars and streamlines over the earth between 50°N and 50°S latitude in January and July. The shaded areas are the regions of least variability of the wind, with wind direction within 45° of the average streamline more than 80 percent of the time. The shading is mostly in the trade winds equatorward of the subtropical HIGHS, showing that the trade winds are the

Figure 12.2   Cumulus and cumulonimbus clouds in ITCZ over Pacific Ocean near the Marshall Islands. [Photo by Joanne Simpson.]

steadiest features of the atmospheric circulation near the ground. Nevertheless, as we shall see, within this degree of steadiness the easterly trades are subject to significant fluctuations.

Notice that the streamlines cross the isobars at large angles near the equator. The convergence zone between the trade winds of the two hemispheres, known as the Intertropical Convergence Zone (ITCZ), is shown by heavy dashed lines. It is interrupted or displaced by the influence of continents. In general, the ITCZ migrates, being in the Northern Hemisphere in the northern summer and the Southern Hemisphere in the southern summer. However, over the Atlantic and eastern Pacific oceans, it is north of the equator in January as well as in July. The ITCZ is characterized by convective cloudiness and showers. The clouds take the form of bands or clusters of cumulus clouds, only a few of which develop into cumulonimbus and thunderstorms. Over the oceans the typical cloud clusters have lifetimes of several hours, but usually less than a day. Over land they undergo a typical diurnal cycle, forming in the morning, with maximum development and showers in midafternoon, and dissipating after nightfall. This convective activity is considered the principal mode of transfer of heat upward in the Hadley circulation cells.

Over the Indian Ocean and the western Pacific the winds are dominated by the Asiatic Monsoon, with continuous flow outward across the equator from the anticyclone over Asia in January, and inward toward the LOW over Asia from south of the equator in July. In January the northeast flow turns to northwest in crossing the equator; in July the southeast flow changes to southwest in crossing it. The eastern Atlantic and the adjacent portion of West Africa are subject to a similar monsoon effect. In July the southeast trade winds are pulled into the LOW over the Sahara and are deflected into southwest winds as they cross the equator. Since Africa does not extend to high latitudes where the temperature is low in winter, the mean flow in January over this area is the unperturbed trades, unaffected by a monsoon effect.

As is usual, the easterly winds increase in speed with height through the layer of frictional influence. This layer is deeper in the tropics than at higher latitudes both because of the up-and-down exchange of air due to convective activity and because the dynamic laws that govern its thickness say that the smaller the Coriolis parameter f the greater the thickness. Even though the rate of temperature decrease with latitude is small in the tropics, there is a small westerly thermal wind effect (Section 8.7). The effect of the slight poleward decrease in temperature is to reduce the strength of the easterlies aloft; combined with the friction effect, which produces an increase with height at low levels, the result is maximum strength of the easterlies in the middle troposphere.

12.2   Temperature structure in the trade-wind region

Since the trade winds are directed equatorward, for the most part they carry air toward regions of warmer water. Thus, the air near the sea surface is heated from below and has moisture added to it. In response to the instability thus produced, convective clouds—trade-wind cumulus—are characteristic of the region. At the same time, as discussed in Section 10.3, the flow around the subtropical anticyclone is divergent and an inversion is present that limits the height to which convective clouds can penetrate.

The trade-wind inversion is lowest at the eastern portions of the oceans, and slopes upward toward the west. It also slopes upward toward the ITCZ. Figure 12.3, which is an average vertical cross section from San Francisco to Honolulu in summer, illustrates the upward slope of the inversion as one follows a streamline southwestward. At the coast the inversion base is at 400 m, and the clouds can

Height m)

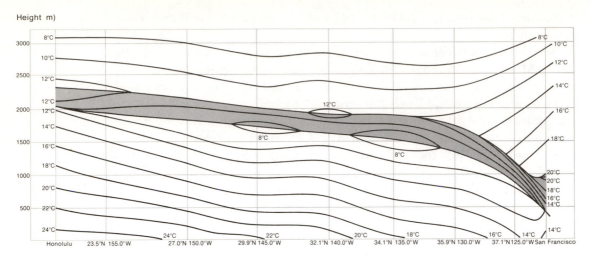

Figure 12.3 Average vertical cross section from San Francisco to Honolulu in summer. Sloping solid lines are isotherms. The inversion layer is shaded. [From M. Neiburger, D. S. Johnson, and C. W. Chien, *Studies of the Structure of the Atmosphere over the Eastern North Pacific Ocean in Summer,* vol. 1:1 (1961). Published by the University of California Press; reprinted by permission of The Regents of the University of California.]

form only in the moist air below that height. Consequently, fog, stratus, or stratocumulus are present much of the time along the California coast. Farther southwest the inversion slopes upward, first rapidly, then less rapidly, and there is room for convective activity below it, permitting the typical trade cumulus to form. Over the eastern portion of the oceans, the clouds still are not deep enough to produce rain, but farther west showers occur, particularly at times when the inversion is raised or dissipated by passing disturbances.

The increase in height of the inversion downwind along the trades is primarily due to transport of heat from the sea as the air moves over warmer water. In a fashion similar to the diurnal heating by solar radiation, the air columns moving westward are rendered unstable to successively greater heights, leading to formation of clouds that penetrate the inversion and mixing air from the inversion layer downward into the boundary layer (the unstable layer of air in contact with the sea).

The motion of the air toward the west on both sides of the equator has a remarkable effect on the temperature of the sea surface. This is a consequence of the effect of the Coriolis force on the motion of the water due to the frictional stress exerted on it by the wind. The Coriolis force acts to propel the water to the right of the wind in the Northern Hemisphere and to the left in the Southern Hemisphere. The easterly winds on the two sides of the equator thus cause the surface water to move northward north of the equator and southward south of the equator. To replace the water that diverges at the equator, water must rise from the depths, where it is cold. The *upwelling* due to the frictional drag or stress of the easterly trades thus produces a belt of cold water along the equator over the eastern part of the oceans. Where this cold belt is present there is no heating of the air from below, and no convective cloudiness and precipitation are present. This part of the equatorial belt, contrary to expectation, is arid.

12.3 Waves in the easterlies

The existence of disturbances in the trade winds was first recognized in the 1930s by tracing the movement of pressure changes across the tropical North Atlantic Ocean. It was found that these fluctuations in pressure moved westward and were correlated with variations in the thickness of the moist layer below the inversion as shown in airplane soundings at the Virgin Islands. When more observations, both at sea level and aloft, became available, the structure of the disturbances was shown to have the form of wave-shaped perturbations in the easterly flow somewhat similar to the waves in the westerlies aloft at high latitudes.

Whereas the instability that leads to the formation of waves in the westerlies is associated with baroclinicity, that is, the existence of horizontal temperature gradients, in the tropics the temperature varies very little. Theoretical investigations have shown that if the speed of an easterly current that is barotropic has a maximum at some latitude with lower speed on either side, it may be unstable and give rise to waves. It has been shown that the small amount of baroclinicity present in some tropical areas also contributes to the instability. However, the theory does not enable forecasting the specific time and place of the occurrence of easterly waves.

Most of the easterly waves that are observed over the North Atlantic Ocean originate over sub-Saharan Africa, where the barotropic instability in the midtropospheric easterly wind maximum is augmented by some baroclinic instability. They move westward over the coast of Africa at the rate of one every three or four days during the months of June to October. They have wavelengths between 2000 km and 3000 km, and move with speeds between 5 m s^{-1} and 10 m s^{-1}, about one-half the speed of the easterly winds at the height of the strongest winds. Some of them intensify into tropical storms, and a few become hurricanes either over the western Atlantic, particularly the Caribbean Sea and the Gulf of Mexico, or over the eastern Pacific after crossing Central America. It is not clear where the easterly waves that are found over the western Pacific originate, but most of the typhoons found there appear to develop from easterly waves or similar disturbances on the ITCZ.

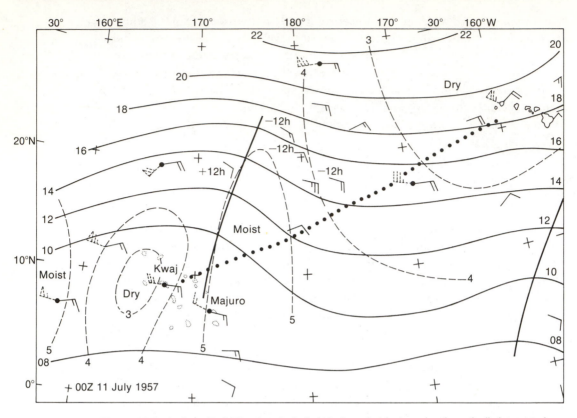

Figure 12.4 Isobars (*solid lines*) and winds (*wind symbols*) at sea level, vertically integrated water-vapor content (g/cm²), and winds at 12 km (*dashed symbols*) over central Pacific Ocean. [From Palmén and Newton (1969).]

Figure 12.4 is an example of a wave in the easterlies over the central Pacific. In addition to the isobars and symbols for surface winds, the figure contains dashed lines drawn through places with the same total water-vapor content in the vertical, and the winds at 12 km are shown by dashed symbols. The trough of the wave is shown by a heavy line; to the east of it the surface winds are from the east–southeast and the moisture content is high, corresponding to a thick, moist layer. To the west of the trough the low-level winds are mostly from the east–northeast, and the integrated vapor content is much less. In the upper troposphere the winds are from the west, as in middle latitudes.

Figure 12.5 shows the vertical cross section along the dotted line in Figure 12.4. Since the wave is moving westward, the western (left) part of the figure represents the weather that would be experienced at a station before the wave trough passes, and the right (eastern) part that after the trough passes. In advance of the arrival

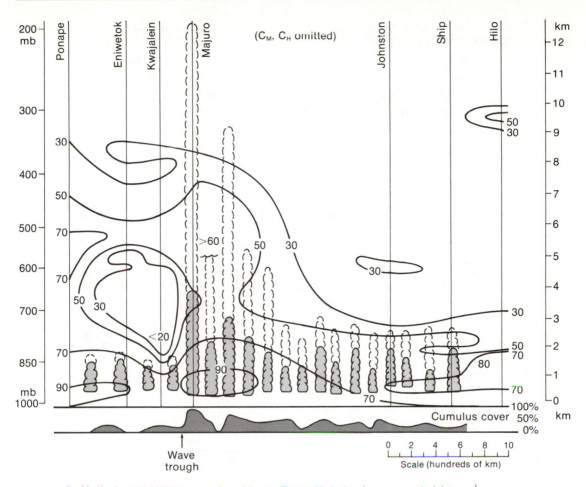

Figure 12.5 Vertical cross section along dotted line in Figure 12.4, showing average heights and maximum vertical extent of cumulus clouds. Solid lines show relative humidity. The graph at the bottom shows the percentage of cumulus cloud cover. [From Palmén and Newton (1969).]

of the trough, the winds would be east–northeast and the clouds would be few and not well developed. After it passes the winds become east–southeast, and the clouds are more plentiful and extend to greater heights, with some showers.

This pattern is typical of easterly waves that move slower than the basic easterly current. If they move faster, the clouds and precipitation are more nearly symmetric or may be predominantly ahead of the trough. Sometimes the cloudiness forms bands in the form of chevrons or inverted V's. In easterly waves that are deepening and becoming vortices, the cloud and precipitation patterns take the form of commas, similar to those in extratropical cyclones.

In addition to waves in the low-level easterlies, waves have been found in the tropical upper troposphere and lower stratosphere. These waves are much longer, 5000–10,000 km in length. They too move westward, at speeds that give them periods of about four or five days. In some instances they appear to be responsible for inducing easterly waves at low levels, or causing them to amplify and become cyclonic vortices. Except for this occasional possibility they have no direct effects on the weather.

12.4   Tropical cyclones, hurricanes, and typhoons

Ordinarily the weather in tropical regions is characterized by only small variations, with dry spells and showery spells in those areas affected by easterly waves, but mostly the weather is the same day after day. Occasionally, however, tropical disturbances develop into cyclones that may achieve such intensity as to cause the loss of many lives, blow down trees, and destroy or severely damage buildings. The intense tropical cyclones are known in various parts of the world by different names: *hurricanes* when they occur in the Atlantic or along the west coast of Mexico, *typhoons* in the western Pacific, *baguios* in the Philippines, and simply *cyclones* in Australia and in the Indian Ocean, including the Bay of Bengal and the Arabian Sea. For convenience we shall refer to all of them as hurricanes.

Tropical cyclones are called hurricanes (or typhoons) only when their winds, in the part of the storm where they are strongest, exceed 32 m s^{-1}. By international agreement, when they have winds between 17 and 32 m s^{-1} they are called *tropical storms,* and when the winds in them are less than 17 m s^{-1} they are called *tropical depressions.* The terminology of the U.S. National Weather Service includes an additional category, *tropical disturbances,* for incipient storms that have a slight cyclonic circulation at the surface but no more than one closed isobar on the weather map. Also, the U.S. National Weather Service uses slightly different values of the wind speed for distinguishing between the other categories. Tropical cyclones undergo development from disturbance to depression to storm to hurricane in a matter of a few days. They may continue as mature hurricanes for periods of as long as two weeks or more, until they move onto land or out of tropical latitudes. When a storm moves from the tropics to higher latitudes over the ocean, it usually maintains hurricane intensity but rapidly changes in structure to become an extratropical cyclone.

Fully mature hurricanes range in diameter from 100 km to more than 1500 km. The winds spiral inward toward the center, where the pressure is usually below 970 mb, with increasing speeds that reach values between 50 and 100 m s^{-1} near the center. At the center itself, however, there is a region of light winds, about

Figure 12.6   View of Hurricane Gladys, taken by astronauts aboard Apollo 7 at 1531 GMT (10:31 EST), October 17, 1968. [Courtesy of NOAA.]

5 m s^{-1} or less, called the *eye of the storm*. Within the eye there is no rain or low clouds, middle clouds are scattered or broken, and cirrus and cirrostratus clouds are visible with sometimes partially clear skies. Surrounding the eye is the eye wall, a circle of towering cumulonimbus clouds from which heavy rain falls, and spiralling inward to the eye wall from the edge of the storm are bands of cumulus and cumulonimbus with rain. The strongest winds occur in the region of intense convection just outside the eye.

The spiral, banded structure is shown in Figure 12.6, which is a photograph of Hurricane Gladys taken October 17, 1968, by the astronauts on the Apollo 7 space flight. The photograph was taken looking southeast when the space vehicle was over the Gulf of Mexico at a height of 185 km and the storm was centered about 200 km west of Florida. The circular cap of cirrostratus near the storm's center is probably associated with the glaciation of the cumulonimbus clouds in the vicinity

Figure 12.7   GOES satellite photograph (1800 GMT August 4, 1980) showing one hurricane in the West Indies and another over the Pacific Ocean west of Mexico. [NOAA–NESS photo. Courtesy of V. J. Oliver.]

of the wall of the eye. The spreading of the cirrostratus prevents the structure near the center from being seen from above. Gladys had formed two days earlier in the Caribbean Sea in a region of widespread shower activity and moved northward at about 5 m s^{-1}, becoming a hurricane shortly before crossing western Cuba. Winds exceeding 35 m s^{-1} were observed in Cuba, and heavy flash floods with severe damage to crops and industrial installations were experienced. Continuing northward and north–northwestward, it crossed the Florida Keys and reached the position to the west of Florida where it was photographed from the spacecraft. It then curved toward the east and passed inland, causing damage to the Florida citrus crop and structural damage estimated to be about $6 million. Although the measured rainfall in some places was as high as 8 inches, flooding from rain was not serious. After crossing Florida it moved northeast along the coast, passing near Cape Hatteras, and gradually became extratropical.

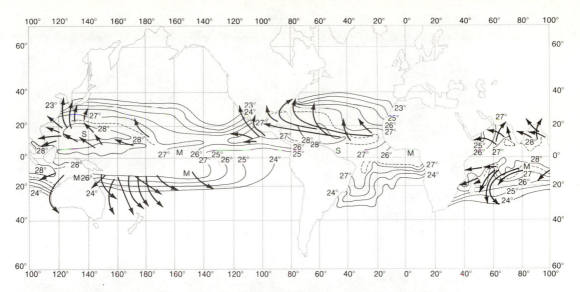

Figure 12.8 Principal tracks of tropical cyclones in relation to sea surface temperature. Isotherms show average sea surface temperature in September for the Northern Hemisphere and in March for the Southern Hemisphere. [From Palmén and Newton (1969).]

The banded structure, together with the eye, enables easy identification of hurricanes on satellite photographs. Since hurricanes form over tropical oceans where surface stations are sparse and ships are infrequent, satellites provide an invaluable forecasting aid in detecting and tracking them. Figure 12.7 gives an example of a satellite photograph of a hurricane.

Tropical cyclones always have their beginnings in regions where the sea surface temperature over large areas is high. In Figure 12.8 the generalized tracks representative of tropical cyclones that reach hurricane intensity are shown. They all originate in areas where the sea surface temperature is higher than 27°C in the season of their occurrence. Furthermore, hurricanes do not form closer to the equator than about 4°N and 4°S, presumably because the Coriolis force at lower latitudes is too small to create the rotational circulation. The sea surface north of 4°N has its highest temperatures from June to November, and these are the months in which hurricanes develop most frequently, with September the month of greatest frequency, although occasionally hurricanes occur in May in the North Atlantic; in the western North Pacific and the North Indian oceans, tropical cyclones have occurred in every month of the year. Correspondingly, the period from December to March is the part of the year when the sea-surface temperatures are highest and tropical storms are most frequent in the Southern Hemisphere. However, there

Figure 12.9   Satellite photograph of hurricane-strength "Cyclone Kerry" at coast of Australia 0000 GMT March 3, 1979. [Photo received in Australia from Japanese Geostationary Meteorological Satellite; furnished courtesy of N. A. Streten.]

have never been any hurricanes in the South Atlantic Ocean, presumably because the sea-surface temperatures do not reach sufficiently high values there.

Over the entire earth, on the average about 80 tropical cyclones that attain winds exceeding 20 m s^{-1} are formed each year. Of these more than one-half reach hurricane strength. More than two-thirds of them occur in the Northern Hemisphere, with about one-half of these over the western North Pacific Ocean, about one-fourth over the eastern North Pacific Ocean, about one-sixth over the North Atlantic Ocean, and about one-eighth over the North Indian Ocean. Of the ones occurring in the Southern Hemisphere, almost one-half form over the waters north of Australia, one-third over the South Indian Ocean, and one-fourth over the South Pacific Ocean.

While the total number of tropical cyclones for the earth varies relatively little from year to year, ranging from a low of 67 to a high of 97 in the 20 years 1958 to 1977, the number in individual ocean basins varies greatly. Thus, the largest number over the Atlantic during that 20-year period was 14, in 1969 and 1971, and the lowest was four in 1972, a factor of 3½ for the ratio of highest to lowest number, compared to less than 1½ for the entire earth.

The frequency of formation increases with latitude up to about 15 degrees, with two-thirds of the tropical cyclones forming at latitudes between 10 degrees and 20 degrees. The increase to 15 degrees emphasizes the importance of the vorticity of the earth's rotation in producing the cyclonic circulation. The decrease poleward of that is due to the less-frequent occurrence of high sea-surface temperatures.

The warmth of the sea surface as a prerequisite for the formation of hurricanes is obviously related to the creation of air that is sufficiently unstable and moist to remain warmer than the surroundings while raising saturation adiabatically to high levels. An adequate theoretical explanation of the formation of tropical cyclones has not yet been developed, that is, a theory that will make it possible to select from the frequent occurrences of high sea temperatures at appropriate latitudes the rare occasions when circumstances are right for the development of an intense cyclonic vortex. Recent advances toward such a theory emphasize the importance of larger scale circulations interacting with the cumulus convection in cloud clusters associated with easterly waves and other tropical disturbances. One theory emphasizes preexisting large-scale low-level convergence in releasing conditional instability that reinforces the convergence in a feed-back mechanism called CISK (Convective Instability of the Second Kind). Another theory shows that amplification of easterly waves due to nonlinear vorticity advection by the large-scale flow can explain development of intense rotational motion. A criterion based on the latter concept has had some success in predicting storm formation.

The large amounts of kinetic energy in the hurricane-force winds come from the latent heat released by the condensation in the rising air currents of the moisture that has evaporated from the warm sea surface. The air at the surface picks up heat and moisture as it moves inward toward the center. The added heat from the warm water compensates for the temperature decrease that would otherwise occur because of the decrease in pressure. When the hurricane moves over land, the absence of evaporation from the sea surface eliminates the source of wind energy, and the increase in roughness of the ground quickly reduces the wind below hurricane strength. Heavy rains continue for a considerable distance inland, frequently accompanied by flooding. While this results in destructive floods at times, the rain from the moist air carried inland by hurricanes is a principal source of moisture beneficial to agriculture.

The general structure of a hurricane is shown schematically in Figure 12.11. The air is shown streaming toward the center at the surface. It gains rotational momentum due to the horizontal convergence, and rises both because of convergence and because of the hydrostatic instability that leads to intense convection, particularly near the eye wall. The upward velocity is greatest near the eye, and smaller farther out, but there is convergence and upward motion in the spiral

Figure 12.10   Satellite photograph of Hurricane Beulah, taken at 1828 GMT (12:28 P.M. CST) September 18, 1967, over the Gulf of Mexico. Beulah subsequently curved northward into the United States and caused so much damage it received the nickname "billion dollar destroyer." [Courtesy of NOAA.]

cloud bands, with relative clearing in the zones between the cloud bands. In the eye there is sinking motion. At the eye wall the descending air tends to mix with the cloudy air around the eye. As a result of the adiabatic heating of the descending air, the eye is warmer than the surroundings right up to the tropopause.

At upper levels the rising air flows outward, and at the outskirts of the storm there is sinking motion. The hurricane is a warm-core cyclone, which must decrease in intensity with height and eventually become an anticyclone. Except near the eye, the outflow in the upper troposphere is anticyclonic.

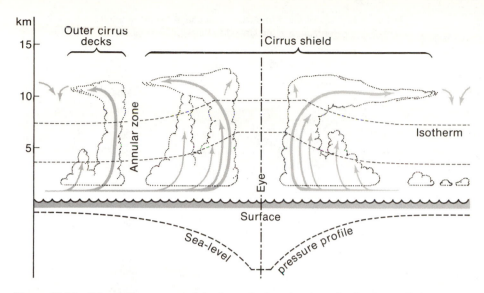

Figure 12.11 Schematic representation of a vertical section through a hurricane. The weight of the streamlines is intended to indicate the strength of the inflow and upflow motion. Lower part of figure shows pressure variation at sea level.

The mature hurricane can be regarded as being composed of four parts: (1) an outer region in which there are cyclonic winds with higher speeds closer to the center and slight convection at low levels, and flow outward from the center with cirrus or cirrostratus clouds at high levels; (2) a belt of squall lines, frequently in the form of spiral cloud bands, with winds reaching hurricane strength and heavy convective showers; (3) the ring-shaped inner region of extremely strong winds and heavy rains; and (4) the eye, inside a transition zone in which the wind drops almost to a calm and the precipitation ceases. In zones 2 and 3 severe thunderstorms and sometimes tornadoes occur. Over the ocean the winds create tremendous waves, the tops of which are carried as heavy spray that merges with the rain "so that the mariner can scarcely tell where the ocean ends and the atmosphere begins."

As a hurricane approaches land it drives water ahead of it in the form of a "tidal wave" or storm surge that inundates coastal lowlands with very deep water. Frequently the storm surge causes more deaths and damage than the winds or flooding due to rain. Thus Hurricane Camille was accompanied by a 24-ft tide when it crossed the Louisiana–Mississippi coast the night of August 17–18, 1969. It is impossible to tell whether the 90 m s^{-1} (200 mi/hr) winds or the tremendous hurricane wave contributed more to the toll of 256 deaths and almost \$1.5 billion in property damage.

Table 12.1 Loss of Life and Property Damage from Some of the Most Severe Hurricanes in the United States

Property damage		
Hurricane	Year	Millions of dollars
Camille	1969	1420.7
Betsy	1965	1420.5
Diane	1955	831.7
Carol	1954	461.0
Donna	1960	426.0
Carla	1961	408.2
New England	1938	306.0
Hazel	1954	251.6
Dora	1964	250.0
Beulah	1967	200.0
Audrey	1957	150.0
Cleo	1964	128.5
Miami	1926	112.0

Loss of Life		
Hurricane	Year	Deaths
Galveston	1900	6000
Louisiana	1893	2000
South Carolina	1893	1000–2000
Okeechobee	1928	1836
Keys and Texas	1919	600–900
Ga. and S.C.	1881	700
New England	1938	600
Keys	1935	408
Audrey	1957	390
Atlantic Coast	1944	390
Mississippi and Louisiana	1909	350
Galveston	1915	275
Mississippi and Louisiana	1915	275
Camille	1969	256

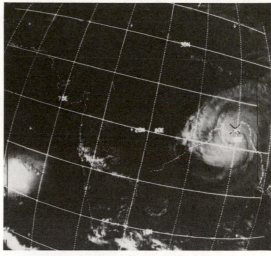

Figure 12.12   Bay of Bengal cyclone, November 11–12, 1970, photographed by ITOS 1 Satellite. [Courtesy of NOAA.]

Table 12.1 gives the cost in lives and property due to hurricanes for a selection of severe storms of the past century. While the property damage in recent years has been great because of increased building of homes and industry in hurricane-prone regions, the loss of lives has been relatively small because of improved forecasting and warning-dissemination methods.

Tropical cyclones in the Bay of Bengal have been responsible for greater losses of life than any other natural disasters. Several times in the past three centuries cyclones have caused the inundation of the low land at the mouths of the Ganges, causing hundreds of thousands of deaths by drowning. The most recent case was the cyclone of November 12, 1970, in which the official death toll, confirmed by the number of burials, exceeded 200,000, and unofficial estimates indicate that as many as 500,000 people died. Figure 12.12 presents satellite photographs of that Bay of Bengal storm on two successive days. Among the previous occurrences, a storm in 1876 is reported to have caused a 40-ft tide that was responsible for 100,000–400,000 deaths, and on October 7, 1737, a storm with a similar tidal wave killed an estimated 250,000 people and destroyed 20,000 boats and ships.

It would seem clear that people should not live in low coastal areas subject to hurricanes. Regrettably, in recent years the attraction of living at the shore in the warm climates in the southeastern United States and on the coast of the Gulf of Mexico has led to thousands of homes and hotels being built with no regard to the danger. In fact, in many places protective dunes, which would reduce danger of flooding by the storm surge, have been bulldozed to give more ready access to the sea. Needless to say, removal of the dunes also gives ready access for the storm surge to flood the structures if and when a hurricane strikes.

In spite of increased understanding of meteorological conditions and improved methods of forecasting them, human activities have yet to be adapted to minimize damage due to the unfavorable aspects of our atmospheric environment. For optimum achievement of this objective, further research is required.

12.5   Investigations of tropical weather. The GATE program

Previous to World War II, relatively few systematic observations were made of the weather in low latitudes, and consequently the knowledge of it was mostly based on general impressions of travelers. In the war, army and navy weather forecasters found their training and experience in middle latitudes inadequate preparation for predicting the weather changes that affected land, sea, and air operations in the South Pacific. Summaries of existing knowledge were rapidly prepared, and these, together with observations made at the various bases that were occupied, quickly multiplied the store of information on the subject. But after the war the places where observations were made in the tropics, particularly upper air observations, again became sparse, although somewhat more plentiful than previous to it.

An additional handicap for analysis and predicting weather in the tropics, besides the sparsity of conventional observations, is the fact that the nature of tropical conditions makes desirable special types of data in order to study them. Furthermore, in recent years it was recognized that improvement of weather forecasting at high latitudes requires more complete understanding of the global circulation, and that knowledge of the weather processes in the tropics is essential to this understanding. These needs have led to a series of research experiments investigating tropical weather phenomena. These included, in the 1960s, the International Indian Ocean Expedition, the Line Islands Experiment (LIE), and the Barbados Oceanographic and Meteorological Experiment (BOMEX). They were followed after several years of planning by the most ambitious of all tropical meteorological investigations, the GARP Atlantic Tropical Experiment (GATE), the observational phase of which was carried out in 1974. (GARP is the acronym for Global Atmospheric Research Program. This program will be described in Chapter 13.) Analysis of the GATE data has already illuminated many aspects of tropical weather and is continuing at the date of this writing.

The observational phase of GATE, which lasted 100 days during June–September 1974, involved the greatest number of satellites, ships, airplanes, ground equipment, and scientists ever assembled for a peaceful program up to that time. Approximately 5000 scientists and technicians from 72 nations carried out observations aboard 39 ships and 13 research aircraft, augmented by about 1000 land stations and six satellites. The phenomena studied ranged in scale from single convective clouds (the "D" scale, 1–10 km) through convective patterns ("C" scale, 10–100 km) and cloud clusters ("B" scale, 100–1000 km) to large flow patterns, including the ITCZ and easterly and equatorial planetary waves (the "A" scale, 1000–10,000 km).

For the A-scale studies, observations were made in a band between 10°S and 24°N, extending from the western Indian Ocean across Africa, the Atlantic Ocean, and Central and South America to the extreme eastern Pacific Ocean. In this 50-million square kilometer area, the routine observational system was augmented by additional rawinsonde stations to fill the gaps over land and by special stationary ships distributed over the Atlantic Ocean, where no observations are otherwise available. For the A-scale, the attempt was to have stations no farther than 500 km–600 km apart. For the B-scale and smaller phenomena, arrays of ships centered about 1000 km west of Dakar were established. The ships were equipped with weather radar in addition to rawinsonde and surface observing equipment. For the structure of the smaller scale phenomena, research aircraft, operating out of Dakar, where the headquarters of the program was established, were equipped with radar as well as instruments to measure wind, temperature, humidity, pressure, and cloud-water content. Sea-surface temperature was observed both by ships and by satellite, and radiation measurements were made at the surface and by satellite. The satellite observations were made by GOES and SMS satellites stationary over the equator at 100°W and 45°W, and by four polar-orbiting satellites, three operated by the United States and one by the USSR.

GATE was designed to obtain data that would provide descriptions of the structure, functioning, and mutual interaction of tropical weather phenomena of various scales, including individual cumulus, cloud clusters, easterly waves, and the ITCZ. In particular, it aimed to provide data for testing numerical models of the tropical atmosphere, and methods of representing in mathematical models of the global circulation the effects of the smaller tropical systems, the specification of which would require a finer network of observations than can be used in practice in global models. The data were collected by the various national and regional processing centers and filed in the two world data centers, in Asheville, N.C. and in Moscow, where they are available to scientists anywhere in the world wishing to do research with them.

A very active program of study of the GATE data is under way. Some of its results provide the basis for the material in previous sections of this chapter. Previous to GATE, little was known about the vertical structure of the ITCZ, the relationship between easterly waves and cloud clusters, and the energy exchange between ocean and atmosphere in the tropics. As the GATE data continue to be analyzed, our understanding of these and other aspects of tropical meteorology will be improved, and the role of the tropics in the global circulation will be clarified. With development of numerical models of the tropical atmosphere, our ability to forecast will be improved, not only for tropical regions, but in temperate latitudes where the influence of the energy input in the tropics must have an effect.

Questions, Problems, and Projects

1. Discuss the similarities and differences between waves in the easterlies and waves in the westerlies.
2. Would you expect to find frontal waves at the zone of convergence between the trade winds of the two hemispheres (the Intertropical Convergence Zone, ITCZ)? Explain.
3. Why don't hurricanes ever occur right at the equator?
4. Why don't hurricanes develop in the Atlantic from December to April?
5. Why do the winds in hurricanes decrease rapidly when the storms pass over land? Why does heavy precipitation continue in them over land?

Weather Forecasting   13

13.1   Methods of weather prediction

In Section 1.3 we saw that we were able to forecast with some accuracy the 24-hour movement of a cyclone from its past movement and to estimate the weather changes that would occur in the area over which it passed by consideration of the weather conditions around the low as shown on the synoptic weather map. With additions and refinements, this empirical method of forecasting was used for the daily forecasts issued by weather services throughout the world until late in the 1950s. Based on the synoptic weather map, it is called *synoptic weather forecasting*.

In the past three decades a more rigorous method, which applies the laws of physics, has been developed. Since the solution of the equations expressing those laws for atmospheric motions requires the use of numerical computations carried out on high-speed electronic digital computers, the method is generally referred to as *numerical weather prediction* (NWP), although it would be more proper to call it *dynamic weather prediction* to distinguish it from other methods that include numerical procedures.

Statistical forecasting is another method in which numerical procedures are used, first, to analyze past data to establish the predictive relationships, and then to utilize the current data to obtain the prediction. A particular kind of statistical relationship that is sometimes sought is *periodicities* or *cycles*. If it could be established that the weather repeated itself at specific intervals, the problem of forecasting would be solved. A related approach is the *analogue method*, in which past weather situations similar to the current one are sought. If a situation identical to the present weather could be found, the future weather should be a repetition of the weather that occurred subsequent to the analogue. In that case the weather would have a periodicity equal to the time between the analogue and the present. In actuality, no well-marked periods other than the diurnal and annual have been found, and while past weather situations can be found in which some of the features resemble the present fairly closely, there are always enough differences to ensure that the development will be different.

The forecasting procedure used at present in the United States involves a combination of numerical, statistical, and synoptic methods. At the National Meteorological Center of the U.S. National Weather Service near Washington, D.C., forecasts of the general flow patterns at various levels are made using NWP. Statistical procedures are then applied to the output of the computers to make specific local forecasts of maximum and minimum temperatures, probability of precipitation and frozen precipitation, precipitation amounts, cloud amounts, ceilings, visibilities, surface winds, and probability of thunderstorms and severe weather for several hundred stations. Forecasts are also made for marine requirements, including surface wave and swell conditions and storm surges for the North Atlantic Ocean, the North Pacific Ocean, the Gulf of Mexico, and the Great Lakes. The forecasts made at the National Meterological Center are distributed as guidance to local forecast offices, where the forecasters adapt them to their localities, utilizing their synoptic experience as well as local indications, and to specialized forecasting units such as the National Severe Storms Forecast Center and the National Hurricane Center.

Similar procedures are used in the weather services of most developed countries.

13.2 Predictability of the weather. Accuracy of forecasts

With all these methods being used, the question arises, why are the daily weather forecasts that are broadcast on television and radio frequently in error? The same question might also arise if we compare prediction of the weather with the prediction of eclipses of the sun and the moon and the movement of planets among the fixed stars. If astronomers can forecast to the minute the occurrence of an eclipse many years in advance, why cannot meteorologists forecast with similar accuracy the occurrence of precipitation just one day in advance?

The answer lies in the complexity of the physical systems involved. For the prediction of eclipses, the motions of the sun, moon, and earth can be treated as a problem involving three bodies, each with its mass concentrated at a point. This three-body problem is not simple, but it is infinitely less complex than keeping track of the movements of the innumerable parcels of air with which one must be concerned in forecasting the weather. The problems of celestial mechanics reduce to the dynamics of a small number of isolated mass points moving in response to the forces they exert on each other. The problems of meteorology are those of the dynamics of a thermally active fluid, and fluid dynamics is innately more difficult than particle dynamics.

If the problem were only to forecast the movement of the high and low centers and if they obeyed simple laws of interaction, it might be no more difficult than the eclipse problem. But in addition to the movement of the centers at sea level it is necessary to forecast their intensity there, their position and intensity at all other levels, the distribution of pressure in between, the distribution of temperature at various heights, the variations of humidity, the occurrence of clouds and precipitation, and the wind velocity. The laws governing the complex interactions of these factors are only partially understood. To the extent that they are understood, they can be represented by mathematical equations, but it turns out that the solution of the equations in their full complexity would require computers of much greater capacity and speed than the largest so far available.

Another limitation on forecast accuracy is due to the fact that the changes in weather conditions at any one place are influenced by the conditions over a large area surrounding it, the size of the area increasing with the length of time for which the forecast is being made. The forecast of tomorrow's weather requires a complete delineation of the present pressure, temperature, humidity, wind, and so forth, throughout the troposphere and lower stratosphere over a large part of a hemisphere. Observations are made only at a discrete number of stations, sometimes so far apart that important weather features may lie between and be missed. At the surface over land, at least in the more advanced countries, a fairly dense network of stations reports the conditions, but over the sea and in less-developed countries the observations are sparse. There are fewer upper-air observations even in the areas where surface observations are adequate, and over much of the earth observations of conditions aloft are absent.

As part of a program called World Weather Watch, the World Meteorological Organization has been assisting the less-developed member nations to fill the gaps in the observational network throughout the world and to communicate the observations promptly to data centers and from them to forecast centers. The recent polar-orbiting satellites (TIROS-N, NOAA-6, and the latest Russian Meteor satellites) are equipped to measure radiation emitted by the atmosphere and the earth in several wavelengths; from these measurements, profiles of temperature and humidity can be evaluated at many positions along their orbits. These activities are gradually reducing the deficiencies in the knowledge of the state of the atmosphere at any instant. In the meantime the current weather conditions are not known accurately enough and in sufficient detail to enable forecasting them with certainty.

In spite of these difficulties and deficiencies weather forecasts at a useful level of accuracy have been made for many years. The beginnings of weather forecasting coincided with the development of the electric telegraph. Previously, there had been precursors of the synoptic weather map, which were based on past data collected by mail. The first of these was a series of charts showing winds and departure of pressure from average for each day during the year 1783. These charts were published in 1820 by Heinrich W. Brandes (1777–1834), who may be regarded as the first German meteorologist. The availability of the observations was due to the foresight of Antoine L. Lavoisier (1743–1794), the famous French chemist. In about 1780, Lavoisier proposed setting up strictly comparable instruments for taking meteorological measurements at a large number of places in France, Europe, and even the entire world. Acting on his own proposal, he distributed several barometers to observatories in Europe, enabling Brandes subsequently to collect the data and prepare his charts. Shortly afterward, James P. Espy (1785–1860), an American meteorologist, organized an observational network in the United States and drew daily weather maps on which the winds, pressure, and precipitation observed at the stations of the network were summarized. From these early charts prepared by Brandes and Espy came the first demonstration that the low pressures associated with bad weather occur in areas that move regularly, usually from west to east. When the telegraph enabled the collection of observations on a current basis, the possibility of using this concept for weather forecasting became a reality.

Beginning in the 1860s, meteorological services in various countries of Europe were established to prepare daily weather maps and issue forecasts based on them. While telegraphic collection of reports and transmission of weather indications in the United States began in 1849 under the aegis of Joseph Henry (1797–1878), the first secretary and director of the Smithsonian Institution, a national weather service was not established and regular forecasting in the United States was not begun until 1870. Initially, the Weather Bureau was part of the Signal Service of the U.S. Army, then successively of the Department of Agriculture and the Department of Commerce. Now called the National Weather Service, it is a branch of the National Oceanic and Atmospheric Administration in the Department of Commerce. Whereas the attention of the public now seems frequently to be focused on the occasions when the forecasts are wrong, in the early years the successful prediction of the changes in something so variable as the weather was so surprising that the misses were ignored or accepted, and, increasingly, reliance was placed on the forecasts by nautical interests, farmers, shippers, and the general public.

How accurate are weather forecasts? The answer depends on the particular weather feature being forecast and the length of the period and the locality for which the forecast is made. Wind direction and speed and temperature are among the elements for which the forecaster can claim greatest success. The occurrence of cloudiness and precipitation is usually forecast fairly well, although the timing may be off. Forecasts of amounts of precipitation are subject to larger errors. The optimum forecasts are those covering periods of 6–24 hours. Forecasts covering periods of 2 or 3 days are fairly accurate, although their accuracy decreases with each day. Beyond 5 days the results of numerical weather prediction usually are no better than climatological expectations, although in some situations the conditions are sufficiently stable or slow-moving that specific forecasts for periods as long as a week can be made.

The length of time for which we can expect to be able to forecast the sequence of weather can be estimated by the methods of NWP by carrying out computations based on two initial weather situations that differ by an amount corresponding to the uncertainty or expected error of the observations and then seeing how rapidly the differences increase. If the forecasts based on the two initial conditions do not diverge, the expectation would be that observational errors would not affect the accuracy of the prediction, and the length of time for which the forecast could be made would depend entirely on the validity of the forecasting method. If, however, the forecasts progressively become more and more different, there would be no way of deciding which would correspond to what would really take place in the atmosphere, and the time it took for them to differ by an amount equal to the acceptable error in a forecast would be the longest time for which a forecast would have any validity. Tests using this procedure have indicated that good forecasts should be possible for periods as long as two weeks, but not for longer periods. Actually, the procedure depends on the degree to which the NWP computations correspond to the way in which the atmosphere behaves, so the conclusion might be changed if improved NWP methods are developed.

There have been many attempts to measure the accuracy of weather forecasts. A fundamental difficulty is the problem of deciding whether a forecast is right or wrong, and if wrong, how wrong. If the forecast says it will rain today in Kansas, and it rains in the western part of the state but not in the eastern part, is the forecast right, wrong, or one-half right? If the forecast says the minimum temperature tomorrow morning will be 40°F, and it turns out to be 41°F, should one say the forecast is as wrong as it would be if it were 50°F? The National Weather Service expresses the accuracy of forecasts as a percentage only for the occurrence of precipitation. For the United States as a whole, the forecasts of precipitation average about 80–85 percent right. The average error in the temperature forecasts

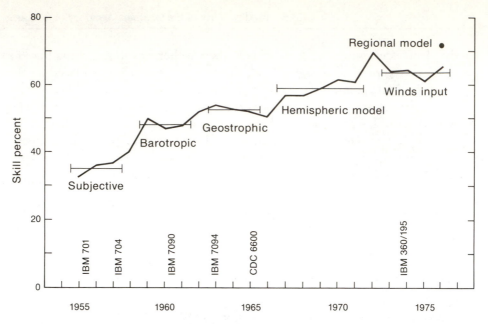

Figure 13.1 Skill, averaged annually, of the forecasts of the wind field at 500 mb 36 hours in advance, made by the U.S. National Meteorological Center. The horizontal lines show the average values for years during which there were no major changes in procedures. The skill percent is designed so that no skill, for instance random choice, would give zero score, and a perfect forecast would yield 100. The words "Barotropic," "Geostrophic," etc. refer to the computational models used during the periods indicated, and the symbols IBM 701, IBM 704, etc. are entered at the years when the progressively larger and faster computers were introduced. [From F. Shuman, "Numerical Weather Prediction." *Bulletin of the American Meteorological Society,* vol. 59, no. 1, Jan. 1978.]

is about 4°F. To rate the success of the forecasts of ceiling and visibility for airway terminals, a point system is used. Scores that estimate the degree to which the temperature and precipitation forecasts are better than reliance on climatology or persistence are used in evaluating the success of the forecasting methods. In addition, statistical measures are computed of the correspondence of the isobaric contours on the prognostic maps prepared in numerical weather prediction to the observed maps. The main purpose of these evaluations is to judge the improvement, if any, in the forecasts when new methods are introduced and when gaps in the observational network are filled.

These measurements show that the forecasts are considerably better than climatological averages, pure chance, or persistence, although they are far from perfect, and that there has been some improvement in recent years as the empirical methods have been replaced by successively more sophisticated NWP models using larger and faster computers and as the observed data have become more plentiful. The improvement has been greatest in the forecasting of the general pressure pattern aloft. Figure 13.1 shows the change in skill of forecasting the pressure gradient at 500 mb 36 hours in advance. In the mid-1950s, before the NWP methods were incorporated into the forecasts, the measure of skill was 35 percent; in the mid-1970s, it was about 65 percent. Based on the pressure pattern, the wind forecasts for aviation are usually very close to the observed. The procedures used in deducing the changes in temperature and the occurrence and amount of precipitation from the forecast of the pressure and wind distribution have not improved as much.

13.3 Analysis of the present weather

The first step in forecasting, whether by empirical methods or by NWP, is to determine the present weather situation and to represent it in a fashion that can be comprehended by the human forecaster or used in the machine computation. This process takes place in three parts: (1) observation, (2) transmission and collection, and (3) analysis.

As discussed in Chapter 1, observations are characterized as surface and upper air. The observations that are made for use in forecasting are called *synoptic observations*. The surface synoptic observations are made simultaneously four times a day (0000, 0600, 1200, and 1800 GMT) at all stations and are standardized by international agreement through the World Meteorological Organization, a specialized agency of the United Nations, with 147 member countries. The regulations of the WMO ensure that the instruments and the procedures used result in comparable accuracy of the observations throughout the world.

Surface weather observations are made on a regular schedule at more than 12,000 weather stations over the earth and aboard many hundreds of naval, merchant, and passenger ships as they ply the oceans. Even with this tremendous number of stations there are large land areas, particularly in tropical and polar regions, and tremendous portions of the oceans away from the principal shipping lanes where there are few observations.

The routine surface observations include the pressure, temperature, dew-point temperature, cloudiness (amount, type, and heights), wind direction and speed, current weather, pressure tendency, and amount of precipitation since the last report. These observations are transmitted by radio, telegraph, or teletype every six hours in a standard code, which consists of five-digit words. The form of transmission of an observation and a sample message are as follows:

IIIii	N dd ff	VV ww W	PPP TT	N_h C_L h C_M C_H
			$T_d T_d$ a pp	7 RR R_t s
72405	83220	12706	14701	67292
			00228	74541

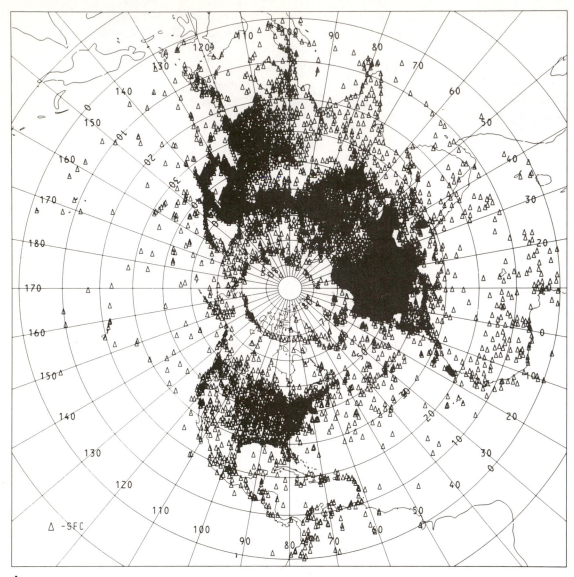

A

Figure 13.2   Surface data coverage received by World Data Center in Washington, D.C., on October 21, 1979. (A) Northern Hemisphere; (B) Southern Hemisphere.

The first five-digit word gives the station identification number (Washington, D.C., in the sample message), and the remainder represents the weather data according to the symbols and tables given in Appendix E.

Figure 13.2 shows the locations from which surface observations were received at the World Data Center in Washington on a day in late 1979.

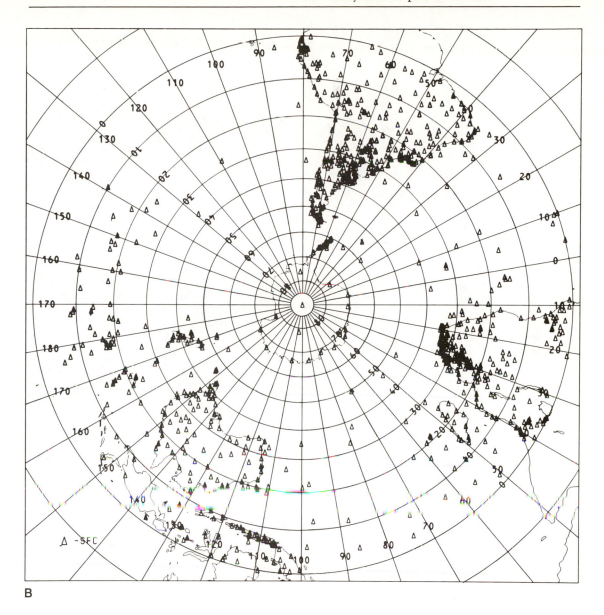

B

The conditions aloft in the troposphere and lower stratosphere are observed principally with radiosondes, rawinsondes, and pilot balloons. Augmenting these are temperature profiles from the polar-orbiting weather satellites and cloud-level winds obtained from successive pictures by the geostationary satellites. The radio-sonde observations (RAOBS) give the temperature, pressure, and humidity at

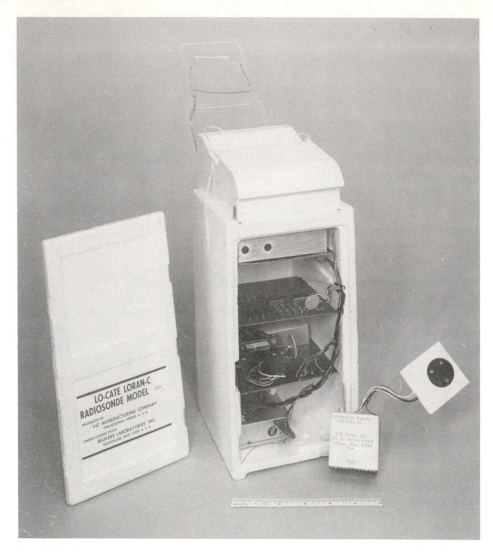

Figure 13.3 Radiosonde for use in rawin observations. This instrument, carried aloft by a balloon, measures and transmits the temperature, humidity, and pressure as the balloon rises and enables determination of the upper-air winds by tracking, using navigational aid transmissions. The temperature is sensed with a thermistor, a ceramic rod whose electrical resistance varies with the temperature. Similarly, the humidity is measured by the electrical resistance of a slide that is coated with a carbon film. The pressure is measured by an aneroid baroswitch with 180 contacts, or points at which the pressure is accurately determined. For wind determination, the transmission from Loran C or Omega navigational aid stations are received by the radiosonde and retransmitted to the ground station. Automatic comparison of the navigational aid signals received directly at ground stations and those received after being transmitted from the balloon enables the position of the balloon to be determined, and from successive positions the wind speed and direction are derived. [Courtesy of Beukers Laboratories, Inc. and Viz Manufacturing Co.]

various levels. In addition, the rawinsondes (RAWINS) give the wind direction and speed as determined by tracking the radiosonde transmitter with direction-finding equipment. The pilot balloon observations (PIBALS) are upper-wind measurements made by tracking small balloons visually, using theodolites (small telescopes similar to surveyors' transits, but with the eyepiece along the axis of vertical rotation). For accurate wind measurements, two theodolites (and thus two observers) separated by a known distance are required, but good approximate values are obtained by the single-theodolite method, which assumes a known ascension rate of the balloon. The data from RAOBS and RAWINS are reported for specified "mandatory" pressures—1000, 850, 700, 500, 400, 300, 200 mb, etc.—as well as at "significant" levels that enable reproduction of the details of the sounding. Like the surface observations, RAOBS and the wind soundings are transmitted in a numerical code consisting of groups of five-digit words.

There are approximately 750 radiosonde or rawinsonde stations throughout the world, the densest network being in Europe. About 100 observations are taken at U.S. National Weather Service stations in the United States and on Caribbean and Pacific Islands, and about 50 more at other stations in the Western Hemisphere.

The RAOBS are taken twice daily, at noon and midnight Greenwich time. PIBALS are usually taken four times daily. In places where RAWINS provide wind data in conjunction with the RAOBS, the winds at the other two synoptic times are obtained using PIBALS. Figure 13.4 shows the radiosonde observations that were received in Washington on a typical day.

Satellites are providing additional wind, temperature, and humidity data, particularly over the oceans and in other areas of sparse observations. The wind measurements are obtained by measuring the displacement of identifiable clouds in successive pictures taken by the geostationary satellites. They are subject to uncertainty of the height of the clouds, as well as to errors because of formation and dissipation processes that may influence the apparent motion of the clouds. The temperature and humidity profiles are computed using the TIROS–N Operational Vertical Sounder (TOVS), which consists of instruments on the polar-orbiting satellites that measure the emissions from the atmosphere at points along the orbits in the infrared and microwave positions of the spectrum, and computer procedures ("software modules") for converting the radiation measurements into temperature and humidity values in layers of the atmosphere. In contrast to the RAOBs, which measure at the positions the balloon passes through and are made at prescribed synoptic times, the TOVS measurements are averages over areas about 250 km in diameter, and the successive soundings are made at different times of day as the satellites progress along their paths. The TOVS profiles are most accurate when made through cloudfree air, but procedures have been developed, using the microwave signals, which are unimpeded by clouds, to provide useful temperature soundings from cloudy areas as well.

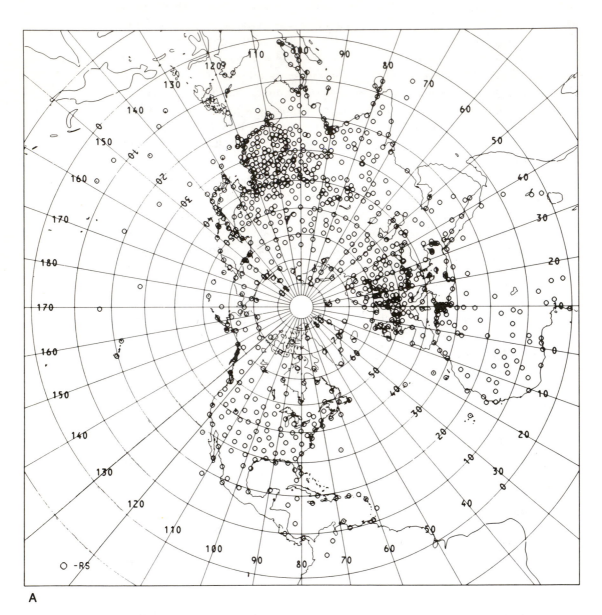

A

Figure 13.4 Radiosonde observations received by the World Data Center in Washington, D.C., on October 21, 1979. (A) Northern Hemisphere; (B) Southern Hemisphere.

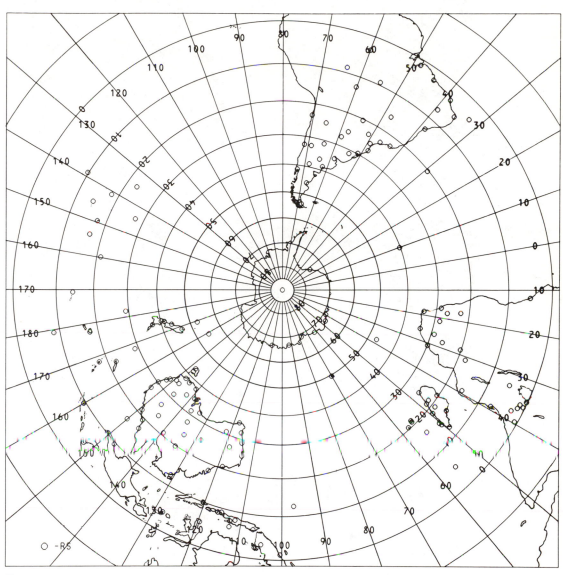

B

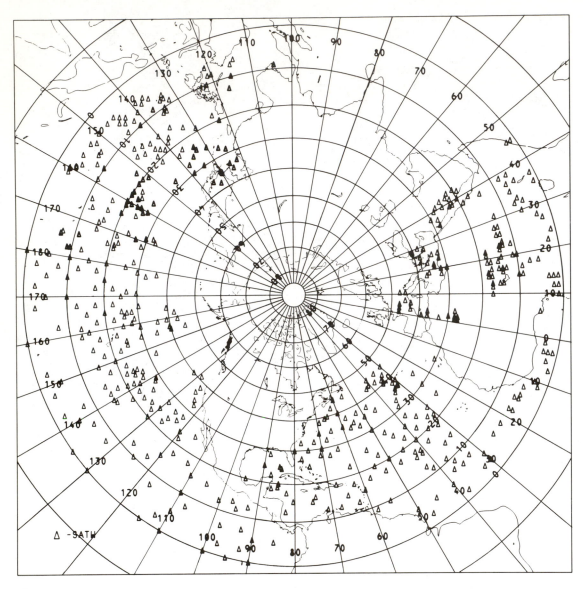

A

Figure 13.5 Satellite measurements of wind received by the World Data Center in Washington, D.C., on October 21, 1979. (A) Northern Hemisphere; (B) Southern Hemisphere. Note that these observations fill much of the regions not covered by surface and radiosonde observations.

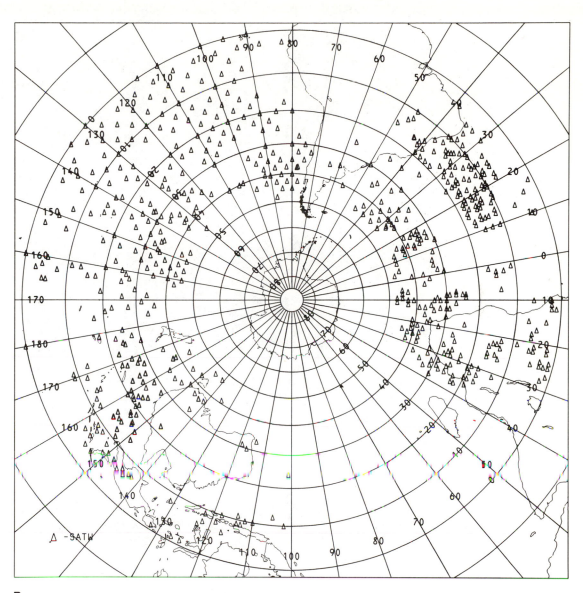

B

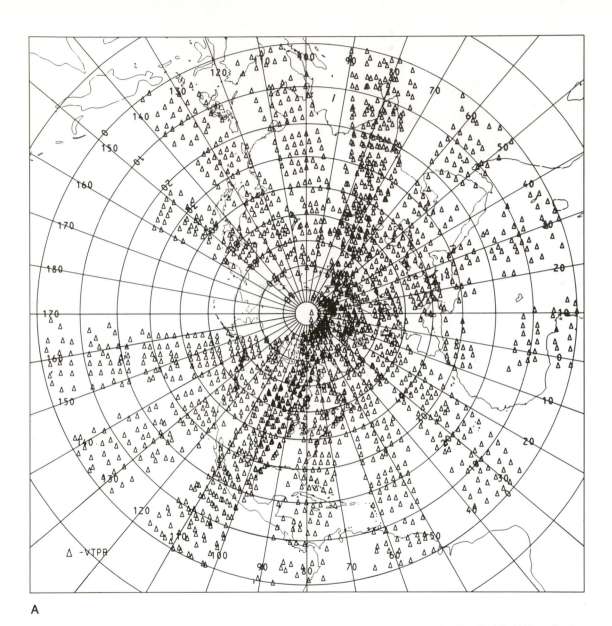

A

Figure 13.6 Locations of temperature and humidity soundings made by satellite received by the World Data Center in Washington, D.C., on October 21, 1979. (A) Northern Hemisphere; (B) Southern Hemisphere. Note that these soundings fill most of the gaps in the radiosonde coverage shown in Figure 13.4.

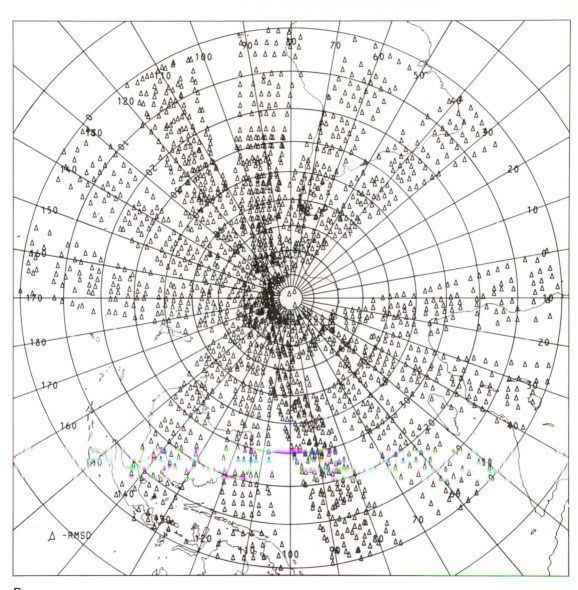

B

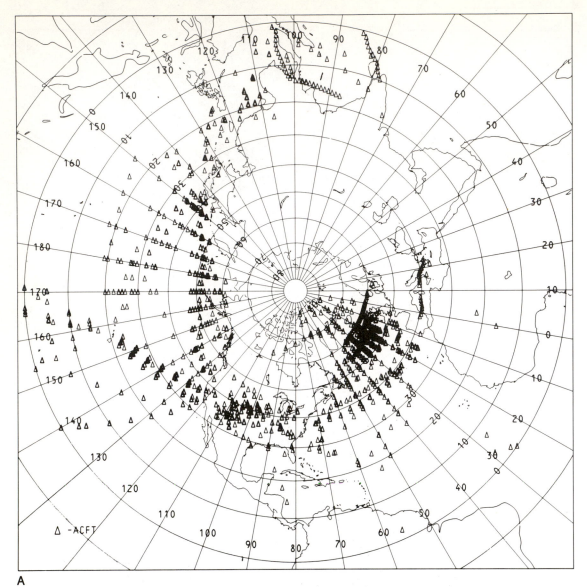

Figure 13.7 Reports from aircraft received by the World Data Center in Washington, D.C., on October 21, 1979. (A) Northern Hemisphere; (B) Southern Hemisphere.

In addition to the regular synoptic observations and the data obtained by satellite, a variety of observations are made to supplement them or for special purposes. Among these are the hourly airways observations that are used to provide more frequent and more detailed information for pilots of aircraft, the pilot reports of conditions experienced by planes in flight (PIREPS or AIREPS), and the radar and sferics observations that enable determination of the location and progress of

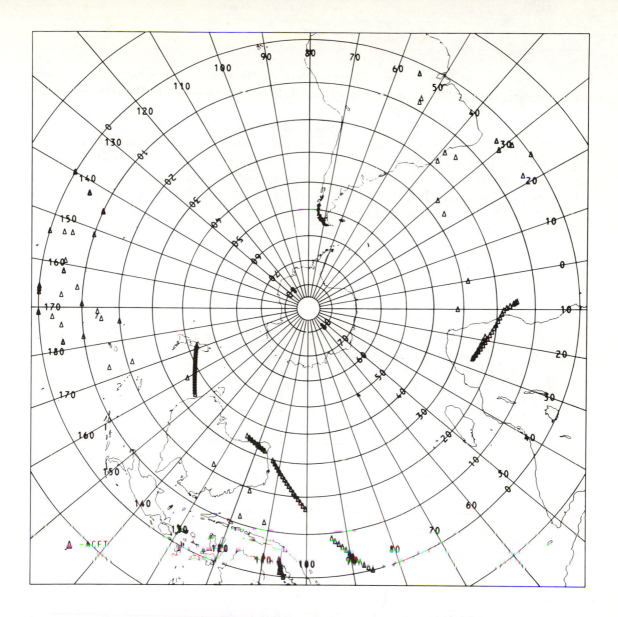

precipitation areas and thunderstorms. The radar echoes show the position and intensity of storms from which precipitation is falling. Sferics is a system of determining the distance and direction of lightning strokes.

The satellite photographs provide additional information to that obtained from the observational network, both by corroborating and completing the data on cloud distribution in areas of dense conventional observations and by filling in the gaps where the conventional observations are sparse.

To determine the location and characteristics of hurricanes and typhoons, specially equipped weather reconnaissance planes fly into the storms at various levels.

The various observations are sent by the observing stations to national and regional data centers. Collections of them are transmitted by these centers to the three World Meteorological Centers at Washington, Moscow, and Melbourne, Australia. In turn, the world centers and regional centers redistribute the collections as needed to the various forecast centers. In addition to the raw (unprocessed) observational data, the centers transmit synoptic charts on which the present situation has been summarized and prognostic charts showing the expected developments. The collected observations are transmitted by teletype or by radio. The charts are distributed by facsimile transmission, either by wire or by radio; in some cases they are broadcast by geostationary satellite. Within the United States, the National Weather Service is replacing current methods of data exchange with a much faster computerized system called AFOS (Automation of Field Operations and Services) (see Figure 13.8). In AFOS the various charts, etc., are transmitted over telephone lines and can be displayed on television screens at the request of the local forecaster operating a computer terminal. Different charts and sequences of observations can be viewed side by side or superimposed for comparison, and if desired "hard copies" can be printed out quickly. In addition to the many charts received from the National Meteorological Center in this fashion, the local forecast offices may also prepare charts of their own for more detailed representation of conditions in the area for which they forecast. Special and supplemental observations are also distributed to forecast centers.

The purpose of the synoptic charts, whether prepared at the weather centrals or at forecast stations, is to provide the forecaster with a three-dimensional picture of the current weather situation in a comprehensible fashion. Clearly, the forecaster cannot hold in his mind the hundreds or thousands of individual observations of conditions that may influence the weather over the area for which he is forecasting. By representing the observations on maps and other charts in an appropriate fashion, the separate data are brought together into patterns that are readily comprehended, such as high- and low-pressure centers, air masses, fronts, wave cyclones, waves in the westerlies, and jet streams. Using the physical principles that govern atmospheric processes, like those that have been discussed in the previous chapters, the forecaster is able to reason about the future behavior of these various features.

Before the advent of large computers and sophisticated computer programs for analyzing data, the weather charts used by forecasters were analyzed by hand. Hand analysis is still used extensively for locating fronts and very detailed small features of the flow and for checking the analyses that are produced by machine.

Preparation of a hand analysis begins with the plotting of the observational data. The model for plotting the observations on the surface synoptic map was presented in Figure 1.15 and in Appendix E, and was discussed in Section 1.3. Once the

Figure 13.8   AFOS, the system being introduced for distribution of weather data, maps, and forecasts by the U.S. National Weather Service. [Courtesy of NOAA.]

data at all stations have been plotted, the map analyst proceeds to locate the fronts and to draw the isobars that summarize the pressure field. In placing the fronts, the principle of continuity is used. Since fronts are the boundaries between air masses, continuity requires that the movement of the fronts from their position on the previous map be consistent with the movement of the air masses on either side. The characteristics of the fronts that help in identifying their position were listed in Section 11.2. The cloud patterns shown in satellite photographs assist in locating the fronts and pressure centers, particularly in regions where the observational data are sparse. For the times when RAOBS are available, the surface analysis is coordinated with the upper-air analysis. The upper-air data, particularly the isotherms at 850 mb and the isopleths representing the thickness of the layer between 1000 and 500 mb, help to locate the positions where significant differences in air-mass temperatures occur.

The upper-level maps are drawn for some or all of the standard pressure values—850, 700, 500, 400, 300, 200, and 100 mb. In the U.S. National Meteorological Center these charts are all plotted and analyzed by the computer. The data are plotted by the machine in accord with the models shown in Figure 13.9. Contours showing the height of the pressure surface are drawn at 30-meter intervals on the

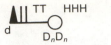

Figure 13.9 Model for plotting rawinsonde observations and reports from aircraft on upper-air charts.

HHH   Height of constant pressure surface, in meters with thousands digit omitted at 850 mb and in decameters at 500 mb.

TT   Temperature (°C).

D_nD_n Dewpoint temperature depression (°C).

d   Wind direction in tens of degrees with hundreds digit omitted. Wind direction is plotted to 36 compass points. Flag denotes 50 knots, barb denotes 10 knots, and half-barb denotes 5 knots. Circle is blacked in when dewpoint depression is 5°C or less.

GGZ   Time of report GMT to nearest hour.

hh   Altitude of aircraft in thousands of feet.

850-mb and 700-mb charts, 60-meter intervals on the 500-mb chart, and 120 meters on the 300-mb, 200-mb, and 100-mb charts. The centers of high and low are labeled H and L, as for the high- and low-pressure centers on the sea-level map. It will be remembered that contours of isobaric surfaces correspond to isobars at a constant level and have a similar relationship to the wind. Isotherms are drawn as dashed lines, at intervals of 5 K. Isotherms on a constant-pressure surface are also lines of equal potential temperature.

For the computer-produced charts, the observational data are fed directly from teletypewriter circuits into a computer that has been programmed to decode the messages, check the data for accuracy and consistency, and list the reports in an organized sequence. The computer then evaluates the height and temperature at fixed grid points (see Figure 13.10) on the basis of the values at the locations of observations, and height contours and isotherms are drawn by machine. In addition to the contours and isotherms, isotachs—lines of constant wind speed—are shown as dotted lines on some of these charts.

A number of other upper-air charts are prepared at the National Meteorological Center and transmitted to the forecasters throughout the nation. These show various derived quantities, such as the 1000- to 500-mb thickness referred to earlier and the 24-hour change in 500-mb heights.

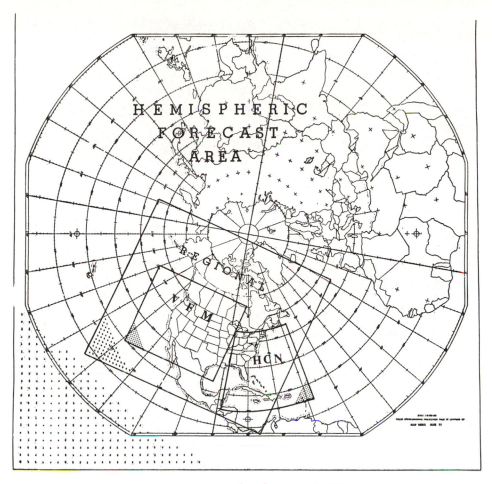

Figure 13.10 Forecast areas and grid distributions (approximate) for various numerical forecasting models used at the U.S. National Meteorological Center. VFM = Very Fine Mesh. HCN = Hurricane, also called Moveable Fine Mesh.

13.4 Numerical weather prediction

Numerical (or dynamic) weather prediction is based on the assumption that weather phenomena obey physical laws expressed in equations that can be solved in the required detail and with sufficient accuracy. As has already been indicated, this is only partially true because of the complexity of the atmospheric system and because of the inadequacy of the observations defining the initial state. It would be hopeless to attempt to predict the motions and properties at all scales from global circulations down to the smallest eddies. However, it turns out that by appropriate treatment of the equations they can be made to represent only the large-scale motions. The solutions of these "filtered" equations can be carried out by using high-speed computers in such a fashion that the main, larger-scale features of flow patterns, such as cyclones, are forecast with considerable accuracy for three or more days in advance.

As an indication of the procedure, consider equation (7.1), which may be written

$$\Delta M\mathbf{v} = (\mathbf{F}_1 + \mathbf{F}_2 + \ldots)\Delta t \qquad (7.1a)$$

This equation gives the change in momentum, $\Delta M\mathbf{v}$, of an air parcel of mass M when forces $\mathbf{F}_1$, $\mathbf{F}_2$, and so forth, act on it for a short interval of time Δt. As shown by equations (7.4) and (8.1), the forces acting depend on the present values of the pressure gradient, the density (and thus the pressure and temperature), and the velocity. If we know these quantities at the present time, which we represent by t_0, we can use the equation to compute the velocity at time $t_1 = t_0 + \Delta t$. Analogous equations give the change in temperature, pressure, and so forth. With the values we have obtained for time t_1, we can repeat the procedure to obtain the velocity, pressure, and temperature at time $t_2 = t_1 + \Delta t = t_0 + 2\Delta t$. By repeated iterations we could find the distribution of these quantities for all parcels of air at any time in the future.

In addition to the magnitude of the task if we were to attempt to divide the atmosphere into small enough parcels to show all the small-scale motions and to keep track of the position and motion of each, there are computational problems that arise when the time scale, as indicated by the choice of Δt, and the distance scale [for instance, Δn in equation (7.4)] are not properly related. Progressively, during the period since the first application of an electronic digital computer to the weather prediction problem in the late 1940s, these problems have been overcome. Methods have been developed that filter out the effects of small-scale ("subgrid") motions and eliminate the possibility of explosive accumulation of computational errors. From the computational standpoint, the forecast could be extended for several months without "blowing up." It appears that the present limitation of accuracy is due to deficiencies in the equations specifying the physical processes and in the observations specifying the initial state of the atmosphere, and to the coarseness of the grid of points at which the initial state is specified and the subsequent developments are computed.

The early numerical weather predictions gave the flow pattern over a limited portion of the globe at a single level representative of the middle of the troposphere, which was taken to be the 500-mb pressure surface. In the NWP procedure presently used at the U.S. National Meteorological Center, solutions representing

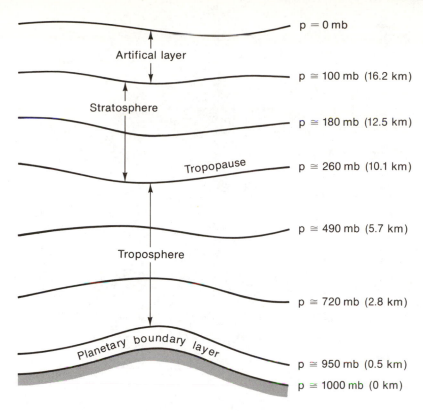

p = 0 mb

Artifical layer

p ≅ 100 mb (16.2 km)

Stratosphere

p ≅ 180 mb (12.5 km)

Tropopause

p ≅ 260 mb (10.1 km)

p ≅ 490 mb (5.7 km)

Troposphere

p ≅ 720 mb (2.8 km)

Planetary boundary layer

p ≅ 950 mb (0.5 km)

p ≅ 1000 mb (0 km)

Figure 13.11   Vertical layout of seven-layer primitive equation model used in Numerical Weather Prediction at the U.S. National Meteorological Center.

the flow in seven layers are computed routinely for the regional and hemispheric grids shown in Figure 13.10. On an experimental basis, when the weather situation suggests need for finer resolution, computations are also carried out for the hurricane (moveable fine mesh) or the very fine mesh grids. Figure 13.11 shows the vertical structure used in these "seven-layer primitive equation" models. Computations of a twelve-layer global model covering most of the earth are also carried out daily.

The initial data are interpolated from the irregularly distributed observations to the equally spaced grid points at the various layers. In order to utilize the data, such as those from satellites and commercial aircraft, that are not made at synoptic hours, a four-dimensional data assimilation scheme is used. Using these initial values of the quantities appearing in the equations analogous to equation (7.1a)—the "primitive equations"—their values at later times are computed. The computations are carried out for small intervals of time, for example, 10 minutes, during which the rate of change can be assumed to be constant, and the process

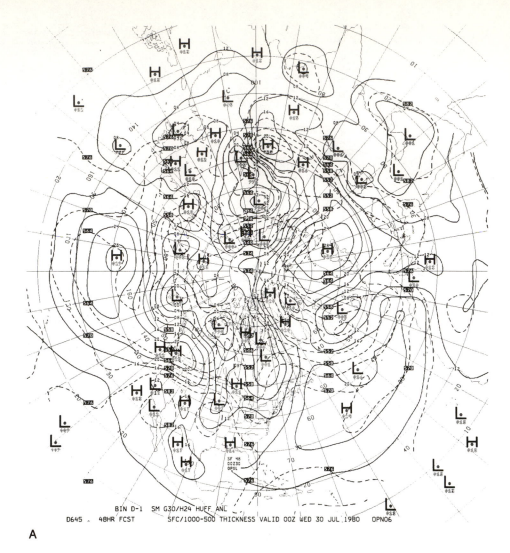

A

Figure 13.12 (A) Prediction of sea-level pressure (*isobars as solid lines*) and 1000-mb–500-mb thickness (*dashed lines*) for 0000 GMT July 30, 1980, made 48 hours earlier. (B) Observed sea-level pressure and 1000-mb–500-mb thickness 0000 GMT July 30, 1980.

is repeated through the forecast period. The computations for the thousands of grid points, the seven or twelve layers, and the many intervals of time make for tremendous numbers of arithmetic and logical operations. The computers presently in use, the IBM 360–195, carry out 10 million to 15 million such operations per second, and it takes about an hour to perform all the calculations required for a 48-hour forecast.

Just as the current NWP procedures represent a large advance from the early single-level method, so may rapid changes be expected to continue in response to research and to increased computer capabilities.

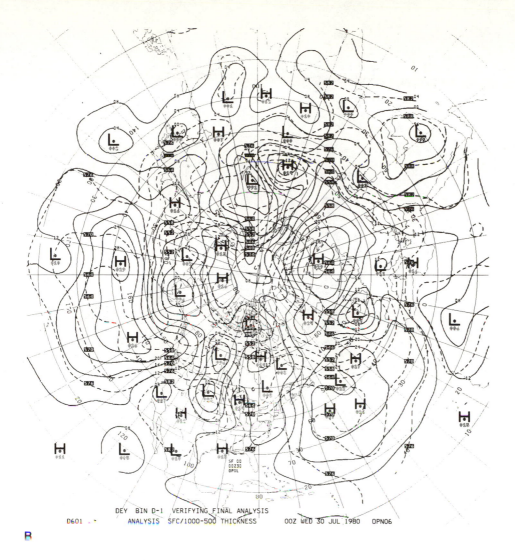

DEY BIN D-1 VERIFYING FINAL ANALYSIS
D601 ANALYSIS SFC/1000-500 THICKNESS 00Z WED 30 JUL 1980 OPN06

An example of the results of an NWP prediction is shown in Figures 13.12 and 13.13. The forecast was produced at the U.S. National Meteorological Center by the global prediction model. Figure 13.12A shows a forecast made July 28, 1980 of the sea-level pressure and 1000-mb–500-mb thickness over the Northern Hemisphere at 0000 GMT July 30, 1980, 48 hours later, and Figure 13.12B shows the data actually observed on July 30. It will be seen that most of the features in the forecast chart correspond well to the observed. For instance, in the United States the low-pressure trough over the center of the country was predicted to be only slightly east of its observed position, and that over the Atlantic states was forecast in almost exactly the correct position, although not extending far enough southward. In Europe the high-pressure center over Scandinavia was located well, but its extension southward to the Mediterranean was held slightly too far to the

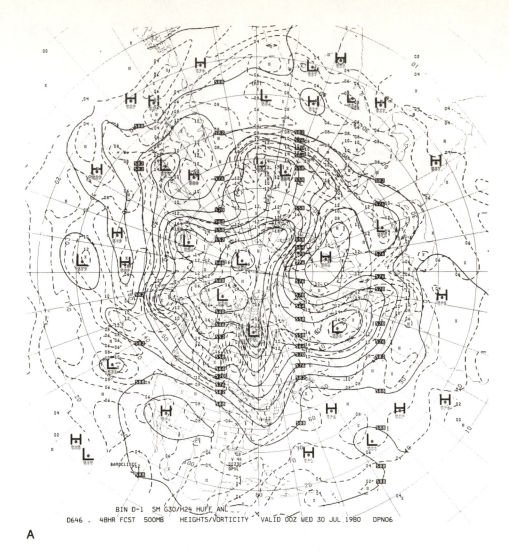

BIN D-1 SM G30/H24 HUFF ANL
D646 . 48HR FCST 500MB HEIGHTS/VORTICITY VALID 00Z WED 30 JUL 1980 OPN06

A

Figure 13.13 (A) Prediction of 500-mb heights (*contours as solid lines, 60-m interval*) and absolute vorticity (*dashed lines*) for 0000 GMT July 30, 1980, made 48 hours earlier. (B) Observed 500-mb heights and absolute vorticity 0000 GMT July 30, 1980.

southwest. The low-pressure area west of the British Isles was correctly placed. Considering that a few years ago forecasts beyond 24 hours were highly speculative, this forecast, which is typical of current forecast capability, confirms the evidence of skill shown in Figure 13.1.

Figure 13.13 shows the corresponding 500-mb 48-hour forecast and verification. At this level the features of the forecast appear to fit the observed very well. As examples, compare the positions of the troughs over eastern North America and the eastern Atlantic Ocean and the highs over the western United States and Scandinavia.

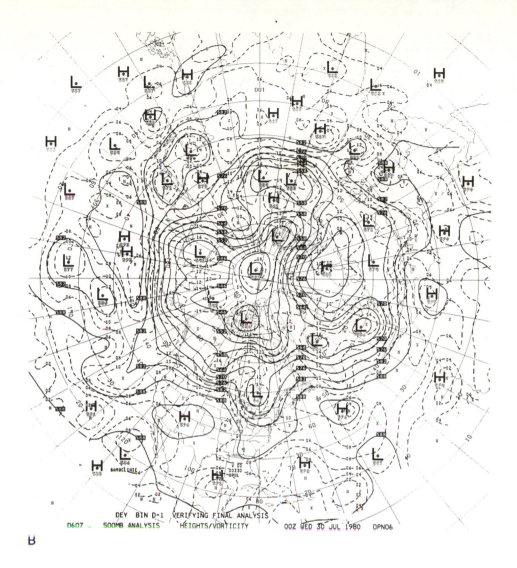

DEY BIN D-1 VERIFYING FINAL ANALYSIS
D607 500MB ANALYSIS HEIGHTS/VORTICITY 00Z WED 30 JUL 1980 OPN06

The NWP output includes predictions of the vertical velocity and the humidity distribution, from which quantitative precipitation forecasts can be made. Figure 13.14 shows a sample forecast of the total precipitation expected between 12 and 24 hours after the initial time, compared with the observed amounts. While the agreement is far from perfect, the areas for which precipitation was forecast are generally correct, and the amounts are of the right order.

While the NWP gives estimates of the flow pattern that are better and can be made for longer periods than those made by the subjective and empirical methods in use previously, forecasts of temperature and precipitation based on these flow patterns are improved when empirical or statistical procedures are used to supplement the NWP results. The statistical procedures consist of finding the relationship between some of the quantities that are predicted numerically (for

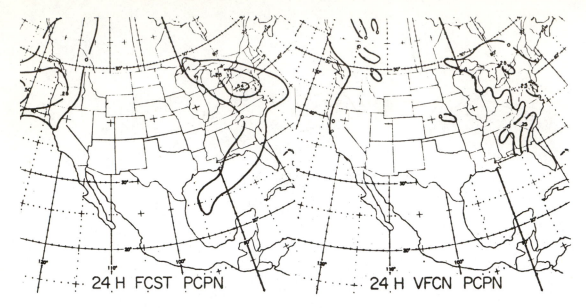

Figure 13.14　Forecast and observed precipitation for period 12–24 hours after initial time. [From F. G. Shuman and J. B. Hovermale, "An Operational Six-Layer Primitive Equation Model," *J. Appl. Meteorol.* vol. 7, no. 4 (1968), p. 528.]

instance, the height of the 700-mb surface over a station or the difference in sea-level pressure between two stations) and the quantity to be predicted (for instance, the probability of precipitation in the next 12 hours or the maximum temperature) on the basis of past records. The quantities that are assumed to be known are called *predictors* and the quantity that is to be predicted is called the *predictand*. The problem is to find those predictors that determine with the least uncertainty the predictand and to find the relationship that enables the evaluation of the predictand for any given values of the predictors.

At the Techniques Development Laboratory of the National Weather Service in Washington, the relationships for the probability of precipitation and the probability of frozen precipitation (sleet or snow) over the United States and the minimum and maximum temperature at 244 cities in the United States and Southern Canada have been developed. The predictors, which were selected for physical as well as statistical reasons, include measures of the circulation pattern as given by the predicted heights of constant pressure surfaces, the humidity at various levels, and the past temperatures and precipitation at various locations. Figure 13.15

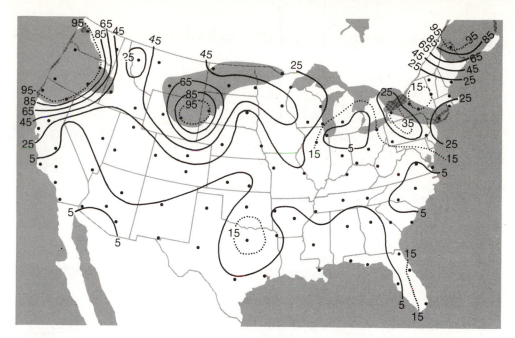

Figure 13.15 Example of forecast of probability of precipitation in period from 1200 GMT, May 8, to 0000 GMT, May 9, 1970, obtained by applying statistical relationships to NWP output based on initial conditions at 1200 GMT, May 8. Shading shows area in which precipitation was observed during the forecast period. [From W. H. Klein, "Computer Prediction of Precipitation Probability in the United States," *J. Appl. Meteorol.* vol. 10, no. 5 (1971), p. 913.]

shows an example of the forecast of the probability of precipitation. This forecast was for the first 12 hours after the time of forecast. Similar charts are made and transmitted for every 12-hour period up to 48 hours after the time of the forecast. Figure 13.16 shows the forecasts of the minimum and maximum temperatures, made for periods of 24 to 60 hours in advance.

These prognostic charts, prepared at the National Meteorological Center, are transmitted to the local forecast offices for their guidance. In applying them, the forecaster at each local office must modify their indications in the light of his own interpretation of the weather situation, local topographic influences, and the special needs of the public in his community. While the forecasts he issues are usually consistent with the NMC products, they contain considerably more detail, including specification of smaller-scale features, such as sea breezes or thundershowers, which the NWP filters out. The computer has not yet displaced the human from the weather forecasting process!

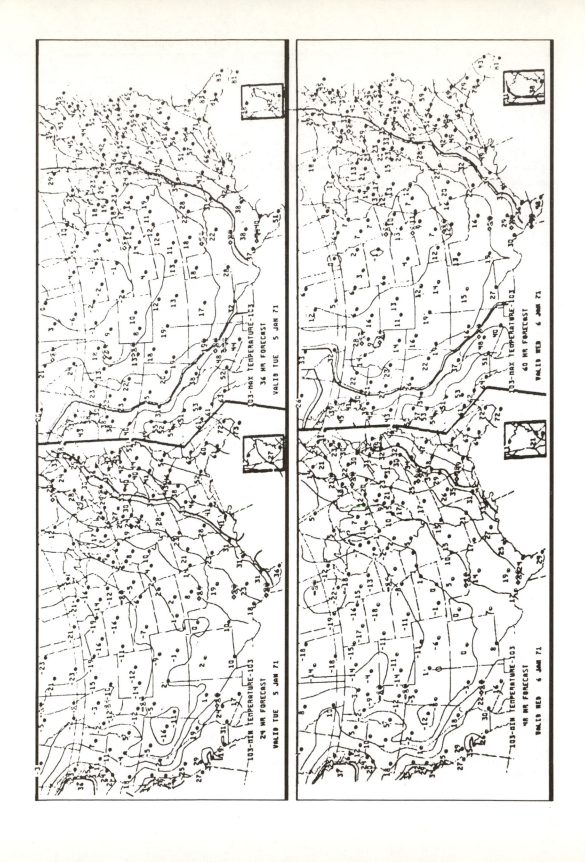

13.5   Special forecasts

In addition to the general forecasts issued for everyday use by the public, the National Weather Service makes various specialized predictions. The forecasting for air navigation has already been referred to. Frost warnings for agriculture, flood warnings, hurricane warnings, warnings of severe thunderstorms and tornadoes, fire-weather forecasts, and air pollution forecasts are other important services provided to meet the needs of special groups or the requirements of special meteorological circumstances.

These special forecast services are made by forecasters with expertise in the particular subjects, using supplemental observations. Thus, for flood forecasting, a special river and rainfall reporting network is maintained, as well as a snow survey that indicates the water content available in case of sudden thaws. The forecaster has available the past records of stream flow in response to various amounts of precipitation, snow melt, and soil and river conditions. Based on the river-stage measurements and the forecasts of precipitation and temperature, he predicts the changes in height of the river at various points along its course. For rivers with large drainage basins, the response is sufficiently slow to permit the forecast to be made with considerable accuracy well in advance. When flooding is expected, it can usually be predicted far enough in advance for protective measures to be taken—in particular, for residents of low-lying areas to move to higher ground, reducing or eliminating loss of life. Small rivers are subject to flash floods, in which the response to the heavy rain is so rapid that unless the amount of precipitation is forecast well in advance there is little time available for protective action. The Weather Service has set up a flash-flood warning procedure in a number of river basins, utilizing volunteer networks of rainfall and river observing stations that enable rapid determination of the trend of the river while heavy rain is still falling.

The forecasts of river stages are increasingly being based on computations using computers. However, the final decision on issuing a flood warning is made by a National Weather Service hydrologist or meteorologist.

The hurricane warning service is headquartered at Miami, Florida. During the hurricane season for the Atlantic Ocean—June 1–November 30—a constant watch for tropical disturbances is maintained. Their early detection has been facilitated

Figure 13.16   Facsimile transmission of maximum and minimum temperatures (°F) for 24–60 hours in advance, forecast by computer using statistical relationships based on initial data for 1200 GMT, January 4, 1971. Inset in lower right-hand corner is northern New England. [From W. H. Klein, F. Lewis, and G. A. Hammons, "Recent Developments in Automated Max/Min Temperature Forecasting," *J. Appl. Meteorol.* vol. 10, no. 5 (1971), p. 917.]

by meteorological satellite photographs. When there are indications of a developing storm, a reconnaissance aircraft, the "hurricane hunter," is sent to determine its location and intensity. As the storm approaches the coast, it comes within range of the weather radar network. The forecasting is a combination of pure extrapolation of past movement and prediction based on the influence of the large-scale circulation as forecast by NWP. Warnings of the strength of winds, the height of the storm tide, and the intensity of the precipitation (which may cause flooding) are issued for the locations that will be affected as the storm moves inland.

Forecasts of severe thunderstorms and tornadoes consist of two stages: the severe storm watch bulletin, designating the areas of probability of severe storms; and the storm warning, giving the actual position and probable movement of a damaging storm or tornado. The watch is based on the general forecast for the area. The tornado warning is based on actual visual or radar detection of the storm. For this purpose the Weather Service maintains a network of radars throughout the central and southeastern parts of the United States, where tornadoes are most frequent, and Severe Local Storm Spotter Networks of volunteer observers and communication facilities for rapid reporting and broadcasting of the sighting of storms.

When a tornado has been sighted, persons in the immediate vicinity should take cover immediately, preferably in a storm cellar or reinforced building, and stay away from windows. A basement or an interior hallway or small closet on the ground floor is best. If possible, one should curl up under a sturdy piece of furniture, which will protect against flying or falling debris.

The fire-weather service provides information about the fire danger to guide the deployment of protective forces by the Forest Service and makes wind and humidity forecasts when fires are burning, enabling the efficient use of fire-fighting men and equipment. Frost warnings enable protection of crops with orchard heaters, by flooding, or by early harvesting. In these and other applications the forecaster must take account of the topography on a very local scale and make use of his knowledge of the behavior of weather systems in relation to the specific phenomena under consideration.

13.6   Extended-range forecasting

The NWP tests of predictability discussed in Section 13.2 suggest that detailed day-by-day predictions are possible for two weeks at the longest. In practice, at present the accuracy of the predictions decreases from day to day, so that after five days they are little better than climatological expectations. At first sight it would seem that weather forecasts cannot be extended beyond a week or so at present.

However, there is valuable information we might be interested in other than the day-by-day sequence of weather. For various purposes we may find it useful to know whether next week or next month will be rainy or dry, or warmer or colder than normal for the season. The attempts at extended-range forecasting have been focused on these questions, rather than on the daily weather.

There have been many methods attempted: weather typing, statistical correlations, analogues, periodicities, and the influence of sunspots. Recently, two procedures based on physical processes have been dominant. In one, the average flow for a week or a month is studied from the standpoint of the movement of the long, slow-moving waves in the flow. This method, combined with the results of statistical studies, has been the basis for the weekly, monthly, and seasonal outlooks issued by the National Weather Service. The other procedure is based on the tendency for certain anomalies of the characteristics of the earth's surface—its temperature and snow cover—to produce persistent effects on the circulation of the atmosphere in subsequent months and seasons. This is particularly true when the surface anomaly, for instance, abnormally low sea-surface temperature, tends to be reinforced by its effect on the atmosphere. Some "feed-back" mechanisms of this type have been found, but so far the procedure has been applied only experimentally.

The value that long-range weather forecasts would have is widely recognized. If abnormally wet or dry seasons or unusually warm or cold months could be predicted in advance, for instance, the planting and harvesting of crops, construction activities, distribution of fuels, and other economic activities could be planned in a more rational fashion. Methods of seasonal forecasting are in a preliminary stage of development and have had slight or limited success. In economic planning, climatological records must be relied on for weather information beyond a week in advance.

13.7   Programs for the improvement of forecasting

Both within the weather services and in the general scientific community, a large amount of research is being conducted for the purposes of improving the accuracy and extending the range of weather forecasting. Much of this effort is aimed at further increasing our understanding of the physical laws governing the structure and behavior of weather systems. For a full understanding, the observations of the global atmospheric structure must be more complete and more extensive. Recognizing the tremendous scope of the task of obtaining a complete picture of the structure of the atmosphere even for one time, and of understanding the complex interrelationships of the processes occurring within it at all scales, meteorologists have launched two international programs. One of them, the World Weather

Watch (WWW), has concentrated on filling the gaps in the routine observational network and improving the system for prompt communication of the observations to the data centers and from them to the forecast centers. It is being conducted by the World Meteorological Organization and the weather services of its member states. The other is the Global Atmospheric Research Program (GARP), which has been organized jointly by the International Council of Scientific Unions (ICSU) and the WMO. GARP's broad objectives are (1) to improve our understanding of the processes that determine the day-to-day changes of the weather, as a basis for better weather forecasting, and (2) to improve our understanding of the processes that control the statistical properties of the atmospheric circulation and their longer-term variations, as a basis for predicting changes in climate. As part of the program to achieve these objectives, the First GARP Global Experiment (FGGE, also called the Global Weather Experiment) set out to accumulate for relatively short periods much more complete observational coverage throughout the world than is routinely available, even with the improvements that the WWW has already achieved. These data will be used to develop and test improved procedures in order to extend the NWP forecasts to the theoretical limit of predictability.

Implementation of the World Weather Watch has been under way since 1968. It has been gradually developing a fully coordinated Global Observing System of surface, upper-air, and satellite observations. Besides the establishment of additional manned surface and upper-air stations by member countries, particularly in tropical regions and in the Southern Hemisphere, automatic weather stations have been established in remote, uninhabited areas and on buoys at sea. The upper-air network has been augmented by reports from aircraft in flight. The routine satellite system has included the two geostationary satellites and two polar-orbiting satellites operated by the United States, and one or more orbiting satellites operated by the USSR. The cloud pictures from the geostationary satellites, in visible light in the daylight hours and in infrared both day and night, enable determination of upper winds from cloud movements in addition to giving a continuous record of the development and movement of weather systems. The orbiting satellites are equipped to make temperature soundings by means of infrared and microwave radiation measurements, but these have not been processed in time for transmission to forecast centers throughout the world as yet.

Equally as important as the Global Observing System, the WWW includes the Global Data-Processing System (GDS) and the Global Telecommunications System (GTS). It is obvious that unless the observations reach the forecasting centers in a timely fashion they may as well not have been made. The GDS includes World Meteorological Centers at Washington, Moscow, and Melbourne, and 23 Regional Meteorological Centers in various parts of the world. These data centers are connected through the Main Trunk Circuit and Regional Telecommunication Hubs by high speed, usually 3000-words-per-minute, transmission systems. The observations are collected at regional centers, organized into a form prescribed by the WMO, and transmitted to forecast centers throughout the world as needed.

The original stimulus for the GARP came from the recognition that meteorological satellites, the first of which was launched April 1, 1960, provided the opportunity to deal with forecasting the weather in new ways. This recognition, together with awareness of the need to emphasize peaceful uses of outer space, led President John F. Kennedy, in a speech before the United Nations in 1961, to propose international collaboration to apply these new tools to the solution of the weather-prediction problem. Responding to this challenge, the United Nations passed two resolutions requesting the WMO and ICSU to develop and carry out international programs that would achieve this objective. The following years saw the development of plans for the WWW to improve world weather services and for GARP to develop the underlying knowledge that forms the basis for scientific weather prediction. Conferences and studies led to the design of a number of experiments investigating various regional aspects of the problem and planning for FGGE, the Global Weather Experiment.

From the start the unique character of tropical processes and their importance in the global circulation were recognized. GATE, which has been discussed in detail in Chapter 12, was conducted to elucidate these processes. Similarly, the important role of the Asiatic Monsoon, both as it affects a large part of the earth and as it interacts with the world-wide circulation by carrying air between the Northern and the Southern Hemispheres, led to the Monsoon Experiment (Monex), the observational phases of which were conducted during FGGE. The Polar Experiment (Polex) is another regional program whose observational phase was carried out at the same time as that of FGGE. These regional programs are largely concerned with the transfer of heat, moisture, and momentum between the atmosphere and the earth's surface. Other programs designed to study specific instances of this interaction include the Joint Air–Sea Interaction Experiment (JASIN), which was carried out over an area of the North Atlantic Ocean in 1978, the Air–Mass Transformation Experiment (Amtex), conducted over the East China Sea in February 1974 and February 1975, and the international mountain experiment (ALPEX), which is planned to be conducted in the Alps in the early 1980s, and possibly in other mountain ranges later.

While these regional investigations form essential parts of GARP, the central feature of the program is FGGE, the Global Weather Experiment. After more than a decade of preparation, the observational phase of this experiment was carried out in 1978–1979. It consisted of a "buildup year," December 1, 1977–November 30, 1978, during which the special observational aspects were put in place and tested, and the operational year, December 1, 1978–November 30, 1979, during which as complete coverage of the weather systems of the entire earth as the concerted efforts of all the nations of the world could muster was obtained. Within

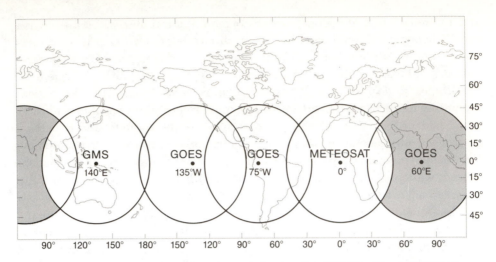

Figure 13.17   Geostationary satellite coverage during the Global Weather Experiment. [From R. J. Fleming et al., "The Global Weather Experiment, I. The Observational Phase Through the First Special Observing Period." *Bulletin of the American Meteorological Society,* vol. 60, no. 6, June 1979.]

the operational year, two 2-month periods—January 5–March 5, 1979 and May 1–June 30, 1979—were designated as Special Observational Periods (SOPs), during which procedures that could not be sustained for an entire year would be employed to make the coverage more complete. Within the SOPs there were specified 1-month periods of intensive observations when extra efforts were extended and/or still more frequent observations were made.

The Global Weather Experiment is probably the largest international scientific experiment conducted to date. In addition to the normal participation by the 147 member countries of the WMO in the WWW system of surface and upper-air observations, 70 of these countries and five intergovernmental organizations provided funds, equipment, and personnel for special observing and data-management systems. These special systems include satellites, an aircraft dropwindsonde program, tropical-wind observing ships, constant-level balloons, drifting buoys, and automated aircraft-observing systems. Additional geostationary satellites were launched, so that a system of five surrounded the earth at approximately equal intervals over the equator. The location of these satellites and the approximate coverage by their equipment is shown in Figure 13.17. The third-generation polar-orbiting satellites, TIROS-N and NOAA-6, provided thousands of soundings and sea-surface temperatures per orbit, as well as serving as a data-collection system for drifting buoys and constant-level balloons. In the aircraft dropwindsonde program, operated by the United States, airplanes flying at about 13-km elevation over the tropical Atlantic and Indian Oceans released instruments—dropwinsondes—every 350 km. The dropwinsondes transmitted reports of temperature, humidity, and wind as they fell to earth on parachutes. There were about 40 Tropical Wind Observing Ships, oceanographic research vessels that filled gaps in the tropics not covered by the aircraft tracks, WWW land stations, or island stations. The United States activated for the experiment four upper-air stations on islands in the equatorial Pacific and increased from one to two observations a day the schedule for eight others. More than 300 constant-level balloons were launched from Ascension Island, in the tropical Atlantic, and Canton Island and Guam, in the tropical Pacific, during the SOPs. Designed to float at about 14.3-km elevation, they provided winds in the tropics above the level of the aircraft. To provide surface data over the Southern Hemisphere, which is primarily ocean little travelled by ship, more than 300 drifting buoys were deployed. Eight nations provided the equipment and 14 deployed them. The data from them were transmitted by radio to the TIROS-N satellite and relayed to France from U.S. ground stations for processing. Figure 13.18 shows the distribution of these buoys on a day during SOP-II. In addition to the usual AIREPS that are observed manually and communicated through the GTS, two systems of automatic reports from aircraft were introduced for the Global Weather Experiment. In the Aircraft Integrated Data System (AIDS), wind and temperature data were recorded on cassette recorders aboard about 80 wide-bodied airplanes of various airlines and processed by a center established by the Netherlands. In the Aircraft-to-Satellite Data Relay System (ASDAR), the data were transmitted from the instruments aboard planes to ground stations via the geostationary satellite data collection systems in real time, and were thus available for current use as well as for research purposes. The AIDS data were delayed and used only in research.

The Global Weather Experiment provides a set of data defining the state of the atmosphere more completely and in more detail than ever before. This set of data can be used to provide initial values for NWP models on a global scale, and to test the accuracy of the results of computations using these models. It will enable determination of the extent to which deficiencies in present models have been due to inadequacy of initial data or to inadequacy of the models themselves.

More fundamentally, it and the several regional experiments are providing data for better understanding the structure of atmospheric motions on all scales. This improved understanding will enable optimum design of numerical models, defining the smallest scale that must be included explicitly and enabling valid parameterization of smaller-scale systems. This improvement of numerical models should enable more specific evaluation of the limit of deterministic day-by-day predictability and the extension of practical forecasting closer to the theoretical limit.

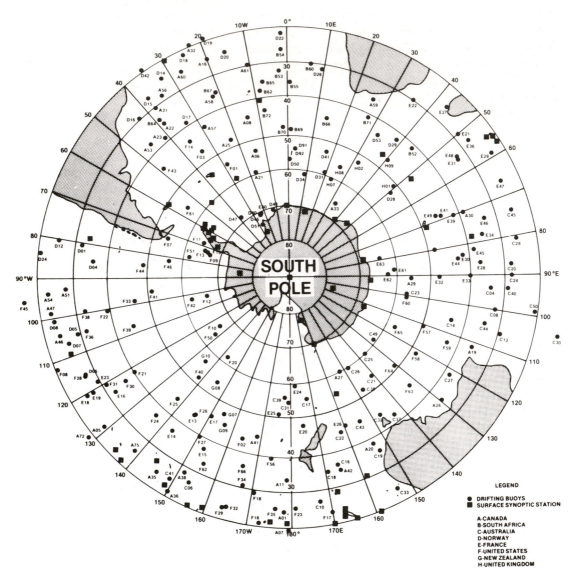

Figure 13.18 Positions of drifting buoys and surface land stations over the Southern Hemisphere oceans and Antarctica reporting on one day (May 30, 1979) during the Global Weather Experiment. [From R. J. Fleming et al., "The Global Weather Experiment, II. The Second Special Observing Period." *Bulletin of the American Meteorological Society*, vol. 60, no. 11, Nov. 1979.]

Studies using these data are being carried out at universities and laboratories, governmental and private, throughout the world. The process of designing the experiments in itself provided the stimulus for much fruitful research. The application of data from the experiments doubtless will produce insights that will bring closer the complete understanding of our atmospheric environment, and thereby the ability to predict the weather further into the future and with greater accuracy.

Questions, Problems, and Projects

1. Discuss the reasons why a prediction of tomorrow's weather cannot be carried out with the same accuracy as the prediction of eclipses of the sun and moon years ahead.

2. In most localities the observed maximum temperature, minimum temperature, and precipitation for each day and the forecast of these quantities for the following day are published in the daily newspaper or broadcast on television and radio.

   a. Keep a record of the observed values and the forecasts for a month.

   b. As measures of the accuracy of the temperature forecasts, evaluate the average difference between the observed and forecast values, separately for the maximum temperature and the minimum temperature.

   c. As a measure of the accuracy of the precipitation forecasts, count the number of days for which the precipitation forecast was right. This should comprise the days precipitation was forecast on which it did occur and the days precipitation was not forecast and none occurred. Express this as a percentage of the total number of forecasts.

14   Air Pollution

14.1   Types and sources of air pollution

Almost all human activities introduce contaminants into the atmosphere. Natural processes also introduce other substances than those we regard as the constituents of pure air. Besides nitrogen, oxygen, argon, carbon dioxide, water in its various phases, and the trace gases that form a permanent part of the atmosphere, the atmosphere always contains emanations from growing or decaying vegetation, salt from sea spray, dust from blowing soil and sand storms, smoke from lightning-caused fires, and gases and fumes from volcanic eruptions. Except locally or temporarily, these natural substances are in low concentrations and have no harmful consequences. The artificial injection of contaminants by human activities, however, has caused illness and death to humans, impaired agricultural productivity, and damaged property. The levels that pollution has reached have already aroused worldwide concern, and the possibility that irreversible consequences will occur unless the emissions of pollutants are reduced or eliminated has been suggested.

Figure 14.1   Air pollution from an industrial complex. [Courtesy of NOAA.]

Air pollution may be defined as the presence in the atmosphere of substances that are toxic, irritant, or otherwise harmful to people or damaging to vegetation, animals, or property. The combustion of fuels to produce energy is the principal source of air pollution. Until the middle of the twentieth century, the problems of air pollution were concerned principally with the products of coal combustion—namely, soot, sulfur dioxide, and fly ash—and the effluents from such industrial operations as smelters and steel mills. At high humidities water drops condense on coal smoke to form the very dense "pea soup" fogs that were characteristic of such cities as London, Pittsburgh, and St. Louis before they enacted legislation requiring the use of clean fuels and efficient means of combustion. The word *smog*, formed by combining the words *smoke* and *fog*, was originally used to designate this type of air pollution. It now has come to be used as a synonym for general air pollution, whether or not it involves smoke or occurs at high humidity.

In recent years petroleum derivatives have replaced coal as the principal source of energy in many places: fuel oil and natural gas for heating and electric power generation; and gasoline, kerosene (for jet airplanes), and diesel oil for transportation. The use of these fuels has led to a new type of air pollution, in which photochemical reactions play an important role. A photochemical reaction is one in which the absorption of radiation initiates or facilitates chemical changes. In photochemical smog, nitrogen dioxide absorbs solar radiation to initiate a chain of reactions. In the presence of hydrocarbons the process results in the formation of ozone and other highly reactive, irritating, and toxic substances. Because it was first observed in Los Angeles, photochemical smog is frequently referred to as *Los Angeles-type smog*. The coal-smoke smog is called *London-type smog*.

Estimating the total emissions of pollutants into the atmosphere involves many uncertainties. There is no strict census of the amounts emitted even in industrialized nations, such as the United States. The latest estimates made by the U.S. Environmental Protection Agency of the amounts of the principal categories of pollutants emitted in the United States are shown in Table 14.1. Only the total annual values are given, as the seasonal variation of the figures on which they are based, such as the consumption of coal, are not available. The sulfur oxides are principally sulfur dioxide (SO_2). The nitrogen oxides include nitric oxide (NO) and nitrogen dioxide (NO_2). Nitrous oxide (N_2O), a natural product of the decay of vegetation, is not regarded as a pollutant and is therefore omitted. The volatile organic compounds are principally hydrocarbons and other compounds occurring in petroleum or resulting from chemical reactions, including partial combustion, in which they take part. It is seen that the main contributor of sulfur dioxide and particulates in the atmosphere is industry, including electric power generation, whereas carbon monoxide comes chiefly from vehicles used in transportation. Industry and transportation emit large amounts of oxides of nitrogen and volatile organic compounds. Until control devices were required on automobiles, the emissions of hydrocarbons from vehicles far exceeded that from industry.

After the pollutants have been emitted, they spread by being carried away by the winds and being mixed horizontally and vertically by turbulent diffusion and convection. Ultimately, they are removed from the atmosphere by processes we shall discuss in a later section. The average length of time that pollutants remain in the atmosphere varies. For large particles and for reactive gases, it may be only minutes or at most hours. For sulfur dioxide it is estimated to be a few days, but for carbon monoxide it may be several months. If the pollutants were emitted uniformly over the earth's surface and mixed throughout the atmosphere, even those with relatively long lifetimes would have concentrations so low as to be inconsequential at present average emission rates. It is because the emissions occur in small areas and it takes time for them to spread that objectionable concentrations accumulate.

Table 14.1 Estimated Amounts of Emissions of Air Pollutants from Various Sources in the United States in 1977 (in millions of metric tons)

Source category	Particulates	Oxides of sulfur	Oxides of nitrogen	Volatile organic compounds	Carbon monoxide
Transportation	1.1	0.8	9.2	11.5	85.7
Highway vehicles	0.8	0.4	6.7	9.9	77.2
Nonhighway vehicles	0.3	0.4	2.5	1.6	8.5
Stationary fuel combustion	4.8	22.4	13.0	1.5	1.2
Electric utilities	3.4	17.6	7.1	0.1	0.3
Industrial	1.2	3.2	5.0	1.3	0.6
Residential, commercial, and institutional	0.2	1.6	0.9	0.1	0.3
Industrial processes	5.4	4.2	0.7	10.1	8.3
Chemicals	0.2	0.2	0.2	2.7	2.8
Petroleum refining	0.1	0.8	0.4	1.1	2.4
Metals	1.3	2.4	0	0.1	2.0
Mineral products	2.7	0.6	0.1	0.1	0
Oil and gas production and marketing	0	0.1	0	3.1	0
Industrial organic solvent use	0	0	0	2.7	0
Other processes	1.1	0.1	0	0.3	1.1
Solid waste	0.4	0	0.1	0.7	2.6
Miscellaneous	0.7	0	0.1	4.5	4.9
Forest wildfires and managed burning	0.5	0	0.1	0.7	4.3
Agricultural burning	0.1	0	0	0.1	0.5
Coal refuse burning	0	0	0	0	0
Structural fires	0.1	0	0	0	0.1
Miscellaneous organic solvent use	0	0	0	3.7	0
Total	12.4	27.4	23.1	28.3	102.7

Note: A zero indicates emissions of less than 50,000 metric tons per year.
Source: National Air Quality, Monitoring, and Emissions Trends Report, 1977. U.S. Environmental Protection Agency, Research Triangle Park, N.C., December, 1978.

14.2   Factors affecting the concentration of air pollution

The primary determinant of air-pollution concentration is, of course, the amount of contaminants emitted into the air. But our experience tells us that even though the same sources pour out pollutants day after day, sometimes the air is relatively clean and sometimes it is very polluted. The concentration of pollution depends on the weather conditions. In addition, for the same amount of emission and the same meteorological conditions, the air-pollution concentration is influenced by the geometric configuration of the sources, including the height above ground of the emission and the area over which the sources are distributed. If the emission is at a considerable height, the pollutants become diluted by the time they are transported to the ground by eddy diffusion. If the emission is spread out over a large area, the downwind concentration will be smaller than if it all comes from a small area, equivalent to a point source. The factors that affect the concentration of air pollution are thus (1) the total amount of pollution emitted; (2) the meteorological conditions; and (3) the configuration of the sources. As an aside we may note that for pollution control we can alter (1) and (3). Although weather modification has been proposed as a means of reducing air pollution, all such proposals made up to the present time have been shown to be impractical.

The amount of pollution emitted in industrial, commercial, and domestic activities and the problem of reducing it are essentially subjects for engineering studies. However, meteorology enters into these questions insofar as the estimation of the capacity of the air to dilute the pollutants determines the allowable emissions to which the engineers should limit the equipment they design. Similarly, the height of smokestacks, the shape of factories, and the layout of commercial areas, housing developments, and cities are the provinces of architecture, engineering, and urban planning, but meteorological factors should be taken into account if the pollution coming from these structures is to be minimized.

The various types of sources are for convenience treated as point sources, line sources, and area sources. The smokestack of an electric power generating plant is an example of a point source. The exhaust pipe of an automobile similarly constitutes a point source; however, because motor vehicles move in rapid succession along them, streets and freeways act essentially as line sources. The chimneys of houses and apartment buildings, regarded individually, are also point sources, but they are so close together that they merge into a continuous area source. The emissions from a point source are expressed in terms of mass per unit time, for instance, kg s^{-1}; from line sources as mass per unit length per unit time, for example, $\text{kg m}^{-1} \text{s}^{-1}$; and from area sources as mass per unit area per unit time, $\text{kg m}^{-2} \text{s}^{-1}$.

The meteorological factors that affect the pollution concentration are (1) the stability of the air as determined by the vertical variation of temperature and

(2) the direction and strength of the wind. These factors determine how fast the contaminant is diluted by mixing with environmental air after it leaves the source. The hydrostatic stability controls the rate at which up-and-down motions mix the pollutants with clear air from above. The wind speed determines into how much air the contaminants are initially mixed, and the irregularities of the wind speed and direction govern the rate at which the pollutant spreads horizontally as it is carried downwind.

The effect of stability has already been discussed in Chapter 4. If there is a superadiabatic lapse rate, vertical mixing occurs readily and pollutants are diluted rapidly. If there is an inversion, the great stability completely suppresses vertical mixing and dilution can occur only to the extent that there is horizontal admixture of clean air.

The effect of stability is rendered strikingly visible by the behavior of smoke issuing from a stack. Figure 14.2 shows schematically the various types of smoke plumes associated with various types of variation of temperature with height. In diagram A the smoke is emitted into an inversion layer. The stability prevents diffusion up and down, so that the only spreading of the smoke is sideways. Since the plume is thin in the vertical and is V-shaped in the horizontal, it has been likened to a lady's fan; hence the name "fanning." In diagram B there is an adiabatic lapse rate in the lower layer, topped by an inversion. Downward mixing goes on readily, but the inversion limits the upward mixing. This configuration arises when there is an inversion at the ground at sunrise and the heating from solar radiation produces an adiabatic or slightly superadiabatic lapse rate through a progressively thicker layer until the level at which the smoke plume has been confined until then is reached. After this level is reached, the pollution mixes downward rapidly, "fumigating" the ground, which has until this time been protected from the pollution by the inversion. This process may result in an abrupt increase of the pollution concentration at the ground to a high value.

When there is a superadiabatic lapse rate through a deep layer (diagram C), the smoke is carried up and down by convective currents, forming a *looping* pattern, and is rapidly diluted by the intense vertical mixing. With a deep adiabatic layer (diagram D), vertical mixing also goes on freely, but the turbulent motions that are induced by irregularities of the ground and shearing of the wind are not amplified by instability. The vertical spreading and lateral spreading are about equal, and the smoke plume resembles a cone. In the remaining possible configuration, diagram E, the smoke is emitted at the top of an inversion layer, where it is kept from mixing downward but spreads upward. This tendency to be carried aloft but not to the ground has been termed *lofting*.

The vertical diffusion of emissions from line and area sources is influenced in the same way by stability. With the increased horizontal dimensions of these

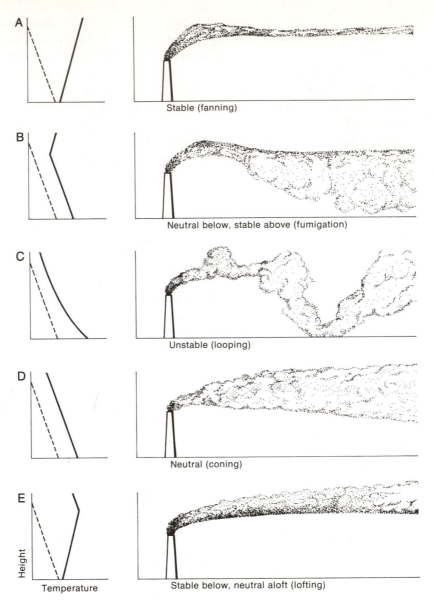

Figure 14.2 Smoke-plume patterns characteristic of various types of variation of temperature with height. The dashed curves in the diagrams on the left represent the adiabatic lapse rate. [After D. H. Slade, ed., *Meteorology and Atomic Energy 1968* (Oak Ridge, Tennessee: U.S. Atomic Energy Commission, 1968).]

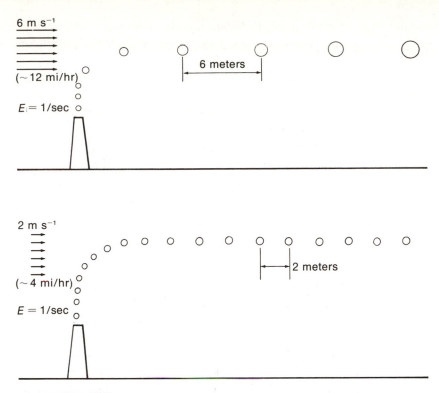

Figure 14.3   Effect of wind speed on pollution concentration. The size of the "bubbles" is intended to show the amount of clean air with which the pollutants in them is mixed. The stronger wind in the upper diagram leads to lower concentrations both because the bubbles of pollution are spread farther apart and because they are diluted more by mixing.

sources, the dilution due to horizontal mixing is reduced. Over an area source, for instance, lateral stirring mixes the pollution from one part of the source with that from other parts of it, rather than with clean, unpolluted air. The presence of an inversion over a large area source leads to very high concentrations.

The way in which the wind speed affects the concentration of pollutants is illustrated in Figure 14.3. The pollution is assumed to be entering the atmosphere at the rate of one unit per second, each unit being represented for convenience by an individual bubble. In the upper diagram the wind is 6 m s^{-1}, and the bubbles are therefore spread six meters apart by the wind, whereas the wind in the lower diagram is one-third as strong and the bubbles are one-third as far apart. Due to the direct effect of wind speed, the concentration is three times as great with the 2 m s^{-1} wind as with the 6 m s^{-1} wind.

A second effect of wind speed, turbulent mixing, is represented by the size of the bubbles as they move downstream. The stronger wind is more turbulent, so that the polluted air mixes more rapidly with the air around it and becomes more dilute; with the light wind there is less turbulence and the concentration remains high. Turbulence manifests itself in fluctuations of wind direction as well as speed. The variations in speed produce mixing along the average wind direction, and the variations in direction lead to side-to-side mixing and to up-and-down mixing. As has been discussed, the amount of vertical mixing is largely controlled by the hydrostatic stability. However, for the same stability, vertical mixing varies with the wind. In light winds the mixing in all directions is much less than in strong winds.

From the foregoing discussion, we see that for strong winds and large lapse rates the diffusion of pollutants is rapid. Under these conditions high concentrations of pollution will not occur except very near intense sources. When the wind is very light and an inversion is present, however, the pollution diffuses slowly and high concentrations develop if there are sources of sufficient magnitude.

Light winds and inversions occur together occasionally everywhere in the world, but there are some places where they are much more frequent than others. The frequency with which they occur, and in particular the length of time they persist, are measures of the *air-pollution potential* of a place. The evaluation of air pollution potential and its distribution in the United States will be discussed in the next section.

In addition to the effects of mixing with clean air due to turbulence and diffusion, the concentration of pollutants is affected by chemical reactions, which change their composition, and by processes that remove them from the air. The photochemical reactions that take place in Los Angeles-type smog have already been mentioned. The nitric oxide (NO) emitted by vehicles and by industry combines with the oxygen in the air to form nitrogen dioxide (NO_2), and this reacts with sunlight in the presence of hydrocarbons that are also emitted by vehicles and industry to form ozone, aldehydes, and other compounds. Pollutants that are in the form they are emitted are called *primary pollutants*. The forms they change to as a result of chemical reactions are called *secondary pollutants*. As a result of the reactions the concentration of the primary pollutants is reduced and that of the secondary pollutants is increased. Other chemical reactions take place more slowly, so that the change in character of the pollutants occurs at longer distances from the sources. Thus, sulfur dioxide (SO_2) is gradually oxidized further to sulfur trioxide (SO_3), which has a strong affinity for water molecules, combining with them to form sulfuric acid drops (H_2SO_4). The sulfuric acid droplets may combine with ammonia, which is frequently present in the air due to decay of vegetation, to form ammonium sulfate particles.

The removal processes are of two types, *wet removal*, in which the pollutants are collected in cloud or precipitation particles, and *dry removal*, in which particulate pollution settles out due to gravity and gaseous pollutants are removed by being adsorbed on or reacting with the substances in the surface of particles or on the ground. Because the pollution particles are very small, so that turbulence offsets the tendency for them to settle out, and because relatively little of the gaseous pollution comes in contact with vegetation and other reactive surfaces at the ground, dry removal proceeds very slowly. On the other hand, when air is cloudy or favorable to the formation of clouds, wet removal can clean the air rapidly.

As discussed in Chapter 5, particles serve as nuclei for formation of cloud drops or crystals. In addition, when pollution particles are in the air between cloud drops, air molecules colliding with them tend to propel them toward the drops. The result is that particulate pollution in cloudy air is taken up by the drops, and when the drops grow and precipitate the pollution is carried out of the air and deposited on the ground. Gaseous pollutants also are removed in this way. They may be adsorbed on the particles before the particles are collected by the cloud drops, or they may be adsorbed on or dissolved in the drops. Either way clouds are usually quite effective in removing gaseous as well as particulate pollution. This process of wet removal, in which the pollution is collected in clouds which then precipitate, is called *rainout*. The process in which raindrops or snowflakes collect the pollutants as they fall through polluted air is called *washout*. Because of the low collision efficiency of precipitation particles for the very small pollution particles (see Section 5.5), washout is relatively ineffective, and rainout is the principal method by which polluted air is cleansed.

As the wind carries polluted air long distances from a source, the concentration becomes lower and lower, but unless the air passes through an area of cloud and precipitation it does not return to its original cleanliness. If it encounters a new source, for instance a large city or industrial center, before being cleansed, the new pollution will produce higher concentrations than it would if the air were initially "clean." In places where sources are close together, such as the cities of the eastern United States and Europe, the cumulative effect results in much higher concentrations than would occur if they were widely separated.

The precipitation that falls when highly polluted air has been cleansed by clouds frequently contains large amounts of sulfuric and/or nitric acid. This *acid rain* has deleterious effects on fish in lakes where it accumulates, on vegetation, on soils, and on structures and monuments. Many lakes in Scandinavia and in the northeastern United States and eastern Canada cannot support fish life because they have become so acidic. There are indications that agricultural yields and forest growth in some areas have been reduced because of the effects of acid rain on foliage and soils. Damage to the Parthenon in Athens and buildings in Venice has been attributed to acid rain, as well as to the direct effects of pollution in the air.

Thus, not only are high concentrations of pollution of concern when they are in the air; their serious effects continue as they are being removed.

14.3   Evaluation of air-pollution potential

Although the frequency with which the combination of light winds and inversions occurs is different for various parts of the world, there is no place where it does not occur. Even the windiest regions on earth would occasionally have high concentrations of pollution if sources were present. The climate of a place determines how much of the time there will be severe smog if sources are present. The intensity of the sources determines whether it will ever be severe. Unless the pollution sources are sufficient, even the most adverse meteorological conditions will not lead to adverse concentrations. The severity of smog depends on both the sources and the meteorological conditions.

Persistent light winds and inversions occur together principally during periods when a stagnant or slow moving anticyclone is present. In temperate and high latitudes the fall and winter are the seasons when such periods are likely. Winds are light near the center of an anticyclone, and the horizontal outflow from the high-pressure center produces sinking motion and adiabatic heating aloft. At the same time, the temperature of the ground and the air immediately above it falls because of radiation into clear skies. This cooling below and heating above produces an inversion so strong that it is not destroyed by the heating of solar radiation during the short days of autumn and winter.

Similarly, the semipermanent subtropical high-pressure areas (see Sections 9.1, 12.1, and 12.2) have light winds and inversions associated with them. The subsidence is concentrated on the eastern sides of the anticyclones so that the continental areas affected by the inversion produced by it are the west coasts, at latitudes ranging from 40° or more in both hemispheres almost to the equator. In winter, invasions of storms, fronts, and cold, unstable air masses interrupt the anticyclonic influence, and periods of high smog potential are less frequent and shorter in duration. But in summer the trade-wind inversion and light winds are present continuously at subtropical west coasts. From the standpoint of pollution potential such places as Casablanca, Morocco, Capetown, South Africa, Santiago, Chile, and Los Angeles, California, are among the worst possible locations for cities.

The pollution potential of a city can be estimated in terms of its size, the wind speed, and a quantity called the mixing height, representing the height through which the pollutants will be completely mixed in crossing the city. If the pollution is thoroughly mixed up to H, the concentration C of pollution within a city at a distance L downwind from its outskirts, according to a simple "box" model, is given by

$$C = QL/UH \qquad (14.1)$$

where Q is the emission rate per unit area, assumed to be uniform, and U is the average wind speed. The formula agrees with our expectations that the concentration is higher the more intense and more extensive the sources over which the air has passed, the slower the movement of the air across the sources, and the smaller the depth of air into which the pollution mixes. In this simple model it is assumed that lateral mixing does not produce a decrease in concentration and that there is no removal of pollutants. Although the model is a gross oversimplification, it focuses attention on the factors that would enter a more accurate expression for computing the concentration of pollution, namely, the properties of the pollution sources, Q and L, and the meteorological factors, U and H.

In the daytime the mixing height is determined by the depth of the layer through which the sun's radiation has established an adiabatic lapse rate. On clear nights the radiational cooling might be expected to establish an inversion and reduce H almost to zero. However, it has been found that over cities the heat retained by the buildings or added by human activities maintains a lapse of temperature through the lowest 100–200 m. Thus, even at night there is some dilution of pollutants over cities because of vertical mixing. The values of H range from 100 or 200 m at night and on some winter days to 3 km or more on summer afternoons.

The wind in most cities undergoes a similar diurnal and seasonal variation. Usually, it is lightest at night and in the early morning and has a maximum in the afternoon, when vertical mixing transfers momentum downward from higher levels where the wind is stronger because it is not impeded by surface friction. In regions such as sea coasts and valleys, the diurnal wind regime will be dominated by topographic influences, especially when the general flow is light. Usually, these influences have the same pattern—light winds at night and maximum winds in the afternoon.

Most primary pollutants have their diurnal maximum concentration an hour or two after sunrise, when human activities have increased Q to a high value but U and H are still small. As indications of the daily pollution potential, the annual median values of the concentration ratio C/Q for the United States (except Alaska and Hawaii) are given in Figure 14.4 for two values of L. The values in this figure were obtained by using a somewhat more sophisticated model than that represented by equation (14.1); for this reason the values for the 100-km city are not exactly ten times those for the 10-km city.

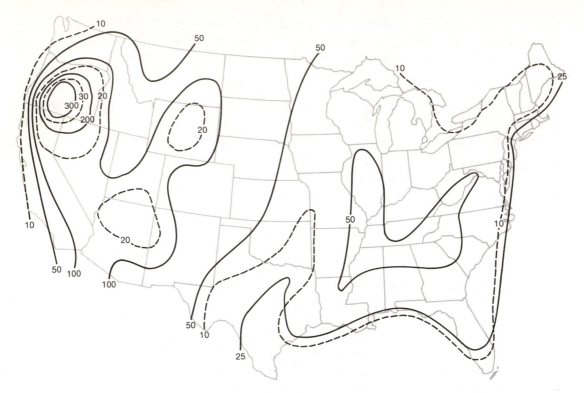

Figure 14.4 Values of median annual concentration ratio C/Q in the morning for cities of 10-km *(dashed lines)* and 100-km *(continuous lines)* length in direction of wind. [After G. C. Holzworth, paper presented at the Second International Conference on Air Pollution, Washington, D.C., December, 1970.]

Of more interest than the average values is the frequency of extremely high values of concentration. Figure 14.5 shows the values of C/Q that would have been exceeded on 10 percent of the mornings each year. Here we see that, whereas a 100-km city would experience values of C/Q greater than 100 one-half of the time only over the western quarter of the country, it would have values in excess of 400 during 10 percent of the time in much of the East as well as the West.

It appears from these maps that the area of the worst pollution potential is southern Oregon, where the smog in a 10-km city would be almost as bad one-half of the time as that which Los Angeles experiences on the worst 10 percent of the days. In turn, Los Angeles has more than twice the pollution potential of New York.

It should be noted that these charts give morning values. Only if the potential remained high long enough for the air to move across the city would the actual pollution concentration reach these values; and in any case if in the afternoon the

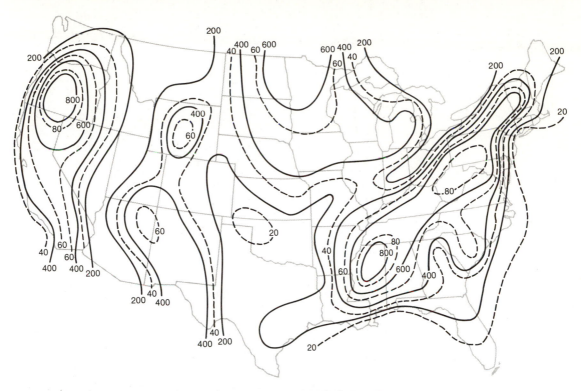

Figure 14.5 Values of concentration ratio C/Q exceeded on 10 percent of the mornings annually, for cities of 10-km *(dashed lines)* and 100-km *(continuous lines)* length in direction of wind. [After Holzworth (1970).]

wind is strong enough and the layer of turbulent mixing is deep enough the concentrations will rapidly decrease. The effects on health are usually dependent on the dosage, that is, on the concentration multiplied by the duration that the concentration has the particular value. If the concentration has a very high value but is sustained only for a short time, its effects will not be as bad as those of concentrations having a lower peak value that is sustained for a long period. Figure 14.6 presents an estimate of the frequency of sustained high concentrations. It gives the number of "episode days" in a five-year period, where an episode is defined as a period of at least two days in which the mixing height and wind speed were sufficiently low to lead to high pollution-concentration ratios. The area of the highest number of episode days is in the West, from the Rocky Mountains westward. San Diego has the largest number. Los Angeles, which presently has much more pollution because of its larger size (greater Q and L), has less than one-half as many, primarily because its afternoon winds tend to be a little stronger. It is clear that, in many parts of the western United States, if cities grow to the size of Los Angeles, they also will be subject to severe smog problems.

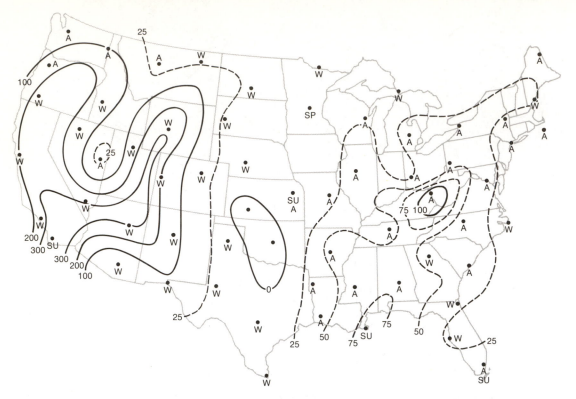

Figure 14.6 Number of days in five years for which conditions favored reduced dispersion of pollution continuously for at least 48 hours. Seasons in which most such days occurred at each station is shown by W (winter), SU (summer), SP (spring), or A (autumn). [After Holzworth (1970).]

14.4   Effects of air pollution

At the beginning of this chapter some of the adverse effects of air pollution were mentioned. It would seem almost unnecessary to go into details about these effects in order to justify the imposition of measures to reduce it. However, the reduction of pollution can be achieved only at considerable expense, and the justification of these expenditures requires an analysis of the benefits to be achieved.

The effects of greatest concern are those on human health. There have been several instances of large numbers of deaths attributed to air pollution. The largest number occurred in the London disaster of December 5–9, 1952. During this period an anticyclone stagnated over England, producing a strong inversion and nearly calm conditions. The temperature was lower than normal, leading to augmented emissions from the millions of domestic fires and industrial chimneys. A persistent dense smog in the true sense of the word developed, with the coal smoke providing hygroscopic nuclei on which the condensation took place. Measured concentrations of particulates and sulfur dioxide were many times the values usually found in the polluted London air. Before the smog was cleared out by the approach of a frontal system, a great many people became ill, and there were over 4000 more deaths than the number normally taking place during a five-day period at that time of year.

The public reaction to this terrible smog "incident" led to the passage by Parliament of the British Clean Air Act of 1956, which required the use of "smokeless" fuels and improved combustion methods. Similar procedures had previously led to a reduction of smokiness of some American cities, notably, Pittsburgh and St. Louis; and since the adoption of this law, the visibility in London has improved remarkably. No longer does London experience the "pea soup" fogs that used to typify it.

In terms of mortality, the first important occurrence of smog in the United States was the Donora disaster of October 26–29, 1948. Donora, Pennsylvania, is a small industrial town in the valley of the Monongahela River, about 20 miles southeast of Pittsburgh. A persistent inversion developed on October 26 because of the stagnation of an anticyclone. Into the calm, foggy air the zinc plant, steel mill, and other industrial establishments poured their effluents for three days, producing increasing concentrations of sulfur dioxide and other contaminants. By the third day almost one-half of the 14,000 people in the town became ill and twenty died.

There had been many occurrences of high pollution concentrations before the disastrous ones in London and Donora, and there have been many since. In some of them it has been established that deaths due to the smog occurred, but in most cases it is uncertain whether any increase in mortality could be ascribed to the pollution. Nevertheless, the few instances in which pollution has been shown definitely to have caused death, together with evidence that some of the contaminants in the atmosphere are toxic, clearly show that air pollution is a hazard to human health.

Many substances commonly found in polluted air are known to be responsible for illness when they are present in high enough concentrations. Usually, the concentrations at which they have harmful effects in laboratory tests are much higher than those observed in the atmosphere. Sulfur dioxide, carbon monoxide, hydrocarbons, nitrogen dioxide, lead, sulfate particles, and ozone are among the pollutants that are known to have adverse effects on humans. Epidemiological studies—that is, studies in which populations that are subject to high concentrations of polluted air are compared with similar populations that live in relatively pollution-free areas—have shown greater incidences of various respiratory illness attributable to high concentrations of sulfur dioxide, sulfate particles, oxides of nitrogen, and photochemical oxidants. Carbon monoxide, which combines with hemoglobin more readily than oxygen and thus reduces the capacity of the blood to carry oxygen, increases the danger of death due to heart disease, reduces the ability to do physical exercise, and affects mental processes, including alertness and visual acuity. The poisonous nature of lead, which has been introduced into urban atmospheres principally from vehicles using leaded gasoline, but is also emitted by industries, particularly smelters, is well known. The increasing use of nonleaded gasoline is reducing this hazard for urban populations, but it remains a serious danger for persons living near or working in lead smelters.

The effects of pollutants on vegetation have long been recognized. Sulfur dioxide has received the most attention, but in recent years the products of photochemical smog in very low concentrations have been shown to cause injury to plants. Photochemical smog damage has been found in the vicinity of most of the major metropolitan areas of the United States. In the San Bernardino Mountains of California, the pine and fir trees of the national forest are dying because of smog from the Los Angeles area. The damage to vegetation includes the burning or marring of leaves, the stunting of growth, a decrease in the size and yield of fruits, and the destruction of flowers.

The total annual cost of damage to vegetation in the United States by air pollution, including the losses to farms, forests, and domestic gardens, has been estimated to be more than $2 billion.

The effects of pollution on materials and structures also have been known for a long time. Corrosion of metals, soiling and erosion of building surfaces, discoloration of paints, soiling and weakening of textiles, deterioration of sculpture, paintings, and other works of art, and weakening of rubber goods, including automobile tires and hoses—all are caused or accelerated by air pollution. The total cost of cleaning, protecting, and replacing materials damaged by smog is not known, but estimates indicate that it amounts to several billion dollars a year in the United States.

One of the most conspicuous manifestations of air pollution is the reduction of visibility. This reduction is due to the scattering of light by particles. It is an effect of particulate pollution only and not of gaseous contaminants. Thus, the striking improvement in visibility that has taken place in Pittsburgh, St. Louis, and London since the adoption of measures to reduce the amount of coal smoke must not be interpreted as a complete solution to their air-pollution problems. The gaseous contaminants were not necessarily reduced thereby, and they may cause more serious effects than the particulates.

The reduction of visibility by pollution may be sufficient at times to interfere with transportation. Airplanes may have to operate on instrument flight rules instead of visual flight rules, with consequent costly slowdown of landing operations.

Pollution also reduces the amount of solar radiation reaching the ground. This reduction, particularly in the ultraviolet, may contribute to the adverse effects of pollution on human health.

Other effects of pollution on the weather will be discussed in the next chapter, in which we shall consider both the intentional and the unintentional influences of humans on the atmosphere.

14.5   Control of air pollution

The obvious way to control air pollution is to keep the contaminants from entering the atmosphere. There is no way to do this completely. All human activities produce wastes, some of which automatically enter the air. The production and use of energy, the refining of minerals, the various modes of transportation, and such commonplace actions as house-painting and dry cleaning—all use the air to carry away waste products. Since complete elimination of these activities is unreasonable, we must seek ways to carry them out that lead to acceptably low concentrations of pollutants.

Because the objective of commerce and industry is to yield a profit for owners or stockholders, no individual company will purchase expensive equipment to reduce its emissions of pollutants unless its competitors are required to do likewise. Thus, control of pollution cannot be expected on a voluntary basis. It must be imposed by law, so that all businesses are required to meet the same standards. Furthermore, since pollution is carried by the wind, which does not respect municipal boundaries, and since businesses in one city or state compete with those in others, the laws controlling air pollution must be at a sufficiently high level of government. Measures passed and enforced by state and national governments have a better chance of producing the desired air quality than those imposed by municipal or county agencies. On the other hand, because situations may be more acute in some cities than in others or in rural areas, there must be provision for stricter standards in some localities than apply to the general region.

There are various different legal approaches to the control of air pollution. Laws may specify the allowable concentrations of pollutants in the air in the vicinity of industrial plants, leaving it to the industries to choose how to achieve this goal, or they may specify the allowable rates of emission of the pollutants from the sources. Often, instead of specifying the limits in the legislation, enforcement agencies are empowered to do so after determining on the basis of expert investigations and hearings what the limits should be and whether their attainment is feasible. Sometimes the legislation simply provides that industries and other sources should employ the best available techniques to reduce emissions. An alternative procedure that has been proposed but seldom utilized is to assess taxes based on the amount of emissions, on the premise that this will provide an economic incentive to reduce emissions. A procedure that has been very successful in reducing emissions by industry in the Los Angeles area is the permit system, in which the installation and operation of equipment that produces pollution is allowed only after it has been demonstrated that the emissions from the equipment will meet the standards that have been set. Initially it was the policy of governmental agencies not to establish requirements that cannot be met with existing technology at reasonable expense. In some instances this policy was abandoned in favor of one which bases rules on the emission levels needed to attain acceptable air quality, independent of whether means are available to attain these levels, in order to stimulate development of more effective ways to reduce emissions. Recently, because of problems related to the energy shortage, there has been pressure to relax regulations limiting effluents emitted during production and use of fuels.

In the early years of pollution control, measures were left almost exclusively to municipalities and were concerned principally with establishing tolerable levels of pollution in the vicinities of large industrial plants. These levels were usually expressed in terms of the maximum allowable concentration at the ground of such contaminants as sulfur dioxide. To meet these requirements, factories and electric generating plants built higher and higher smokestacks as the amount of their emissions increased. Another criterion that was used was the opacity of the effluent as it left the stack. By limiting the darkness of the smoke, the amount of particulate matter entering the atmosphere was controlled to some extent.

As the problem became more general, municipalities and counties joined to form air pollution control districts, and the control measures have attempted to limit the emissions from all sources. With the spread of pollution from one area to another, and with the recognition that such ubiquitous sources as automobiles contribute a large proportion of community air pollution, state governments and the federal government have launched air pollution control programs. At present virtually every one of the fifty states in the United States has pollution control agencies. The federal government, which began with a feeble program emphasizing the exchange of information under the Clean Air Act of 1955, has undertaken vigorous control and enforcement activity under the Air Quality Act of 1967 and the Clean Air Amendments of 1970 and 1977. Under these laws the Environmental Protection Agency (EPA) is empowered to establish air quality control regions, set ambient air quality standards, require local and state authorities to devise plans to bring the air quality in their regions up to the standards, and, in those instances in which the local and state programs are inadequate, to inaugurate measures of its own to enable the achievement of the standards.

Table 14.2   Trend in Air Pollutant Emissions in the United States, 1970–1977 (in millions of metric tons)

Year	Total suspended particulates	Oxides of sulfur	Oxides of nitrogen	Volatile organic compounds	Carbon monoxide
1970	22.2	29.8	19.6	29.5	102.2
1971	20.9	28.3	20.2	29.1	102.5
1972	19.6	29.6	21.6	29.6	103.8
1973	19.2	30.2	22.3	29.7	103.5
1974	17.0	28.4	21.7	28.6	99.7
1975	13.7	26.1	21.0	26.9	96.9
1976	13.2	27.2	22.8	28.7	102.9
1977	12.4	27.4	23.1	28.3	102.7

Source: Same as Table 14.1.

For motor vehicles the 1970 Clean Air Amendments required that emissions of carbon monoxide, hydrocarbons, and oxides of nitrogen from 1975 model year automobiles be reduced to less than 10 percent of those permitted for 1970 cars. The automobile industry succeeded in having this requirement delayed, and the 1977 Amendments provide that, beginning with the 1983 model year, emissions of carbon monoxide and hydrocarbons be reduced by at least 90 percent, and beginning with the 1985 model year, emissions of oxides of nitrogen be reduced at least 75 percent from those emitted by vehicles before federal standards were established. If and when these requirements are met, the amounts of emissions listed for transportation in Table 14.1 will undergo a rapid decrease as new automobiles replace old ones, even with a continued increase in the total number of cars.

Table 14.2 shows the estimated emissions of various pollutants for the period 1970–1977. The controls that have been required of industry have resulted in a considerable decrease in particulate pollution and a small decrease in oxides of sulfur. Control of automobiles resulted in a small decrease in emissions of carbon monoxide and volatile organics (hydrocarbons) in 1974 and 1975, but the increase in number of vehicles offset this decrease in subsequent years.

The continuing rise in population constitutes a threat that pollution will continue to increase in spite of the control measures. Not only might the growing demands for materials and energy offset the reduction in per-capita contribution to air pollution because of control, but the need to use less-rich ores and less-pure fuels may make it impossible to meet the demands and the emission limits at the same time. The energy shortage that has been increasingly recognized, and is emphasized by the many-fold increase in petroleum prices set by OPEC (Organization of Petroleum Exporting Countries), has led to plans for increased use of coal, both directly as a fuel for generation of electricity and other industrial applications, and for conversion to liquid and gaseous fuels for use in vehicles and homes. Control of pollutant emissions from coal is costly, and increasing efforts to relax the requirements for scrubbers and other devices that reduce emissions of sulfur dioxide and particulates will doubtless be exerted.

Nuclear power is almost completely free of atmospheric pollution except for the possibility of accidental release of radioactive material. While the reactor accident at Three Mile Island, near Harrisburg, Pennsylvania, in March 1979, did not lead to release of radioactivity at levels regarded as harmful, its occurrence made the possibility seem more real, even though very thorough measures to guard against accidental release are incorporated in the design of nuclear reactors, and the chances of its happening are considered to be extremely small, much smaller than the chances of plane crashes or railroad accidents. In routine operation, nuclear reactors emit into the air a negligible amount of radioactivity. The amount is so small that someone at the boundary of a reactor site would experience an increase of less than 1 percent of the normal background radioactivity. There are other considerations that need to be evaluated in connection with the decision of whether or not to replace fossil fuel by nuclear reactors as an energy source, such as the problem of disposing of the highly radioactive spent fuel. From the standpoint of air pollution, the shift would be desirable.

Solar energy, in addition to being a renewable source, is essentially nonpolluting. To the extent that its use replaces the combustion of fossil fuel, it will contribute to the elimination of air pollution.

The reduction of emissions and the restoring of tolerable air quality appear technically feasible if we are willing to undertake the costs of doing so. These costs may in the long run include, in addition to higher prices for commodities to cover the expense of their production in a pollution-free manner, the acceptance of a way of life that does not depend as much on energy-consuming appliances and that provides for a stable population level.

It is frequently suggested that it may be easier or less costly to change the meteorological conditions that limit the capacity of the air for pollutants. Various proposals have been made, including large fans to augment the air motion when the wind is too light, "punching holes" in inversions when they are present, using the waste heat of electric generating plants to augment convective mixing, painting the roofs of buildings in alternate city blocks black and white so that differences in solar heating produce convection, and cutting down the Rocky Mountains to permit more rapid flow of air across the United States. As a rule these proposals, except for those depending on heat from the sun or the waste heat from power plants, would require such tremendous amounts of energy that its production, in addition to being very costly, would introduce more pollution into the air than the procedures could remove. The proposals for the use of solar energy or waste energy usually do not take sufficiently into account the magnitude of the problem. Computations have shown that the modification of the temperature structure or air flow to remove pollution by any of the methods so far proposed is impractical. Until a practical procedure for doing so is proposed, the only way to control pollution remains the reduction of emissions at the source.

Questions, Problems, and Projects

1. Make a "census" of the principal sources of air pollution in your community, listing each major type of source and each kind of contaminant that comes from it. What steps could be taken to reduce the emissions from each source?
2. The number of people who become ill or die in a city each day depends on several factors, in addition to the amount of air pollution present. How would you separate out the effects of the other factors in order to tell whether air pollution is responsible for any increase in illness and death?
3. Compare the factors that influence lateral (sideways) mixing with those controlling vertical mixing, and discuss the effectiveness of the two types of mixing in reducing the concentration of pollutants from point sources, line sources, and area sources.
4. Explain why Los Angeles, California, was the first place to have high concentrations of photochemical smog. Why is it less frequent at cities in higher latitudes?
5. Discuss the meteorological factors that should be considered in establishing an air quality control region.
6. How would you interpret the proposal to "punch holes" in an inversion? What would be required in order to eliminate an inversion over an area?

15 Human Influence on the Atmosphere. Inadvertent and Intentional Modification of Weather and Climate

15.1 Effects of human activities

The activities of humans alter atmospheric conditions in three ways: (1) by changing the character of the earth's surface; (2) by adding energy to it from artificial sources; and (3) by adding matter to it. The modification of the surface affects the way in which solar radiation is absorbed and retransmitted to the atmosphere and changes the frictional resistance to the wind. It also influences the absorption or runoff of rain and melting snow, and the evaporation or transpiration of water into the air. The replacing of forests and grasslands by cultivated vegetation or structures, the draining of swamps, the damming of rivers to establish artificial lakes and reservoirs, and the irrigation of arid areas affect evaporation of water and conduction of heat, and thereby change the humidity and the temperature. The combustion of fossil fuels (coal, petroleum derivatives, and natural gas) heats the air and adds particulate and gaseous contaminants. Nuclear reactors likewise add heat and contaminants to the air. Many other agricultural, industrial, commercial, and domestic activities introduce matter (pollutants) into the atmosphere.

In early times when humans subsisted by hunting other animals and collecting seeds and berries, the effects were minimal, being confined to the immediate vicinity of the fires maintained for warmth and cooking. With the beginning of agriculture 8000–10,000 years ago, the effects started to become more extensive both because of the substitution of cultivated crops for natural fields and forests and because of the rapid increase in human population, which it made possible. The effects of agriculture presumably were felt principally in the immediate vicinity of the modified terrain, but since forests were chopped down over tremendous areas the changes that occurred in the exchange of heat, moisture, and momentum between the ground and the atmosphere were on a very large scale. As these transformations took place mostly before the beginning of systematic observations, the amount of change in the resulting weather is not known.

Following the spread of agriculture, trade and commerce led to the development of larger and larger towns and cities, which had their own characteristic effects on the weather. The industrial revolution accelerated the concentration of people in cities, and the increased use of fossil fuels aggravated the influence of cities on the atmosphere. Since most of the growth of cities has taken place since the beginning of weather records, and since the effect of cities on the weather can be estimated by comparison with surrounding rural areas, the extent of their effect is fairly well known.

Already in 1818 Luke Howard wrote a book on the climate of London, in which he described the differences between the world's first large city and the surrounding countryside. In recent years there have been many studies of the weather and climate of cities. One of the most intensive of these was METROMEX (Metropolitan Meteorological Experiment), an investigation that concentrated on the influence of St. Louis, Missouri on the properties of the atmosphere in its vicinity, and particularly the effect of the city on precipitation.

15.2   The weather in cities

Cities influence practically every aspect of the weather. The pollution concentration, both particulate and gaseous, is many times that in rural areas. Consequently, the visibility and the intensity of solar radiation, particularly in the ultraviolet, are reduced. Other elements, including temperature, humidity, wind speed, and occurrence and distribution of cloudiness and precipitation, are all affected.

The most definite influence of cities is on the temperature. Cities are on the average at least 1°C–2°C warmer than surrounding rural areas both day and night and in all seasons of the year. Several factors contribute to this phenomenon, with different ones dominating at different times. The main influence during the day in the growing season is the relative absence in cities of evaporation and transpiration of water from soil and plants. These processes use a large part of the insolation in rural areas, and in their absence the greater amount of solar energy raises the city's temperature. At night the city is kept warmer than its surroundings by the re-emission of heat that was absorbed by streets and buildings because of the greater conductivity and heat capacity of concrete, asphalt, and brick than of soil and vegetation. The energy used for industrial processes, heating, cooling, and illumination of buildings, other industrial and domestic purposes, and transportation ultimately is added to the outside air and raises its temperature. The energy of human metabolism also contributes a little. The dark and irregular surfaces of buildings, with vertical walls that permit multiple reflections of incoming sunlight, reduce the amount of solar radiation reflected back toward space, thereby increasing the energy available to heat the air over cities.

The area affected by the warmth of a city is referred to as an *urban heat island* because it is surrounded by rural areas with lower temperatures. The urban heating effect is greater under clear skies than under cloudy conditions and usually disappears in strong winds and storms. It varies through the day, being greatest in the late evening. Because the air moving from rural areas to cities on clear nights is heated, the surface inversions found at night in the surroundings usually do not occur over the urban heat island. Instead, there is a normal lapse rate through the lowest 100 meters or so, above which the usual nocturnal inversion occurs. Thus, the urban heat island increases the thickness of the layer through which pollutants are mixed on clear nights.

The intensity of the urban heat island also varies with season. For instance, measurements at Akron, Ohio gave values of 5 K on a clear evening in spring, 3 K in summer, 8 K in autumn, and 4 K in winter.

The intensity of the urban heat island increases with the size of the city and the density of its population. Measurements during the development of a new town, Columbia, Maryland, in a previously completely rural area illustrate the early stages of this effect. When the town had 2000 inhabitants the largest temperature differences between it and the surroundings on clear evenings ranged between 1 K and 3 K. Two years later, when its population was 16,000, the values had increased to 2 K–4.5 K. For large metropolises, with populations in the millions, the difference can exceed 10 K at times.

The additional warmth of the urban heat island has both good and bad consequences. In winter, in addition to reducing somewhat the fuel requirements for heating living spaces, it reduces the frequency, depth, and duration of snow cover, thereby lowering the economic costs of traffic slowdowns and snow removal. During heat waves in summer, on the other hand, it increases energy consumption for air conditioning, adds greatly to the discomfort, and sometimes is responsible for heat prostrations and deaths that would not occur at the lower temperatures of the surrounding countryside.

Both because of the higher temperatures and because of the reduced evaporation, the relative humidity in cities is considerably lower—about 5 percent on the average—than the surrounding countryside. In spite of the lower average relative humidity, fogs are more frequent in cities due to the additional hygroscopic nuclei in the polluted air. This was especially true in places where soft coal was used for home heating or by industry. In London, for instance, the change from inefficient burning of soft coal in domestic fireplaces, in response to the Clean Air Act, resulted in a decrease by one-half in the occurrence of fogs. The "pea-soup" fogs that previously were characteristic of London no longer occur.

The winds in cities are generally lighter than in the country, because the buildings act as obstacles, increasing the frictional resistance of the ground. On the average, wind speeds are about 25 percent less in cities than in the surrounding countryside. On the other hand, when strong winds blow along streets with tall buildings on either side, there is a channeling effect, and in this circumstance the winds may be augmented locally. The irregularity of the surface causes increase in the turbulent fluctuations of wind speed and direction over cities.

When the general wind is light, the urban heat island creates a local wind circulation similar to the sea breeze: slight upward currents of air over the city and converging flow of surface winds from the surroundings to compensate.

The effect of cities on precipitation has been hard to determine, both because of the large natural variability of rainfall and because cities are usually located in topographic situations that lead to differences between them and their surroundings. Statistical studies were carried out in various parts of the world, and wherever a significant difference was found it showed greater precipitation in and downwind of the city than in its surroundings and/or upwind of it, with the average increase about 10 percent.

An interesting example that led to considerable controversy and was part of the stimulus for METROMEX was the so-called La Porte anomaly. La Porte is a town in Indiana about 60 km from the steel mills of Chicago and Gary. A study carried out in the early 1960s showed that the annual precipitation reported for La Porte was 40 percent greater than other stations in the vicinity, and that the precipitation there had increased beginning in 1928. From about 1933 the increase paralleled the increase in the steel production in the Chicago–Gary area, and it was suggested that the precipitation excess was caused by the industrial activity. The validity of the conclusion was challenged on the basis of questions regarding the accuracy of the observations, since a single volunteer observer had made the measurements throughout the period during which the increase took place, and after he was replaced the La Porte anomaly seemed to disappear. However, further studies using maps of precipitation amounts suggest that, rather than being eliminated, the La Porte anomaly shifted southwestward in the period subsequent to the original study, presumably in response to changes in the wind patterns associated with rainstorms and snowstorms.

The La Porte study indicated that the major effect of an urban-industrial complex on precipitation might be a considerable distance downwind rather than in the immediate vicinity. This was also suggested by a study of the records of the precipitation in eight cities, of which six had indications of urban influence on precipitation. The studies pointed up the inadequacy of the number of places at which precipitation was measured and the need for additional types of data if conclusions about the amount of the effect and its causes are to be reached.

METROMEX was designed to fill the need for determination of the effect of a city on the frequency, amount, intensity, and duration of precipitation and related severe weather, and to identify the physical processes responsible for these effects. An extensive system of surface and upper-air observations was established in and around St. Louis, including 225 rain gauges and pads for measuring hailstone size, 70 collectors of rainwater for chemical analysis, three radar sets, 14 pibal stations, four meteorological aircraft, a lidar, an accoustical sounder, equipment for releasing and sampling air tracers, radiosondes, and various standard meteorological instruments. This formidable array was operated for five years. It concentrated on the summer months, since previous studies suggested that urban effects on precipitation occur mainly in summer. It was found that 10–30 percent more precipitation fell in and downwind of St. Louis than occurred upwind of it, with the maximum excess about 30 km downwind of the city. The excess occurred principally in association with frontal or squall-line storms, which account for most of the precipitation. Air-mass storms, though accounting for one-third of the occurrences of rain in summer, result in a much smaller fraction of the rainfall amounts. The investigation of the physical factors responsible for the excess indicated that while the increase in number of active cloud condensation nuclei led to more numerous and more uniform-sized cloud drops, which might delay the development of precipitation, there was a sufficient number of giant CCN introduced to lead to rapid initiation of the warm rain process. The mixing height is increased to 2 km or more by the urban heat island in daytime, and this enabled the giant CCN to be carried to levels where cumulus clouds form and where the instability associated with frontal or squall-line convergence is released.

In addition to furnishing giant CCN that affect precipitation, pollution sources are responsible for a smaller amount of solar radiation reaching the ground in cities than in surrounding rural areas, a reduction in the visibility (visual range), and a greater frequency of fog. The insolation is reduced as much as 20 percent in the center and windward portions of cities on days with inversions and light winds. The effect of pollution in increasing the frequency and intensity of fog has already been mentioned.

As the size of metropolitan areas and the total amount of pollution entering the atmosphere from them have increased, the urban influences on the weather have become more extensive. Urbanization has a meteorological significance as well as a geographical and social meaning. In places where cities are close together, their effects may merge to produce regional rather than local changes in weather and climate.

15.3   Human influences on worldwide weather and climate

In addition to the local and regional influences in places where human activities and the pollution associated with them are highly concentrated, there are global effects due to human activities. These include the direct heating of the atmosphere by fuel burning and conversion of nuclear energy, and the temperature changes produced by increase of concentration of carbon dioxide, other gases, and particulate matter throughout the atmosphere.

The amount of energy released directly as heat by human activities, while adequate to raise the air temperature in the vicinity of cities and large industrial complexes, is not yet enough to produce a significant global effect. The present total world use of energy is about $10 \cdot 10^{12}$W, only $1/7800$ of the $7.8 \cdot 10^{16}$W of solar energy absorbed at the earth's surface. However, if the present rate of increase continues, which has been producing a doubling every 10 years, the world use will increase by a factor of 1000 in a century. This would make the average anthropogenic heating rate more than one-tenth the solar energy absorbed at the ground. Because of the limits on growth of population and availability of fuel, continued increase at the present rate is regarded as impossible, and it has been estimated that a leveling off will occur, with the rate of energy release in the year 2100 reaching a value between 50 and 100 times the present one. This amounts to between .05 and 1 percent of solar energy absorption. Experiments with general circulation models indicate that a 1-percent addition of heat over the earth would result in an average temperature increase of about 2 K. Thus, direct release of heat into the atmosphere by space heating and industrial processes might produce a temperature rise of 1–2 K by the end of the twenty-first century.

The possible climatic changes to be expected from an increase in the concentration of carbon dioxide have received a great deal of attention recently. The increase in atmospheric concentration of CO_2 during the past few decades is now well established. The evidence that shows this most clearly is the record of CO_2 concentrations measured at the observatories near the top of Mt. Mauna Loa (elevation 3400 m) on the island of Hawaii, and at the South Pole. Since these

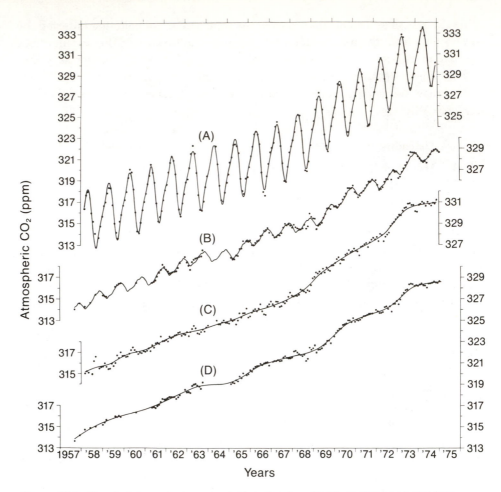

Figure 15.1   Trend of atmospheric concentration of carbon dioxide at stations remote from sources. (A) Monthly average values at Mauna Loa Observatory. (B) Monthly values at South Pole. (C) Twelve-month running mean values at Mauna Loa, to smooth out seasonal variation. (D) Twelve-month running mean values at South Pole. [Courtesy of C. D. Keeling and R. B. Bacastow.]

observatories are remote from local sources and sinks of CO_2, the concentrations measured there are representative of the well-mixed atmospheric background.

Figure 15.1 shows the variation in the concentration of CO_2 at these stations. At both of them there is an annual oscillation superposed on a continuous upward trend. The annual cycle is due to the storage of CO_2 by vegetation during the growing season and its release by their decay during the fall and winter. The annual average rose at Mauna Loa from about 315 ppm in 1958 to 331 ppm in 1975; at the South Pole the increase was from 314 ppm to 329 ppm. The lower values at the South Pole reflect the delay of a year or more for the increase to spread from the sources, which are mainly in the Northern Hemisphere. The increase of 16 ppm in 18 years gives an average increase of 0.9 ppm per year. The increase has continued. In 1979 the average was 334 ppm.

In earlier years measurements were made in places where they may have been influenced by local sources, and the methods of measurement were subject to error, but as nearly as can be determined the CO_2 content of the atmosphere has increased steadily from a value of about 280 ppm a century ago. The increase during the past century corresponds to an increase in amount (mass) of carbon in the atmosphere of $75 \cdot 10^9$ tons. The amount of carbon released by burning fossil fuels (coal, petroleum, and natural gas) during this period has been evaluated to be somewhat more than $150 \cdot 10^9$ tons. Thus, about 50 percent of the carbon put into the air has remained there. Some of the other half may have been taken up by the biosphere, but the most likely sink for the CO_2 not staying in the atmosphere is the oceans, in which it dissolves and gradually makes its way down into the cold water at great depths. There is some evidence that cutting down forests recently, particularly in tropical regions, and burning the resulting wood and paper have contributed to the increase of CO_2 in the air. On the other hand, the extensive reforestation of large areas in China may be tending to offset the increase. The ability of the biosphere to absorb excess CO_2 is limited, and the absorption by the ocean decreases with increased concentration. The indications are that as long as fossil fuel continues to be used, this source of energy will produce a continued increase in the atmospheric concentration of CO_2.

The total fossil fuel resources of the earth are estimated to contain more than $5000 \cdot 10^9$ tons of carbon. About 10–15 percent of this is in the form of oil and gas. The remainder is mostly coal. If this entire amount is burned and the rate of retention in the atmosphere remains the same, the atmospheric concentration of CO_2 will become approximately five times its present value. Estimates of the future rate of fuel consumption vary. The complete exhaustion of the fossil fuel supply would take several centuries, but it is anticipated that the rate of use will be such that the CO_2 concentration will reach twice its present value within the next 100 years. However long it takes for doubling to occur, it is well established that such a large increase will have a major effect on the weather and climate of the world. The reason for this is that the greenhouse effect discussed in Chapter 3 will be larger the greater the amount of CO_2 in the air.

As explained in Chapter 3, the CO_2 and water vapor in the air absorb the outgoing long-wave energy radiated by the ground and radiate some of it back, so that the loss of heat by the ground is less, and the temperature of the ground and the air in immediate contact with it is higher than it would be in the absence of CO_2 and water vapor. The more CO_2 and/or water vapor, the greater the greenhouse effect.

Computations have shown that the effect of an increase of CO_2 concentration to double the present value, if no other consequences took place, would be to increase the average downward long-wave radiation over the earth by 4 W/m^2, somewhat more than 1 percent of the average solar radiation reaching the top of the atmosphere and about 2.6 percent of the average solar radiation reaching the ground. This increase in radiation absorbed at the ground would result in an increase in average surface temperature.

In addition to this direct effect, there are indirect or feedback effects, most of which tend to augment or amplify the increase in surface temperature. The largest is the further increase in the greenhouse effect due to water vapor. The increase in temperature of the sea surface produced by the CO_2 increase would lead to greater evaporation, increasing the water vapor content of the air. Water vapor has a larger effect on the downward radiation than CO_2; this would produce an increase in surface temperature, which in turn would further increase the evaporation and lead to further increase in the temperature. It is estimated that the water vapor effect would multiply the temperature rise by a factor of two or three.

Another positive-feedback influence is the melting of snow and ice because of the rise in surface temperature. The reduction of area covered by snow and ice would reduce the average albedo of the earth's surface and thereby increase the amount of solar radiation absorbed by the ground. On the other hand, if the increase in water vapor together with augmented convection produced by the warmer ground resulted in increased cloudiness, this might reduce the amount of solar radiation reaching the ground and act as a negative feedback.

Computer models of the effect of increasing the CO_2 concentration in the atmosphere have given plausible estimates of the amount of temperature increase that would result, even though none has yet been able to take full account of the various feedback processes. They indicate that at the time the CO_2 content is doubled, the average temperature of the atmosphere near the ground will be between 2 K and 5 K higher than at present, with considerably greater warming at high latitudes, up to 10 K.

The likelihood of such large increases in the global temperature has aroused considerable concern about possible impacts of these changes on human activities. In addition to the possibility of large rises in sea level due to melting of ice in Greenland and Antarctica—a rise of as much as 5 m, which would inundate many inhabited coastal areas—it has been suggested that ocean circulations would be slowed, with consequent adverse effects on yields of fish; that the changes in temperature and precipitation regimes would require large adjustments of agricultural practices, with possible decreases of crop productivity over large areas due to encroachment of deserts and semiarid regions; and that population shifts might be necessitated by the economic and social consequences of the changes in climate. Not all the effects that might result are negative. The increase in CO_2

concentration might increase yields of crops whose growth is limited by present CO_2 availability; increases in yields of cotton and corn should also result from the higher temperatures; and lengthened growing seasons would tend to stabilize yields of those crops limited by the occurrence of frost. Improved models of the effects on climate of the increase in CO_2 are needed before a more definite evaluation of their social consequences is possible.

It might be expected that the temperature increase due to the 15-percent increase in CO_2 concentration that has occurred during the past hundred years— 0.3 K–0.8 K according to the circulation models—would have been observed. It is difficult to determine the trend in the average temperature for the entire earth, both because of lack of observations over large areas and because of natural fluctuations of larger amplitude, but the available data, including recession of glaciers, indicate that up to about 1940 the earth was warming, in accord with the CO_2 expectations. However, since that time the observations indicate a slight cooling, in spite of the continued increase in CO_2 concentration. To explain this it has been suggested that the introduction of particulate matter into the atmosphere by human activities is tending to offset the effect of the increase in CO_2 concentration. Two questions are raised by this suggestion: (1) Is there a continuous increase in particulate content due to human activities? (2) Does increased particulate content produce a cooling?

Whether human activities have led to an increase in the particulate content of the atmosphere has not yet been determined definitely. Direct measurements of particle counts in places away from sources of pollution are not yet available. As an indirect indication, the transmission of solar radiation through the atmosphere has been used as a measure of the *turbidity*, that is, the amount of particulate matter in the entire path of sunlight through the atmosphere. The measurements are affected by the introduction of large amounts of dust into the stratosphere by large volcanic eruptions, such as the one that occurred at Mt. Agung, on the island of Bali, in 1963. Allowing for the effects of volcanic activity, the records of transmission of solar radiation at Mauna Loa show no tendency for the concentration of particulate matter to increase. In contrast, measurements at Washington, D.C., Davos, Switzerland, and several cities in the Soviet Union suggest that there has been an increase of turbidity in recent years. Most of these locations are subject to local pollution, but Davos, for instance, is a high-altitude station at some distance from any metropolitan or industrial center.

The fact that, except for the effect of volcanoes, the average turbidity is not increasing at the Mauna Loa Observatory implies that the particles entering the atmosphere from human activities are removed from the air before it reaches there. As discussed in Chapter 14, the removal processes are of two types: "dry" processes in which clouds and rain are not involved, and "wet" processes in which they are. The dry processes are *fallout*, the settling out of the particles because of gravity, and *impaction*, in which the particles are collected by leaves of trees, wires, buildings, or the membranes of people's eyes and respiratory systems, as they are carried by the wind, or by vehicles or other objects moving through the air. The larger particles are removed rapidly by the dry processes, but the very small ones are carried upward by the turbulent air motions as rapidly as they tend to fall and are swept around objects with the air flow. The removal of the smaller particles is almost entirely due to the wet processes: *rainout*, in which the particle is incorporated into cloud drops, which then grow and fall as precipitation; and *washout*, in which the particles are collected by the raindrops or snowflakes falling from cloud to ground. The length of time that the small particles coming from human activities remain in the air thus depends on the time that elapses before the air into which they are emitted is incorporated into a cloud or a precipitation area. The average period is not known, but it appears from the observations referred to above that the tropospheric air reaching Mauna Loa Observatory has always passed through a cleansing process in or below clouds in crossing the Pacific Ocean, and the air reaching the other stations frequently travels from the pollution sources without doing so. From these considerations one may conclude that in areas remote from sources the particulate concentration will not increase, but at places that can be reached by polluted air before the air is cleansed by rainout or washout, any increase in the amount of emissions will be reflected by an increase in the amount of particulate matter in the air.

The influence of changes in concentration of particulate matter on temperature depends on its height and its absorption coefficient (the imaginary part of its refractive index). The eruption of Krakatoa, an island in the East Indies, in 1883, sent a tremendous amount of fine dust into the stratosphere, where it spread over the entire earth and remained for two or three years, producing a cooling at the ground averaging more than 1 K. Any absorption and reradiation of energy by the particles at the low temperature of the lower stratosphere was clearly not sufficient to offset the reduction of solar radiation due to scattering by the particles. However, particles in the lower troposphere, where the temperature is relatively high, will radiate energy downward and may actually add to the greenhouse effect if their absorption coefficient is sufficiently large relative to their scattering coefficient. Only recently have measurements of the absorption coefficient of some of the kinds of particles in anthropogenic pollution and computations of the effect on temperature of such particles been made. Most of the computations give a cooling throughout the atmosphere, for the initial conditions assumed. However, these conditions are characteristic of polluted air in cities, and not the much smaller particle concentrations found away from sources.

The main point to be made with respect to particulate matter is that volcanic eruptions appear to have much more influence than human activities, both because they put more into the air and because it goes into the stratosphere where it is immune from wet removal processes and can stay for months and years. We must conclude that, except for the accidental intervention of occasional large volcanic eruptions, the temperature of the earth will continually rise because of the increase in carbon dioxide content of the atmosphere and, to a lesser extent, because of the heat added by combustion and by the release of nuclear energy.

15.4   Intentional weather modification

The possibility of purposefully influencing the weather has attracted people from earliest times. In ancient times the attempts to influence it consisted of prayers and sacrifices to placate the gods. Only very recently have proposals or attempts having a scientific basis been made.

Modification of weather or climate on all scales, from the general atmospheric circulation to individual cumulus clouds and to the temperature of a vegetable patch, has been proposed. However, actual attempts have been confined almost completely to the smaller-scale phenomena.

Proposals to modify the general circulation have included changing the radiational budget of the earth by introducing clouds of soot or ice crystals into the atmosphere at particular latitudes and heights, or by coating the snow and ice fields at high latitudes with carbon black, or by eliminating the ice from the Arctic Ocean by pumping warmer water from the Atlantic Ocean into it. Because of the uncertainty of the consequences and the tremendous investment of energy, materials, and money that would be required to carry them out, large-scale programs of this type have remained the subject of vague speculations. The largest meteorological feature for which systematic weather modification experiments have been conducted is the hurricane.

At the other end of the scale, in the control of the microclimate of plants, control measures have been part of standard agricultural practice for many years. Crops are protected successfully from wind damage through the use of rows of trees as windbreaks and from frost by means of orchard heaters or wind machines. In addition to these generally used procedures, there have been less conventional attempts—for instance, covering the ground with carbon black to alter the local radiational balance and improve crop growth.

The control of clouds and precipitation is the aspect of weather modification that has received the most attention in recent years. It received its initial impetus with the discovery by Vincent Schaefer in 1946 that dry ice pellets dropped through a supercooled cloud could initiate the three-phase (Bergeron) process and the finding by Bernard Vonnegut that silver iodide was an effective ice-nucleating agent (see Section 5.5). Since then there have been many tests of cloud seeding

A

B

Figure 15.2 Results of seeding stratocumulus clouds. Dry ice was dropped into the clouds at the rate of 12 lb/mi along three 9-mile-long lines three miles apart. In photograph A, taken 15 minutes after seeding, the change in the character of the cloud top is just beginning to be evident. Photograph B, taken 35 minutes after seeding, shows a marked change in the character of the

C

D

cloud. The ground could be seen from above 40 minutes after seeding, and 75 minutes after seeding (photograph C), small cumulus clouds had formed in the cleared area. Photograph D, taken 80 minutes after seeding when the plane had descended through the cleared area, shows snow showers falling from the clouds in the distance. [Courtesy of U.S. Army.]

to dissipate fog, increase or redistribute precipitation, suppress hail, reduce lightning strokes, which sometimes cause forest fires, or decrease the destructive winds accompanying thunderstorms and hurricanes. In addition to the use of ice-nucleating substances to initiate or modify the three-phase precipitation process in supercooled clouds, there have been experiments attempting to initiate or modify the precipitation process in warm clouds by seeding with salt particles or water spray.

The formation and structure of clouds and the occurrence, form, and amount of precipitation can be influenced in two ways: (1) by altering the dynamic processes, that is, the air currents producing the clouds (see Chapter 6); and (2) by changing the microphysical processes of formation and growth of cloud and precipitation particles (see Chapter 5). To alter the air flow directly would require very large amounts of energy. However, modifying the microphysical processes in some circumstances produces large changes in the dynamic processes. In turn, changes in the dynamic processes will alter the course of the microphysical processes. The total effect on the presence of cloud and the amount of precipitation depends on the interaction of the two types of processes. Thus, in most weather modification efforts, materials (seeding agents) are introduced into clouds to change the growth processes of the cloud particles. This may be done by flying an airplane above the cloud and dropping dry-ice pellets or pyrotechnic devices that release large quantities of silver iodide as they fall, by releasing silver iodide smoke or water spray as a plane passes through the cloud, or by releasing silver iodide smoke or salt particles from ground-based generators and depending on the updrafts that are producing the cloud to carry the seeding agent into the cloud.

The most frequent purpose of cloud seeding is to increase the amount of precipitation in order to increase the crop yield in arid or semiarid regions or regions experiencing a drought, to augment the water supply for domestic or commercial purposes, or to increase the flow of water for hydroelectric power generation. There have been many "rainmaking" projects in various parts of the world. Initially, they were undertaken with a great deal of optimism. Large precipitation increases were attributed to the cloud seeding, particularly by commercial firms, which claimed to have produced increases of several hundred percent. With the passage of time these claims have been modified, and the advocates of cloud seeding usually estimate the average amount of increase to be between 10 and 20 percent.

The reason for uncertainty about the effect of cloud seeding on the amount of precipitation is that if one seeds a cloud one cannot know how much rain would have fallen from it if it had not been seeded. To evaluate the success of cloud-seeding projects, comparisons have been made with control areas—nearby areas that have not been seeded but, according to previous records, have rain amounts correlated with the natural precipitation amounts in the target area. This method of evaluation has been severely criticized by statisticians. To overcome these criticisms, experiments have been designed employing randomization. In a randomized experiment the situations that are considered suitable for seeding are divided into two groups on the basis of pure chance, equivalent to tossing a coin. One of the groups is seeded, and the other is not. The difference in precipitation between the seeded group and the unseeded group is attributed to seeding, the assumption being that the effects of all other factors cancel one another because of randomization.

A number of randomized experiments have been carried out, with the surprising result that although in some of them the seeded group had more precipitation than the unseeded, in many of them there was no significant difference and in some there was significantly less precipitation for the seeded group than for the unseeded group. For example, a randomized experiment, designated Project Whitetop, was conducted in Missouri during the summers of 1960–1964 to test the precipitation effects of seeding summer cumulus with silver iodide by airplane. The days on which to seed were selected by using objective criteria that were designed to identify days on which seeding would be expected to augment the amount of rain. Analysis of the results showed definitely smaller amounts of precipitation on seeded days than on unseeded days. On the other hand, an experiment conducted in Israel from 1961 to 1966, in which winter storm situations were selected for seeding, indicated an increase in precipitation of 18 percent due to seeding.

Actually, this result, that some of the experiments showed smaller amounts of precipitation with seeding, should not surprise us, for it is easy to understand why seeding should have a precipitation-reducing effect under some circumstances. For example, if the number of natural ice nuclei happened to be optimal for the production of precipitation, the addition of artificial nuclei would lead to larger numbers of smaller ice crystals. Overseeding in this fashion has in fact been suggested as a method of combating flood-producing rains. As another example, consider the effect of seeding cumulus clouds. In convective situations rain occurs naturally after the cloud top has risen far enough to reach temperatures at which the nuclei present are effective. If the cloud is seeded before it reaches this level, the precipitation will start when the cloud is not as deep, and further growth of the cloud will be inhibited. The precipitation from the shallower cloud may be considerably less than that which would have occurred had it grown to its full depth before precipitation began. These examples, overseeding and premature development of precipitation, are just two of several ways in which seeding might lead to a reduction in the amount of precipitation.

On the other hand, in some situations the convection is limited to the layer below a slight inversion and the release of the latent heat of freezing when the cloud is seeded gives just enough added buoyancy to the cloud top to enable the cloud to penetrate the inversion. If this occurs, the cloud may grow explosively and a large amount of rain may fall from a cloud that would not have precipitated at all in the absence of seeding.

We conclude that there are some circumstances in which cloud seeding can increase precipitation and some circumstances in which it will decrease precipitation. At present we have only a partial understanding of what these circumstances are. The objective of much of the cloud-physics research currently under way is to find out how to tell in which situations cloud seeding will increase the amount of precipitation.

One of the approaches to determining whether or not a situation would yield more precipitation with seeding than without is numerical modeling. Ideally, given the observations of the general meteorological situation and the measurements of condensation and ice nuclei, it should be possible to compute the rate of development of cloud and precipitation. Then the computation could be repeated with artificial nuclei added, and the result would indicate whether seeding would lead to more precipitation. Considerable progress has been made in developing numerical models that deal with one or another aspect of the overall problem, but so far the models are highly simplified. In spite of the simplicity of the models, they have been applied with some success to the selection of situations suitable for seeding.

Most of the attempts to augment precipitation have been based on the three-phase process and the use of dry ice or silver iodide as seeding agents. There have also been a few experiments to test the possibility of stimulating the warm-rain (collision-coalescence) process. Experiments in Australia, over the Caribbean, and in the central United States have shown a tendency for cumulus clouds to develop precipitation when water was sprayed into their bases. However, the cost of transporting the large quantities of water needed by airplane and the relatively small increase in precipitation make the procedure questionable from the economic standpoint. Seeding by using giant hygroscopic nuclei appears to have more potential for warm clouds. The seeding can be carried out from the ground, greatly reducing the cost. Experiments conducted in India and Pakistan have reported success in producing modest increases in precipitation with this method.

Operational programs and research experiments in precipitation augmentation have continued. Among the important current research experiments are the Precipitation Enhancement Project (PEP), an international experiment under auspices of the World Meteorological Organization, and the Precipitation Augmentation for Crops Experiment (PACE) recently started in the midwestern United States by NOAA. It is expected that these experiments will help identify the circumstances under which precipitation can be increased.

The choice of technique for attempting to eliminate fog likewise depends on the character of the fog. Supercooled fog (liquid drops at temperatures below 0°C) may be affected by seeding by dry ice or silver iodide. Warm fog (at temperatures above 0°C) requires other techniques, such as reduction of the relative humidity by heating, introduction of hygroscopic substances, or mixing with dry air from above in the downwash of helicopters. Ice fog, which usually occurs only at temperatures below -20°C, has not been responsive to modification attempts; the amount of heat that would have to be added is prohibitive.

The dissipation of supercooled fog is perhaps the most successful result of cloud seeding. It has become an operational practice in airports in various parts of the world. The procedure used at some cities in the United States that are subject to frequent fogs in winter is to alert the pilots of small seeding planes when fog is expected, so that they can take off before the fog closes the airport. Dry-ice pellets are released from the seeding plane into the fog at a suitable distance upwind of the runway; then the pilot waits for the development of a hole in the fog as the ice crystals grow and fall out. Unless the wind carries new fog onto the runway, making additional seeding necessary, the plane lands on the cleared runway, which is then available for landing and takeoff by other planes. At the Orly airport near Paris, liquid propane released through expansion orifices is used to seed the clouds. The cooling by expansion as the propane evaporates produces the ice crystals. This procedure has the advantage that it is carried on with equipment at the ground.

From the standpoint of airport operations, the dissipation of warm fog is more important than that of supercooled fog, because a larger portion of the hours of low visibility occurs at temperatures above 0°C at most of the busy airports of the world. During World War II a system called FIDO (Fog Investigation and Dispersal Operations) was used in England to improve visibility at airfields enough to permit landings of military aircraft on days with warm fog. The system consisted of lines of burners on both sides of the approach and runway, which heated the air sufficiently to reduce the humidity and evaporate the fog drops. For safety the burners had to be a considerable distance from the runway and consequently very large amounts of fuel oil had to be used to provide adequate heat. After World War II attempts were made in the United States and England to modify the FIDO system to make it sufficiently safe and economical for use in commercial aviation, but without success. In France, a system called Turboclair has been developed. It uses jet engines to project heat over the runway. This system is used operationally to disperse warm fog at Orly and Charles de Gaulle airports, near Paris. Other techniques, such as seeding with giant hygroscopic nuclei or polyelectrolytes, or flying helicopters over the fog to mix dry air from above into the foggy air, have had some limited success experimentally but have not been shown to be feasible operationally. The Turboclair system is the only system in the world that is used operationally to dissipate warm fog at present.

Hail suppression for the protection of agricultural crops is another objective of weather modification that has received much attention. Attempts to prevent hail were made long before the introduction of seeding with dry ice or silver iodide. For many years, gunfire and rockets were used for this purpose in France, Italy, and Switzerland. Success has been claimed for some of these trials, although no sound explanations why they should work have been put forward. On the other hand, for seeding with silver iodide a reasonable basis has been proposed, namely, that the conversion of the cloud to ice early in its development will prevent the growth of hailstones. There have been tests of seeding to reduce hail in several countries with variable results. The ones reported to have had the greatest success have been a program in France, in which seeding from 1959 to 1966 produced a decrease of 22.6 percent in annual loss due to hail damage, compared with the previous 15 years, and large-scale hail prevention programs in the Soviet Union, in which reductions of damage of 50–90 percent in protected areas, compared with areas that were not protected, were claimed. In the Russian procedure, the seeding is accomplished by rockets or artillery shells that are directed by radar to the precise position in the storm cloud at which the hail production is taking place. Refrangible shells and rockets were developed, which break up into very small fragments when they explode so that the debris falling from the spent projectiles can do no harm. The claims of success of the Russian method stimulated a large-scale research experiment in the United States to test the theories on which their method is based, and several operational programs in the Great Plains.

The National Hail Research Experiment (NHRE) was conducted in the 1970s in northeastern Colorado. It combined detailed physical and microphysical measurements of hail-producing thunderstorms with a statistical analysis of the amount of hail, the size of hailstones, and other precipitation characteristics in seeded and unseeded storms. It had been planned to carry out the seeding on a randomized basis for five years, but the seeding phase of the experiment was suspended after three years because the physical measurements showed that the seeding procedure, which was based on the hypothesis regarding hail formation that the Russian procedure was based on, could not lead to decrease in hail in the type of storms occurring in Colorado. The statistical analysis showed that there was no significant difference between the hail in seeded and unseeded storms, but while the difference that was observed could have occurred by chance, it was in the direction of more hail in seeded storms than in unseeded, the opposite of that desired.

In the operational programs randomization was not used, but comparison of crop damage in seeded and neighboring unseeded areas indicated decreases, though not statistically significant, of 20–60 percent, in all of them. A similar decrease was reported for a hail suppression project conducted during 1971–1974 in South Africa. While the statistical analysis indicates that these results could have occurred by chance, the consistency of them suggests that the seeding was effective in reducing crop damage due to hail.

As lightning is the principal cause of forest fires in the extensive forests of the

United States, the possibility that its occurrence could be reduced by seeding has led to extensive investigations conducted by the U.S. Forest Service and by some state divisions of forestry. There has been some indication that seeding thunderstorms with silver iodide does reduce the number of lightning strokes, but the results are not yet conclusive. Another method that has been tried utilizes chaff— metal-covered nylon needles—which tends to discharge the cloud before the electric potential required for lightning can develop. Tests suggest that the method would be successful if a practical means of dispersing the chaff throughout the cloud could be developed.

In hurricane modification experiments the purpose is to reduce the strength of the destructive winds near the center. The horizontal convergence in hurricanes is concentrated at the eye wall and along the spiral bands radiating out from it. The highest clouds and heaviest precipitation occur there. Associated with the extremely rapid release of latent heat, the strongest winds occur in a ring around the eye wall. Attempts to reduce their severity are based on the idea that broadening the area over which the latent heat is released will increase the distance over which the pressure decreases, reducing the pressure gradient and thus weakening the winds. A few experiments have been conducted in which massive amounts of silver iodide were released into the eye wall of hurricanes by dropping pyrotechnic devices from planes. The winds were lower for a while after the seeding in these experiments, but by amounts that are within the normal variability of winds in hurricanes. After a time they increased again. Since randomized experiments with hurricanes cannot be conducted because there are so few of them, the possibility that the decreases in wind were not due to the seeding cannot be rejected. However, if the same result is obtained in future tests, the probability will be reinforced that seeding is producing the reduction in wind speed.

In reviewing the various weather modification attempts discussed herein, we see that the potential for weather modification exists but the observational data and the theoretical understanding of cloud and precipitation physics are not adequate to enable us to predict with certainty the effect of cloud seeding. Much research remains to be done before we can truly control the weather.

We can visualize the ultimate state of the weather-control process as part of an "ideal" weather service. The data from satellites, balloons, and surface weather stations would be fed automatically into computers, which would process them to forecast the conditions that would develop if no modification treatment is applied. On the basis of criteria for social, economic, and aesthetic benefits, the treatment that would be required to produce optimal weather conditions for the largest number of people would be arrived at by the computer, which would then automatically turn on the nucleus generators and other devices to administer the treatment needed.

We are far from such a state, and may never achieve it. But just like the outcome of improvements in forecasting, the social good and economic benefits that would result from success in weather control justify increased efforts to develop the scientific basis for them, namely, a fuller understanding of our atmospheric environment.

Questions, Problems, and Projects

1. What do you think were the changes in the climate of the United States caused by the settlement by the Europeans, with the accompanying spread of agriculture and development of large cities?

2. What would happen to the concentration of carbon dioxide in the atmosphere if nuclear reactors completely replaced combustion of fossil fuels as a source of energy? What would be the effect on the temperature of the earth?

3. Compare the processes that lead to the removal of particulates, as described in Section 15.3, with those that remove carbon dioxide from the atmosphere. Why is the removal of particulates virtually complete, while about one-half of the carbon dioxide produced by combustion remains in the air? Discuss the processes you consider effective in removing sulfur dioxide, oxides of nitrogen, and other gaseous contaminants.

4. Explain why cloud seeding with dry ice or silver iodide may be expected to increase precipitation. Why does it sometimes decrease precipitation?

5. Why does the introduction of water spray or giant hygroscopic nuclei into warm clouds tend to initiate or increase precipitation? Are there circumstances in which it might decrease precipitation?

6. Discuss the problem of testing the success of a "rainmaking" experiment.

Appendices

A. Units of measurement and conversion factors

The International System of Units (SI), which is an extension of the metric system, is commonly used in science throughout the world. The metric system is also used generally in commerce and everyday life almost everywhere. The main exception is the United States, where the English system is used. Great Britain and the other members of the British Commonwealth converted about a decade ago to the metric system.

The SI is based on seven independent fundamental units. These include the unit of length, the *meter*; mass, the *kilogram*; time, the *second*; temperature, the *kelvin* (equivalent to one Celsius or centigrade degree); electric current, the *ampere*; amount of substance, the *mole*; and luminous intensity, the *candela*. The last three units have not been needed in material presented in this textbook. SI units of other quantities, such as force (the *newton*), pressure (the *pascal*), power (the *watt*), and energy (the *joule*) are defined in terms of the fundamental units.

Multiples or fractions of units, in powers of ten, are designated by prefixes as follows:

p	pico-	one-trillionth (U.S.)	10^{-12}	0.000 000 000 001
n	nano-	one-billionth (U.S.)	10^{-9}	0.000 000 001
μ	micro-	one-millionth	10^{-6}	0.000 001
m	milli-	one-thousandth	10^{-3}	0.001
c	centi-	one-hundredth	10^{-2}	0.01
d	deci-	one-tenth	10^{-1}	0.1
da	deka-	ten	10	10.
h	hecto-	one hundred	10^{2}	100.
k	kilo-	one thousand	10^{3}	1 000.
M	mega-	one million	10^{6}	1 000 000.
G	giga-	one billion (U.S.)	10^{9}	1 000 000 000.
T	tera-	one trillion (U.S.)	10^{12}	1 000 000 000 000.

Table of Units and Equivalents

Metric units	English units	Conversions
LENGTH		
centimeter (cm)	inch (in)	1 cm = 0.3937 in; 1 in = 2.54 cm
meter (m)	foot (ft)	1 m = 3.28 ft; 1 ft = 0.3048 m
kilometer (km)	mile (mi)	1 km = 0.621 mi; 1 mi = 1.61 km
micrometer (μm)		$1\ \mu m = 10^{-6}\ m = 10^{-4}\ cm$
		$= 3.94 \cdot 10^{-5}\ in$
Ångstrom unit (Å)		$1\ Å = 10^{-10}\ m = 10^{-4}\ \mu m$

One degree of latitude = 111.1 km = 69.1 mi = 60 nautical miles

One nautical mile = 1.15 statute miles = 1.85 km

Metric units	English units	Conversions
MASS		
gram (g)	ounce (oz av)	1 g = 0.0353 oz; 1 oz = 28.35 g
kilogram (kg)	pound (lb)	1 kg = 2.20 lbs; 1 lb = 0.454 kg
metric ton	short ton (2000 lbs)	1 m ton = 1.10 short tons
(m ton = 10^{3} kg)		1 short ton = 0.907 m tons

Metric units	English units	Conversions

TEMPERATURE

kelvin (K) or degree Celsius (°C)	degree Fahrenheit (°F)	$1\text{ K} = 1°\text{C} = 1.8°\text{F}$; $1°\text{F} = \frac{5}{9}\text{ K} = \frac{5}{9}°\text{C}$

CONVERSIONS OF TEMPERATURE

Celsius or centigrade (t_c)		$t_c = T - 273.15$ $= 5\dfrac{(t_F - 32)}{9}$ °C
	Fahrenheit (t_F)	$t_F = [1.8\,t_c + 32]$ °F
Kelvin or Absolute (T)		$T = [t_c + 273.15]$ K

AREA

square centimeter (cm^2)	square inch (sq in)	$1\text{ cm}^2 = 0.155$ sq in 1 sq in $= 6.45\text{ cm}^2$
square meter (m^2)	square foot (sq ft)	$1\text{ m}^2 = 10.76$ sq ft 1 sq ft $= 0.093\text{ m}^2$
square kilometer (km^2)	square mile (sq mi)	$1\text{ km}^2 = 0.386$ sq mi 1 sq mi $= 2.59\text{ km}^2$
hectare	acre	1 hectare $= 10,000\text{ m}^2$ $= 2.47$ acres 1 acre $= 43,560$ sq ft $= 0.4047$ hectares

VOLUME

cubic centimeter (cm^3)	cubic inch (cu in)	$1\text{ cm}^3 = 0.061$ cu in 1 cu in $= 16.39\text{ cm}^3$
cubic meter (m^3)	cubic foot (cu ft)	$1\text{ m}^3 = 35.3$ cu ft 1 cu ft $= 0.028\text{ m}^3$
liter (l)	gallon (gal)	$1\text{ l} = 10^{-3}\text{ m}^3 = 10^3\text{ cm}^3$ $= 0.264$ gal (U.S.) 1 gal (U.S.) $= 231$ cu in $= 3.7853$ l

SPEED

centimeter per second (cm s^{-1})	foot per second (ft/sec)	$1\text{ cm s}^{-1} = 0.0328$ ft/sec $= 1.97$ ft/min 1 ft/sec $= 30.48\text{ cm s}^{-1}$ $= 0.592$ kt
meter per second (m s^{-1})	mile per hour (mi/hr)	$1\text{ m s}^{-1} = 2.24$ mi/hr $= 1.94$ kt 1 mi/hr $= 0.447\text{ m s}^{-1}$ $= 0.868$ kt

knot (kt) 1 kt $= 1$ nautical mile per hour $= 0.515\text{ m s}^{-1} = 1.15$ mi/hr

(continued)

Table of Units and Equivalents *(continued)*

Metric units	English units	Conversions
		WORK AND ENERGY
erg		1 joule $= 10^7$ ergs 1 cal $= 4.186 \cdot 10^7$ ergs $\quad = 4.186$ joules
joule (J)	foot pound (ft lb)	1 ft lb $= 1.36$ joules 1 joule $= 0.738$ ft lb
calorie (cal)	British Thermal Unit (BTU)	1 BTU $= 252$ cal $\quad = 1055$ joules 1 cal $= 3.97 \cdot 10^{-3}$ BTU
langley (ly)		1 ly $= 1$ cal/cm^2 $\quad = 4.186 \cdot 10^4$ J/m^2
		POWER
watt (W)	horsepower (hp)	1 W $= 1$ J/sec $= 1.34 \cdot 10^{-3}$ hp 1 hp $= 33{,}000$ ft lb/min $\quad = 746$ W 1 ly/min $= 6.98 \cdot 10^2$ W/m^2
		PRESSURE
pascal (Pa)	lb/sq in	1 Pa $= 1$ N/m^2 $= 10$ dynes/cm^2 $\quad = 1.45 \cdot 10^{-4}$ lb/sq in
dyne/cm^2		1 dyne/cm^2 $= 0.1$ Pa $\quad = 1.45 \cdot 10^{-5}$ lb/m^2
millibar (mb)		1 mb $= 10^2$ Pa $= 10^3$ dynes/cm^2 $\quad = 0.750$ mm Hg $\quad = 2.95 \cdot 10^{-2}$ in Hg
millimeters of mercury (mm Hg)	inches of mercury (in Hg)	1 mm Hg $= 133.3$ Pa $\quad = 1.333$ mb $= 0.03937$ in Hg $\quad = 0.0193$ lb/sq in 1 in Hg $= 3386.5$ Pa $\quad = 33.865$ mb $= 25.4$ mm Hg $\quad = 0.491$ lb/sq in

One normal atmosphere (atm) $= 101325$ Pa $= 1013.25$ mb $= 760$ mm Hg
$= 29.92$ in Hg $= 14.7$ lb/sq in

B.   Numerical constants

Mean solar distance	$1.495 \cdot 10^8$ km
Equatorial radius of earth	6378.4 km
Polar radius of earth	6356.9 km
Angular velocity of earth's rotation	$\Omega = 7.29 \cdot 10^{-5}$ s^{-1}
Acceleration of gravity (at sea level and 45° lat)	9.806 m s^{-2}
Solar constant	1.37 kW/m^2 = 1.96 ly/min
Gas constant for dry air	287 J/(kg K)
Mean molecular weight of dry air	28.964 kg/kmol
Specific heat of dry air:	
at constant pressure c_p	1005 J/(kg K)
a constant volume c_v	718 J/(kg K)

C.   The Greek alphabet

In science the number of quantities that must be represented by symbols in equations is so large that in addition to the Latin alphabet used in English, letters from other alphabets are used—particularly, the Greek. The following list gives the Greek letters, many of which are commonly used in meteorology. The capital letters that are identical with Latin letters are shown in parenthesis.

Small Letter	Capital Letter	Name	Small Letter	Capital Letter	Name
α	(A)	alpha	ν	(N)	nu
β	(B)	beta	ξ	Ξ	xi
γ	Γ	gamma	o	(O)	omicron
δ	Δ	delta	π	Π	pi
ε	(E)	epsilon	ρ	(P)	rho
ζ	(Z)	zeta	σ	Σ	sigma
η	(H)	eta	τ	(T)	tau
θ	Θ	theta	υ	Υ	upsilon
ι	(I)	iota	φ	Φ	phi
κ	(K)	kappa	χ	(X)	chi
λ	Λ	lambda	ψ	Ψ	psi
μ	(M)	mu	ω	Ω	omega

D.   Saturation vapor pressure over pure water and ice

t_c (°C)	Over ice e_{si} (mb)	Over liquid water e_{sw} (mb)	t_c (°C)	e_s (mb)
−40	0.128	0.189	0	6.11
−38	0.161	0.232	2	7.05
−36	0.200	0.284	4	8.13
−34	0.249	0.346	6	9.35
−32	0.308	0.420	8	10.72
−30	0.380	0.509	10	12.27
−28	0.467	0.613	12	14.02
−26	0.572	0.737	14	15.97
−24	0.698	0.883	16	18.17
−22	0.850	1.05	18	20.63
−20	1.03	1.25	20	23.37
−18	1.25	1.49	22	26.43
−16	1.51	1.76	24	29.83
−14	1.81	2.08	26	33.61
−12	2.17	2.44	28	37.79
−10	2.60	2.86	30	42.43
−8	3.10	3.35	32	47.55
−6	3.68	3.91	34	53.20
−4	4.37	4.54	36	59.42
−2	5.17	5.27	38	66.26
0	6.11	6.11	40	73.77

E.   Weather-map symbols

The symbols that are used on surface weather maps to summarize the conditions observed at individual stations are given on the following pages.[1] The data are transmitted by the observation station to the collection center by teletype or radio in the form of a coded message. They are then plotted around a small circle in the position on the map where the station is located, using the arrangement shown in the station model and the symbols shown in the various tables.

---

1. The symbols and map entries were prescribed by the World Meteorological Organization for use by the meteorological services of all countries. The presentation in this appendix is taken from the U.S. National Weather Service publication "Explanation of the Weather Map."

1

SYMBOLIC STATION MODEL   SAMPLE PLOTTED REPORT

$$ff$$
$$TT \; dd \quad \begin{matrix} C_H \\ C_M \end{matrix} \; PPP$$
$$VV \; ww \; Ⓝ \qquad pp \; a$$
$$T_d T_d \; C_L \; N_h \; W \; R_t$$
$$h \qquad RR$$

34   147

¾ ∗   ● +28 /

32 - - - 6 . 4

2   45

2

EXPLANATION OF SYMBOLS AND MAP ENTRIES

Symbols in order as they appear in the message	Explanation of symbols and decode of sample message in block	Remarks on coding and plotting
IIiii	Station number 72405 = Washington	Usually printed on manuscript maps below station circle. Omitted on Daily Weather Map in favor of printed station names.
N	Total amount of cloud 8 = completely covered	Observed in tenths of cloud cover and coded according to code table in block **5**. Plotted in symbols shown in same table.
dd	True direction from which wind is blowing 32 = 320° = NW	Coded in tens of degrees and plotted as the shaft of an arrow extending from the station circle toward the direction from which the wind is blowing.
ff	Wind speed in knots 20 = 20 knots	Coded in knots (nautical miles per hour) and plotted as feathers and half-feathers representing 10 and 5 knots, respectively, on the shaft of the wind direction arrow. See block **9**.
VV	Visibility in miles and fractions 12 = 12/16 or 3/4 miles	Decoded and plotted in miles and fractions up to 3⅛ miles. Visibilities above 3⅛ miles but less than 10 miles are plotted to the nearest whole mile. Values higher than 10 miles are omitted from the map.
ww	Present weather 70 = slight intermittent snow	Coded in figures taken from the "ww" table (block **8**) and plotted in the corresponding symbols same block. Entries for code figures 00, 01, 02, and 03 are omitted from this map.
W	Past weather 6 = rain	Coded in figures taken from the "W" table (block **11**) and plotted in the corresponding symbols same block. No entry made for code figures 0, 1, or 2

2 (con'd)

PPP — Barometric Pressure (in millibars) reduced to sea-level 147 = 1014.7 mb. — Code and plotted in tens, units, and tenths of millibars. The initial 9 or 10 and the decimal point are omitted.

TT — Current air temperature in °C. 01 = 34°F. — Coded in whole degrees Celsius and plotted in nearest equivalent whole degrees Fahrenheit.

N_h — Fraction of sky covered by low or middle cloud 6 = 7 or 8 tenths — Observed in tenths of cloud cover and coded according to code table in block **6** Plotted on map as code figure in message.

C_L — Cloud type 7 = Fractostratus and/or Fractocumulus of bad weather (scud) — Predominating clouds of types in C_L table (block **3**) are coded from that table and plotted in corresponding symbols. "0" is coded if no C_L clouds are observed.

h — Height above ground, of base of lowest cloud· 2 = 300 to 599 feet — Observed in feet and coded and plotted as code figures according to code table in block **7**.

C_M — Cloud type 9 = Altocumulus of chaotic sky — See C_M table block **3**.

C_H — Cloud type 2 = Dense cirrus in patches — See C_H table in block **3**.

T_dT_d — Temperature of dewpoint in °C. 00 = 32°F. — Same as TT above.

a — Characteristic of barograph trace 2 = rising steadily or unsteadily — Coded according to table in block **10** and plotted in corresponding symbols.

pp — Pressure change in 3 hours preceding observation 28 = 2.8 millibars — Coded and plotted in units and tenths of millibars.

7 — Indicator figure — Not plotted.

RR — Amount of precipitation 45 = 0.45 inches — Coded and plotted in inches to the nearest hundredth of an inch. If precipitation less than .005 inch is reported, RR is coded 00 and plotted as T

R_t — Time precipitation began or ended 4 = 3 to 4 hours ago — Coded and plotted in figures from table in block **4**.

s — Depth of snow on ground in inches 1 = 1 inch — Not plotted. (used in Snowdepth analysis)

3

Code No. C_L   DESCRIPTION
(Abridged From International Code)

1   Cu of fair weather, little vertical development and seemingly flattened

2   Cu of considerable development, generally towering, with or without other Cu or Sc bases all at same level

3   Cb with tops lacking clear-cut outlines, but distinctly not cirriform or anvil-shaped; with or without Cu, Sc, or St

4   Sc formed by spreading out of Cu; Cu often present also

5   Sc not formed by spreading out of Cu

6   St or Fs or both, but no Fs of bad weather

7   Fs and/or Fc of bad weather (scud)

8   Cu and Sc (not formed by spreading out of Cu) with bases at different levels

9   Cb having a clearly fibrous (cirriform) top, often anvil-shaped, with or without Cu, Sc, St, or scud

Code No. C_M   DESCRIPTION
(Abridged From International Code)

1   Thin As (most of cloud layer semi-transparent)

2   Thick As, greater part sufficiently dense to hide sun (or moon), or Ns

3   Thin Ac, mostly semi-transparent; cloud elements not changing much and at a single level

4   Thin Ac in patches; cloud elements continually changing and/or occurring at more than one level

5   Thin Ac in bands or in a layer gradually spreading over sky and usually thickening as a whole

6   Ac formed by the spreading out of Cu or Cb

7   Double-layered Ac, or a thick layer of Ac, not increasing; or Ac with As and/or Ns

8   Ac in the form of Cu-shaped tufts or Ac with turrets

9   Ac of a chaotic sky, usually at different levels; patches of dense Ci are usually present also

Code No. C_H   DESCRIPTION
(Abridged From International Code)

1   Filaments of Ci, or "mares tails," scattered and not increasing

2   Dense Ci in patches or twisted sheaves, usually not increasing, sometimes like remains of Cb; or towers or tufts

3   Dense Ci, often anvil-shaped, derived from or associated with Cb

4   Ci, often hook-shaped, gradually spreading over the sky and usually thickening as a whole

5   Ci and Cs, often in converging bands, or Cs alone; generally overspreading and growing denser; the continuous layer not reaching 45° altitude

6   Ci and Cs, often in converging bands, or Cs alone; generally overspreading and growing denser; the continuous layer exceeding 45° altitude

7   Veil of Cs covering the entire sky

8   Cs not increasing and not covering entire sky

9   Cc alone or Cc with some Ci or Cs, but the Cc being the main cirriform cloud

4

R_t **TIME PRECIPITATION BEGAN OR ENDED**

0   No Precipitation

1   Less than 1 hour ago

2   1 to 2 hours ago

3   2 to 3 hours ago

4   3 to 4 hours ago

5   4 to 5 hours ago

6   5 to 6 hours ago

7   6 to 12 hours ago

8   More than 12 hours ago

9   Unknown

5

Code No. N **SKY COVERAGE** (Total Amount)

0   No clouds

1   Less than one-tenth or one-tenth

2   Two-tenths or three-tenths

3   Four-tenths

4   Five-tenths

5   Six-tenths

6   Seven-tenths or eight-tenths

7   Nine-tenths or overcast with openings

8   Completely overcast

9   Sky obscured

6

N_h **SKY COVERAGE** (Low And/Or Middle Clouds)

0   No clouds

1   Less than one-tenth or one-tenth

2   Two-tenths or three-tenths

3   Four-tenths

4   Five-tenths

5   Six-tenths

6   Seven-tenths or eight-tenths

7   Nine-tenths or overcast with openings

8   Completely overcast

9   Sky obscured

7

h **HEIGHT IN FEET** (Approximate)

0   0–149

1   150–299

2   300–599

3   600–999

4   1,000–1,999

5   2,000–3,499

6   3,500–4,999

7   5,000–6,499

8   6,500–7,999

9   At or above 8,000, or no clouds

8

WW PRESENT WEATHER

(Descriptions Abridged from International Code)

	0	1	2	3	4
00	Cloud development NOT observed or NOT observable during past hour	Clouds generally dissolving or becoming less developed during past hour	State of sky on the whole unchanged during past hour	Clouds generally forming or developing during past hour	Visibility reduced by smoke
10	Light fog	Patches of shallow fog at station, NOT deeper than 6 feet on land	More or less continuous shallow fog at station, NOT deeper than 6 feet on land	Lightning visible, no thunder heard	Precipitation within sight, but NOT reaching the ground
20	Drizzle (NOT freezing) or snow grains, NOT falling as showers, during past hour but NOT at time of observation	Rain (NOT freezing and NOT falling as showers) during past hour, but NOT at time of observation	Snow (NOT falling as showers) during past hour, but NOT at time of observation	Rain and snow or ice pellets (NOT falling as showers) during past hour, but NOT at time of observation	Freezing drizzle or freezing rain (NOT falling as showers) during past hour, but NOT at time of observation
30	Slight or moderate dust storm or sand storm, has decreased during past hour	Slight or moderate dust storm or sand storm, no appreciable change during past hour	Slight or moderate duststorm or sandstorm has begun or increased during past hour	Severe dust storm or sand storm, has decreased during past hour	Severe dust storm or sand storm, no appreciable change during past hour
40	Fog or ice fog at distance at time of observation, but NOT at station during past hour	Fog or ice fog in patches	Fog or ice fog, sky discernible, has become thinner during past hour	Fog or ice fog, sky NOT discernible, has become thinner during past hour	Fog or ice fog, sky discernible, no appreciable change during past hour
50	Intermittent drizzle (NOT freezing) slight at time of observation	Continuous drizzle (NOT freezing) slight at time of observation	Intermittent drizzle (NOT freezing) moderate at time of observation	Continuous drizzle (NOT freezing), moderate at time of observation	Intermittent drizzle (NOT freezing), heavy at time of observation
60	Intermittent rain (NOT freezing), slight at time of observation	Continuous rain (NOT freezing), slight at time of observation	Intermittent rain (NOT freezing) moderate at time of observation	Continuous rain (NOT freezing), moderate at time of observation	Intermittent rain (NOT freezing), heavy at time of observation
70	Intermittent fall of snow flakes, slight at time of observation	Continuous fall of snowflakes, slight at time of observation	Intermittent fall of snowflakes, moderate at time of observation	Continuous fall of snowflakes, moderate at time of observation	Intermittent fall of snowflakes heavy at time of observation
80	Slight rain shower(s)	Moderate or heavy rain shower(s)	Violent rain shower(s)	Slight shower(s) of rain and snow mixed	Moderate or heavy shower(s) of rain and snow mixed
90	Moderate or heavy shower(s) of hail, with or without rain or rain and snow mixed, not associated with thunder	Slight rain at time of observation; thunderstorm during past hour, but NOT at time of observation	Moderate or heavy rain at time of observation; thunderstorm during past hour, but NOT at time of observation	Slight snow or rain and snow mixed or hail at time of observation; thunderstorm during past hour, but not at time of observation	Moderate or heavy snow, or rain and snow mixed or hail at time of observation; thunderstorm during past hour, but NOT at time of observation.

8 (con'd)

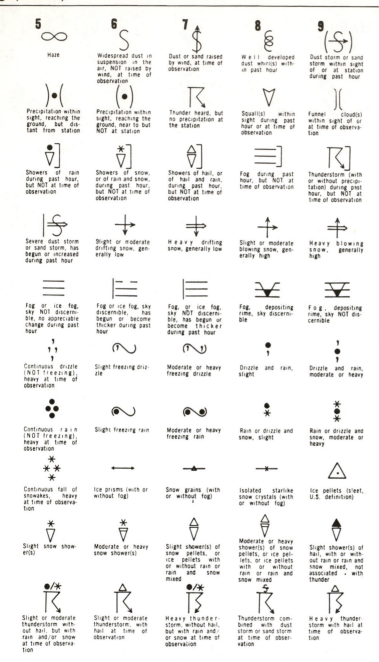

5 Haze	**6** Widespread dust in suspension in the air, NOT raised by wind, at time of observation	**7** Dust or sand raised by wind, at time of observation	**8** Well developed dust whirl(s) within past hour	**9** Dust storm or sand storm within sight of or at station during past hour
Precipitation within sight, reaching the ground, but distant from station	Precipitation within sight, reaching the ground, near to but NOT at station	Thunder heard, but no precipitation at the station	Squall(s) within sight during past hour or at time of observation	Funnel cloud(s) within sight of or at time of observation
Showers of rain during past hour, but NOT at time of observation	Showers of snow, or of rain and snow, during past hour, but NOT at time of observation	Showers of hail, or of hail and rain, during past hour, but NOT at time of observation	Fog during past hour, but NOT at time of observation	Thunderstorm (with or without precipitation) during past hour, but NOT at time of observation
Severe dust storm or sand storm, has begun or increased during past hour	Slight or moderate drifting snow, generally low	Heavy drifting snow, generally low	Slight or moderate blowing snow, generally high	Heavy blowing snow, generally high
Fog or ice fog, sky NOT discernible, no appreciable change during past hour	Fog or ice fog, sky discernible, has begun or become thicker during past hour	Fog, or ice fog, sky NOT discernible, has begun or become thicker during past hour	Fog, depositing rime, sky discernible	Fog, depositing rime, sky NOT discernible
Continuous drizzle (NOT freezing), heavy at time of observation	Slight freezing drizzle	Moderate or heavy freezing drizzle	Drizzle and rain, slight	Drizzle and rain, moderate or heavy
Continuous rain (NOT freezing), heavy at time of observation	Slight freezing rain	Moderate or heavy freezing rain	Rain or drizzle and snow, slight	Rain or drizzle and snow, moderate or heavy
Continuous fall of snowflakes, heavy at time of observation	Ice prisms (with or without fog)	Snow grains (with or without fog)	Isolated starlike snow crystals (with or without fog)	Ice pellets (sleet, U.S. definition)
Slight snow shower(s)	Moderate or heavy snow shower(s)	Slight shower(s) of snow pellets, or ice pellets with or without rain or rain and snow mixed	Moderate or heavy shower(s) of snow pellets, or ice pellets, or ice pellets with or without rain or rain and snow mixed	Slight shower(s) of hail, with or without rain or rain and snow mixed, not associated with thunder
Slight or moderate thunderstorm without hail, but with rain and/or snow at time of observation	Slight or moderate thunderstorm, with hail at time of observation	Heavy thunderstorm, without hail, but with rain and/or snow at time of observation	Thunderstorm combined with dust storm or sand storm at time of observation	Heavy thunderstorm with hail at time of observation

9

ff	KNOTS	(MILES) (Statute) Per Hour
◎	Calm	Calm
—	1–2	1–2
⌐	3–7	3–8
\\	8–12	9–14
\\	13–17	15–20
\\\	18–22	21–25
\\\	23–27	26–31
\\\	28–32	32–37
\\\\	33–37	38–43
\\\\	38–42	44–49
\\\\\	43–47	50–54
◣	48–52	55–60
◣	53–57	61–66
◣\	58–62	67–71
◣\\	63–67	72–77
◣\\\	68–72	78–83
◣\\\\	73–77	84–89
◣◣	103–107	119–123

10

Code No.	a	PRESSURE TENDENCY	
0	⌒	Rising, then falling same or higher than 3 hours ago	
1	⁄	Rising, then steady; or rising, then rising more slowly	Barometric pressure now higher than 3 hours ago
2	⁄	Rising steadily, or unsteadily	
3	✓	Falling or steady, then rising; or rising, then rising more rapidly	
4	—	Steady, same as 3 hours ago	
5	＼	Falling, then rising, same or lower than 3 hours ago	
6	＼	Falling, then steady; or falling, then falling more slowly	Barometric pressure now lower than 3 hours ago
7	＼	Falling steadily, or unsteadily	
8	⌒	Steady or rising, then falling; or falling, then falling more quickly	

11

Code No.	W	PAST WEATHER	
0		Clear or few clouds	
1		Partly cloudy (scattered) or variable sky	NOT PLOTTED
2		Cloudy (broken) or overcast	
3	⭙⁄	Sandstorm or dust-storm, or drifting or blowing snow	
4	≡	Fog, ice fog, thick haze or thick smoke	
5	،	Drizzle	
6	•	Rain	
7	✳	Snow, or rain and snow mixed, or ice pellets (sleet)	
8	▽	Shower(s)	
9	⏀	Thunderstorm, with or without precipitation	

F.   Meteorological instruments and observations

F.1   Temperature measurement

The thermometers used in operational meteorology are of three types: liquid-in-glass, deformation, and electrical resistance. The liquid-in-glass type is the most commonly used indicating thermometer; the deformation type is used in recording instruments (thermographs); and the electrical-resistance type is used for remote sensing.

For measurement of the surface temperature, defined as the temperature near the observer's eye level, about 1.5 m above the ground, the liquid-in-glass thermometer is commonly used. It operates on the principle that when the temperature rises the liquid expands more rapidly than the glass container. By having the container in the form of a tube, the relative expansion shows as a rise in the level of the liquid. Because the rate of expansion is small—mercury, one of the two most commonly used liquids, increases in volume only 0.0182 percent per kelvin—it is necessary to use a method that in effect magnifies the distance the liquid rises in the tube. This is accomplished by having a relatively large bulb connected to a long thin stem, with the amount of liquid such that its volume exceeds that of the bulb alone at the lowest expected temperature, but not the combined volume of the bulb and stem at the highest temperature expected. In this way the top of the liquid will always be in the stem. By making the inside diameter of the stem sufficiently small and/or the volume of the bulb sufficiently large, the thermometer can be made sensitive enough to enable it to be read to any desired accuracy. The mercury-in-glass thermometers in the surface observational network are designed so that temperatures can be estimated to the nearest 0.1°C.

Another liquid, alcohol, is also used in thermometers. It has the advantage that it expands with increasing temperature at about six times the rate at which mercury does. It is well suited for use in minimum thermometers because stems with a larger bore can be used and still provide the required sensitivity. Minimum thermometers require a stem with a larger bore so that a double-ended piece of dark-colored glass will fit easily inside the stem and act as an index that can move back and forth along the stem (see Figure F.1). The thermometer in the operational mode is mounted with the bulb end five degrees above the horizontal. The glass index is confined inside the alcohol column. When the top of the alcohol column moves toward the bulb end (temperature decreasing), the glass index is dragged along by the surface tension at the free surface of the alcohol. When temperature increases, the alcohol moves freely past the index leaving the index at the position indicating the lowest temperature reached. To reset the instrument after reading, the thermometer is put in a vertical position, bulb up. Gravity pulls the index

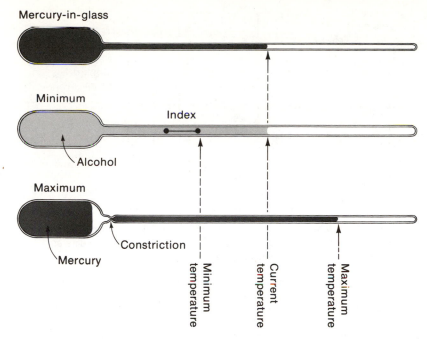

Figure F.1 Liquid-in-glass thermometers.

down to the free surface of the alcohol. Then the thermometer is returned to its near horizontal orientation and is ready to measure the next minimum.

Another advantage of alcohol is that it has a much lower freezing point than mercury. Mercury freezes at $-38.87°C$. There are areas on the earth where surface temperatures reach lower values than this, in particular the Arctic and Antarctic. Alcohol is still liquid at $-100°C$ and can be used safely in thermometers there.

There is one other type of liquid-in-glass thermometer that should be discussed, the maximum thermometer. In contrast to the minimum thermometer, no index is used in this instrument. Instead the bore has a very narrow constriction between the bulb and the stem of a mercury-in-glass thermometer, so narrow that a force greater than gravity is required to push the mercury past the constriction. The thermometer is mounted in a horizontal position. When temperature is increasing, the expansion of the mercury in the bulb pushes mercury past the constriction into the stem, but when temperature decreases there is no corresponding force to push the liquid back past the constriction into the bulb. Consequently, the mercury column above the constriction remains undiminished, indicating the highest temperature that was reached. To reset the instrument it is spun around, the centrifugal force pushing the mercury back into the bulb. It is then returned to its horizontal position and is ready to record the next maximum temperature.

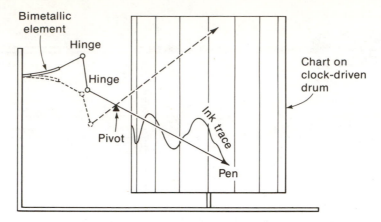

Figure F.2 Schematic diagram of thermograph using bimetallic strip as sensor. Figure 5.3 shows a photograph of a thermograph using a different kind of deformation type sensor.

It is necessary sometimes to have continuous records of the temperature; for example, when you want to know exactly the times when yesterday's maximum and minimum temperatures were attained. Liquid-in-glass thermometers cannot provide this information since they are read only periodically, usually once an hour. The deformation thermometer is one which can readily be adapted to continuous recording.

There are a variety of mechanical objects that can be devised whose shapes change measureably with changes in temperature. Such objects can be used as thermometers. As an illustration we will describe here the bimetallic thermometer. The sensor is made up of two strips of different metals bonded together. The types of metal used are chosen such that they have very different coefficients of thermal expansion. The result is that one metal experiences a much greater expansion with a given increase in temperature than the other one. If the strip was perfectly straight at the initial temperature, after heating it will warp into a curved strip with the metal of least expansion on the inside of the curve. In a recording thermometer (thermograph) using such a strip as the sensor (Figure F.2), one end of the strip is fixed and the motion of the other is magnified by means of linkages (levers) that terminate in a pen arm. The pen rests on a clock-driven rotating drum and traces out a continuous line, providing temperature as a function of time. The instruments most commonly found in weather stations have drums that rotate once per week, thus providing weekly charts.

We frequently need to know the temperature in places where it is either inconvenient or impossible to have an observer or a recording device. In such cases a remote sensing thermometer is required. In such situations it is often the electrical resistance thermometer that is put to use. Most electrical conductors—copper, iron, aluminum, etc.—are better conductors of electricity when they are cold than when they are hot. In other words, their electrical resistance increases when temperature increases. One can determine the temperature of a copper wire by measuring its resistance, but good conductors such as copper show such a small change in resistance for a 1°C change in temperature that they usually are not used in meteorological applications. Instead semiconductors (ceramic materials, for instance) called *thermistors* are used. They show changes in resistance with temperature that can be orders of magnitude larger. Unlike metals, their resistance decreases with increasing temperature.

The electrical resistance thermometer consists of a thermistor at the remote position where the air temperature is desired and a recording device that measures and records its resistance at the location convenient to the observer. For measurement of air temperature near the ground, a wire connects the thermistor to the recorder. For temperatures aloft the sensor is carried by free balloon, its varying resistance changing the frequency of a radio signal that is detected at ground level by a radio receiver. Similarly, the thermistors on free-floating buoys govern radio signals transmitted to satellites that relay the data to collection centers.

F.2 Humidity measurement

Three types of humidity-measuring instruments will be described: the *psychrometer*, an indicating device; the *hair-hygrograph*, a recording instrument; and an *electrical hygrometer* used for measuring humidities aloft.

The psychrometer uses two thermometers (Figure F.3). In one of them, called the *wet-bulb thermometer*, the bulb is covered with a moistened muslin or linen wick. When the psychrometer is ventilated with a fan or by spinning it through the air, the rate at which evaporation takes place from the wet bulb depends on how moist the air is. If the air is saturated, 100-percent relative humidity, no evaporation at all takes place from the moistened wick of the wet-bulb, and both wet- and dry-bulb thermometers register identical temperatures. For unsaturated air, less than 100-percent relative humidity, the evaporation from the wet-bulb cools it to a temperature below that of the dry-bulb thermometer. The difference in temperature is called the *wet-bulb depression*. The greater it is, the drier the air must be. To obtain a value for the relative humidity, the observer refers to psychrometric tables, which relate air temperature and wet-bulb depression to the relative humidity.

A different sort of humidity sensor is one whose physical dimensions change when the humidity changes. Many substances swell up when the humidity

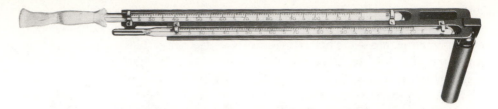

Figure F.3 Photograph of a sling psychrometer. To take an observation the wick is moistened with distilled water and the thermometers are spun around by swinging the handle until the wet-bulb thermometer has cooled to a stable value. The temperatures of the dry-bulb and wet-bulb thermometers are read, and the relative humidity, dew point temperature, and vapor pressure obtained from tables, using the temperature and wet-bulb depression. [Courtesy of Science Associates, Inc.]

increases. In our house some of the doors jam in wet weather, doors that swing free in dry weather. Dresser drawers stick for the same reason—wood increases in size in the presence of air with a high water-vapor content. Ivory displays the same behavior. In fact, an early hygrometer, built by J. A. Deluc in 1773, had an ivory sensor. It was carved in the shape of a delicate (thin-walled) bulb and was attached to a glass tube. It looked very much like a mercury-in-glass thermometer. In dry weather the ivory bulb shrank and forced mercury up the tube. A long mercury column was associated with low humidity, a short one with high humidity.

The physical-deformation type of sensor is used in most dial hygrometers and in recording hygrographs. The material used as the sensor in the modern hygrograph used in most weather stations is the human hair (Figures F.4 and 5.3). A strand of hair 12 in. long will change length by about one-third of an inch when the relative humidity varies from zero to 100 percent. Before it will do this, however, it must be carefully cleaned and all oil removed (by treatment with alcohol and ether). The sensor is usually made up of a harp of many strands to provide enough power to move the magnifying levers that drive the pen across the chart, which is mounted on a clock-driven drum. Hair hygrometers and hygrographs are not as accurate as the psychrometer. The hair gets dirty and as a result its response to humidity does not remain constant. To be useful regular cleaning and recalibration is required.

Measuring humidities aloft requires the use of a humidity sensor that can affect a radio signal. Just as with the remote temperature sensor described above, the one commonly used consists of an element whose electrical resistance varies with humidity. It is a small strip of insulating plastic upon which an electrical conducting layer has been applied, the resistance of which varies with changing humidity. Two types of layers are commonly used. One contains lithium chloride, a highly

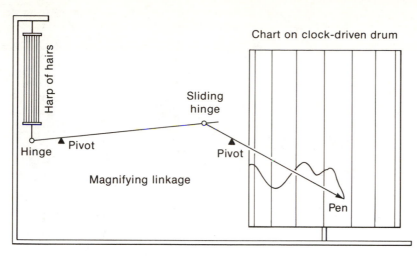

Figure F.4 Schematic diagram of a hair hygrograph. See Figure 5.3 for a photograph of a combined hygrograph and thermograph.

hygroscopic substance. Its great affinity for water causes it to absorb water from the air when the humidity is high, thereby reducing the electrical resistance of the coating. The other type of coating used is a mixture of very finely divided carbon particles and a special cement. This cement swells and shrinks with changing humidity much as wood does. The electrical resistance of this black coating depends on how closely packed the carbon particles are. Electrical current passes through this mixture of cement and carbon more easily when the highly conducting carbon particles are close together. Both the lithium chloride and carbon types of humidity elements are in common use in radiosondes. The balloon-borne radio transmitter sends back signals whose frequencies depend upon the resistance of the coating on the plastic strip.

F.3 Pressure measurement

There are two types of pressure measuring instruments, the *mercurial* and the *aneroid barometers*. The mercurial barometer has the great sensitivity and accuracy required to provide data from which the sea-level pressure field can be constructed. The pressure must be known to the nearest tenth of a millibar. This means that errors must be held to 0.01 percent.

Essentially the mercurial barometer balances a column of mercury against "the column of air" (see Figure F.5). To get an impression of the way the device works,

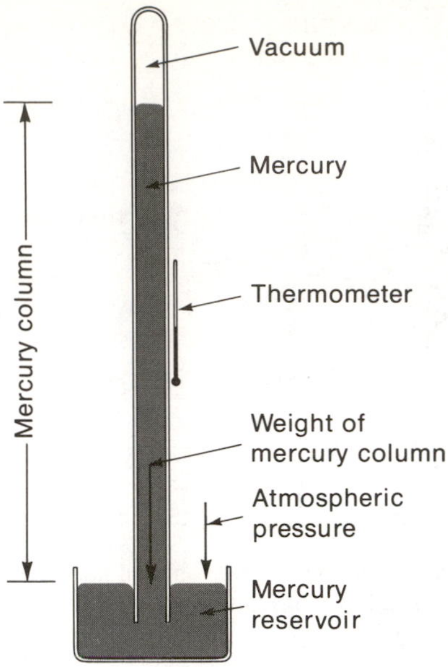

Figure F.5 Schematic diagram of a mercurial barometer. See Figure 4.2 for a photograph of one.

you can perform an experiment in which you slowly lift an upside-down glass out of a basin of water. As the glass emerges bottom-first from the water in the basin, the water inside the glass rises with it. It feels heavy. You are aware of lifting both the inverted glass and the water inside it. This continues until the open end of the glass finally rises above the water level and suddenly air rushes in as the water in the glass drops out.

It is unfortunate for our purposes here that drinking glasses are not 35 feet tall so that you could perform this experiment on a larger scale. Imagine that your wash basin is big enough and that you are strong enough to lift a 35-ft inverted glass out of the water. The water in the glass will continue to rise with the glass for quite a while, but finally, when the closed end of our inverted glass is at about 33 feet above the water level in the basin, the water in the glass will stop rising. It will remain at a fixed level even though you continue to lift the glass, a vacuum developing between the water column and the closed end of the glass. (For simplicity we will ignore any water vapor that might form in this space.)

It seems clear that the glass does not lift the water above 33 feet. The water gets pushed up the glass tube by the water in the basin beneath the open end. The water in the basin is under pressure, the pressure exerted by the over-burden of the atmosphere pushing down on its surface. Pressure equal to atmospheric pressure is exerted in all directions in the water and it is this pressure directed upward into the open end of the glass that holds up the 33-ft column of water. Only an increase in atmospheric pressure can force more water into the tube. Now we are in a position to determine the atmospheric pressure because we can calculate the weight of water that it is supporting. The downward force of gravity on this column of water divided by the cross-sectional area of the glass tube equals that pressure. The measurement we need to make for the calculation is the length of the water column along with the density of the water and the local value of the acceleration of gravity.

Torricelli, the inventor of the barometer, chose mercury, not water, for his barometer. Since mercury is more than 13 times as dense as water, only about 30 in. (76 cm) of it is required to balance the average atmospheric pressure at sea level. To get atmospheric pressure to the desired accuracy (0.1 mb), it is necessary to measure the length of the mercury column to the nearest one-thousandth of an inch and to know the temperature of the mercury to the nearest half-degree Fahrenheit (for determination of the density of the mercury) and the local elevation and latitude (for gravity determination).

Although the mercury barometer represents quite an economy in size compared to our water barometer, it is still a rather bulky piece of plumbing contrasted with most meteorological instruments. For purposes of making continuous recordings of the atmospheric pressure aloft, a smaller, more rugged device is required, the aneroid barometer.

The pressure sensor in the aneroid barometer ("aneroid" means "without liquid") is a flexible air-tight container, usually in the form of a metal disk-shaped can from which most of the air has been removed and a spring inserted. Were it not for the opposing action of the spring, atmospheric pressure would collapse the container. The spring, however, balances the compressional force provided by the atmospheric pressure on the circular ends of the can. If the pressure increases, the spring gives a little and the can gets thinner in the middle. If it decreases the can gets thicker. To make a recording device (Figure F.6), one side of the cell is fixed; the other, the moveable side, is attached to magnifying mechanical linkages that move a pen back and forth when the side moves in and out. This change in thickness of the aneroid cell is so small that even the magnification provided by the mechanical linkages is not enough to allow pressure to be determined to the nearest tenth of a millibar. Consequently, most recording barometers in weather stations have as the pressure sensor a stack of aneroid cells fastened together (looking much like a cylindrical accordion). This increases the sensitivity of the instrument to the point where pressure can be read easily to about the nearest one-half millibar. For balloon-borne measurements, such a stack would usually be too big and heavy in that application, so single aneroid cells are used with, of course, a sacrifice of sensitivity.

Aneroid cells have their inaccuracies and to be useful must be checked regularly against a mercurial barometer. As an example, they are affected by temperature

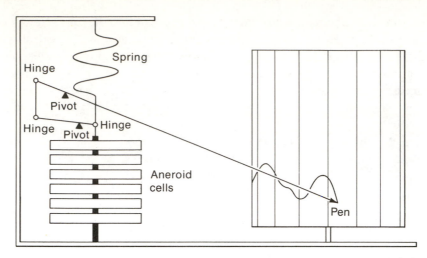

Figure F.6 Schematic diagram of an aneroid barograph.

changes. You would suspect that changes in temperature would change the pressure of whatever gas remains inside the partially evacuated aneroid cell and that they would also expand and contract the mechanical linkages that provide the magnification. These effects, however, turn out to be small compared to another temperature effect, the weakening of the spring with increase in temperature. A good aneroid barometer will have a bimetallic element built into the linkages which is designed to oppose and correct for this effect.

F.4   Upper-air measurements

Almost all upper-air observations of temperature, humidity, and pressure made on a regular operational basis use a free balloon rising through the atmosphere as the instrument platform. The airborne instrument package, consisting of a thermometer, hygrometer, barometer, and radio transmitter, is suspended below the balloon and is called a radiosonde. The radio provides a signal that is received at the ground as a series of alternating audible tones, the frequencies of which correspond to various temperatures and humidities.

The temperature sensor, a thermistor, is an integral part of the audio circuit of the radio transmitter. When its resistance changes, the audio frequency changes. At frequent intervals the humidity sensor, either a lithium chloride or a carbon strip, is switched into the audio circuit in place of the thermistor. A change in its resistance also changes the tone of the audio signal. The switching back and forth between the temperature and humidity signals is accomplished by a pressure-operated switch, the *baroswitch*. It consists of a conducting pen arm moved by an aneroid cell. The pen moves across a tiny keyboard consisting of alternating electrical contacts and insulators. When the pen is in contact with an insulator, the thermistor is in the circuit and the radio is transmitting the temperature tone. When the pen touches one of the keyboard's electrical contacts, an electromagnetic switch disconnects the thermistor in the circuit and replaces it with the humidity strip and the humidity tone is transmitted. By counting the number of times since launch that the signal has switched back and forth between temperature and humidity, one can determine the pressure, since the baroswitch's keyboard is calibrated, each contact corresponding to a given pressure value.

The balloon, about 5 or 6 ft in diameter at ground level, rises to the 100,000-ft level in less than 2 hours. The flight is terminated when the balloon bursts, having expanded to its elastic limit at these heights, where the pressure is but a few millibars. The instrument package, weighing only a few kilograms, returns to the ground by parachute. No attempt is made to retrieve it, although a few are found by accident and some get mailed back to the government for possible repair and reuse. The winds aloft can also be obtained from the radiosonde. (Then it is called a *rawinsonde*.) This is done by tracking its path as it rises using a radio direction finding antenna. The method of wind determination by tracking free balloons is described in the next section, which deals with wind measurement at ground level and aloft.

F.5   Wind measurement

Two types of wind-speed sensors are in common use at weather stations for measuring the surface wind speed, defined as the wind at about 10 m above the earth's surface. In one, the cup anemometer (Figure F.7), a number of cups, usually three or four, are symetrically arranged around a vertical shaft in such a fashion that the open end of each cup faces the closed end of the cup in front of it. Wind striking this array from any direction causes the vertical shaft to rotate, the force exerted by the wind on the open ends of the cups being much greater than that exerted on the backs of the cups. The other sensor, the vane anemometer (see Figure 7.1), rotates about a horizontal shaft and looks like a windmill or an airplane propellor. To measure the wind correctly this sensor must always face into the wind (horizontal shaft parallel to the wind). It therefore requires a wind vane to continually keep it oriented into the wind.

Two different sorts of systems can be used to translate the speed of the rotating anemometer shafts into wind speeds, one mechanical, the other electrical. The mechanical system is the old one and has pretty much been replaced in the field today. The shaft drives a mechanical counter. The observer reads the counter before and after a prescribed time interval, say 10 minutes, and determines the

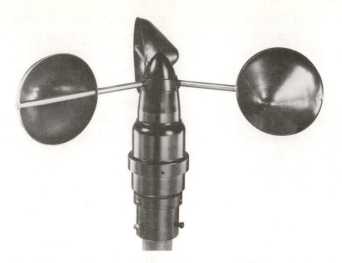

Figure F.7 Photograph of a three-cup anemometer. See Figure 7.1 for picture of windmill-type anemometer. [Courtesy of Science Associates, Inc.]

number of rotations per unit time made by the anemometer. This corresponds to a certain wind speed as determined from the instrument's calibration. The electrical system uses the shaft rotation to generate an electric current whose magnitude is related to the wind speed. This system, of course, is readily adapted to remote indicating and recording, something not true of the mechanical system.

That other characteristic of the surface wind, its direction, also must be measured. The sensor in this case is the wind vane, an asymmetrical object, often resembling an arrow with a large tail, pivoting on a vertical shaft, so that it points into the wind. The orientation of the vane on the shaft with respect to true north can be transmitted either mechanically or electrically (almost always electrically nowadays) to an indicator or recorder in the weather station.

As mentioned earlier, winds at levels above the ground are determined by tracking balloons, usually radiosonde balloons. This is the way it works: The balloon rises vertically through the air but moves horizontally with the air. If we can determine this horizontal component of the balloon's motion at various elevations along its path, we will have determined the wind at these elevations. A convenient way to get the horizontal component of the balloon's motion is to keep track of the point at the surface that lies vertically beneath the balloon (like following the balloon's shadow). The wind speed and direction at the elevation of the balloon is equal to the speed and direction of this "shadow" over the ground. We can determine the location of this "shadow" with respect to the observer (place where

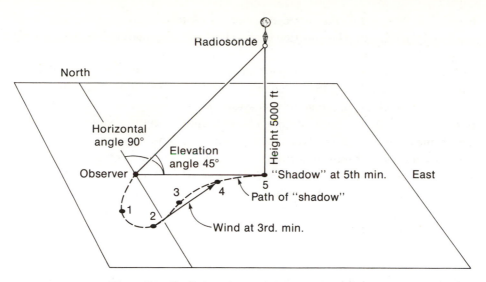

Figure F.8 Illustration of upper-level wind computation.

balloon was launched) if we know the line of sight from the observer to the balloon and the height of the balloon. We get the line of sight by tracking the radiosonde with a radio direction-finding antenna and recording the horizontal and vertical angles of the antenna beam. We get the height of the balloon by using the radiosonde's pressure, temperature, and humidity readings. (The aneroid barometer is an excellent altimeter when corrected for the temperature and humidity of the air column.)

As an example, suppose that 5 minutes after balloon launch, the radio direction-finding antenna is pointing toward the east, upward at an angle above the horizon of 45 degrees (that is, horizontal angle of 90 degrees, elevation angle of 45 degrees) (Figure F.8). And suppose that our elevation computation based on the radiosonde data puts the balloon at 5000 ft above the ground. Then the balloon's "shadow" is going to be due east of the observer. To figure out how far east, imagine yourself moving up this imaginary line that slants from the launch position upward to the east at a 45-degree angle. As you move up the sloping line you keep reeling out a tape measure that tells you how high you are. When the tape measure reads 5000 ft you are at the balloon's position and the other end of the tape measure is at the "shadow's" position. In this case, because the elevation angle is 45 degrees, the horizontal distance from observer to "shadow" is the same as the height of the balloon, 5000 ft. Smaller horizontal angles would give us horizontal distances larger than 5000 ft and larger angles distances less than 5000 ft. If the angle should increase to 90 degrees, the distance would shrink to zero and the balloon would be directly overhead. (The equation used is: horizontal distance to "shadow" equals height of balloon times cotangent of the vertical angle.)

That is how we determine the position of the balloon's "shadow." It is done at 1-minute intervals during the balloon's ascent. To determine the "shadow's" (balloon's) horizontal speed, we merely measure the distance between two successive positions and divide by 1 minute, the elapsed time. To get the wind direction we merely measure the angle between the line connecting these consecutive "shadow" positions and a true south-to-north line. Do this for consecutive pairs of positions and you have the winds at all the heights reached by the balloon. Actually the official procedure is to determine the wind over 2-minute intervals and assign the calculated wind to the height of the balloon at the intermediate point. Obviously the result is not an instantaneous wind but an average wind representing the mean condition over a 2-minute interval and over a height interval given by the balloon's 2-minute vertical displacement.

When radiosonde equipment is not available, winds aloft can be determined by tracking balloons visually, using a tracking telescope called a theodolite. Without the radiosonde data it is necessary to determine the balloon height by some other means. This is done by timing the balloon during ascent and knowing its ascent rate. This works well enough if there are no large updrafts or downdrafts and, of course, just so long as there are no obscuring clouds.

F.6   Precipitation measurement

The rain gauge, used for measuring the accumulation of rain over a 6-, 12-, 18-, or 24-hour period, is a very simple device, a can with an open top (see Figure 5.13A). Its only complication, added so that the precipitation can be measured to the nearest 0.01 in.,[1] is a funnel mounted on top of the can that has a collecting area ten times the cross-sectional area of the can in which the water accumulates. So when 0.01 in. of rain falls into the funnel, 0.1 in. of water collects in the can. It would be impossible for the unaided eye to detect a change of 0.01 in. on a measuring stick, but it easily sees the 0.1-in. change, which thanks to the funnel corresponds to a 0.01-in. accumulation of rain.

It is often of interest to know the rate at which the rain is falling. The simple rain gauge is not read often enough to provide such information with the desired precision. Two types of gauges in current use have this capability. One, the tipping bucket rain gauge, does it by allowing only 0.01 in. of rain to accumulate in a

---

1. In the United States precipitation measurements are recorded in inches; in other countries, in millimeters.

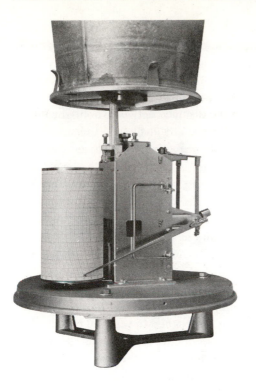

Figure F.9   Detail of weighing rain gauge.
[Courtesy of Science Associates, Inc.]

carefully balanced little bucket before it tips over and spills its contents. At the same moment an identical bucket tips into place and begins to accumulate the next 0.01 in. of rain. In tipping, the buckets make an electrical contact that causes a marker to be printed on a clock driven chart. This then produces a record of the times when each 0.01-in. increment of rain fell. The other type is the weighing rain gauge (Figures F.13B and F.9). It is simply a recording weighing scale that makes a continuous record of the weight of rain water accumulated in the rain gauge as a function of time. For convenience, of course, it reads in inches (not weight units).

Precipitation is notoriously variable from place to place, particularly showery precipitation. One part of a small town may get drenched while another remains completely dry. The official rainfall recorded depends very much on the location of the town's rain gauge. To provide information on the actual spatial coverage of the rainfall, radar can be used. By scanning the atmosphere just above ground level with radar of the proper wavelength (10 cm is fine), a good approximation of the amount of water that is falling in the form of raindrops can be made as a function of horizontal position. In this way the horizontal distribution of rainfall over hundreds of square miles surrounding the radar set can be determined.

F.7   Measurement of the height of cloud bases

Heights of clouds are frequently estimated by eye. However, it is sometimes essential to have more accurate determinations, such as over busy airports during periods of low overcast. Instruments called ceilometers are in use, which continuously record the height of the base of low and middle clouds. The easiest of these to describe is the *fixed-beam ceilometer*. A vertically pointing searchlight puts a bright spot on the bottom of the overcast. At ground level some distance away from the searchlight, usually about 1000 ft, a light-sensitive detector scans the shaft from bottom to top and records the elevation angle at which the bright spot is detected. Large angles denote high cloud bases; small angles, low bases. The readings on the instrument's record chart are conveniently given in terms of height, not angle. The scanning rate of the detector is of the order of a minute or two so that the chart provides a virtually continuous record.

One complicating feature of the ceilometer not mentioned above allows it to function in the full daylight as well as at night. During the day, of course, the clouds are so well illuminated by diffuse sunlight that the searchlight's spot would be undetectable were it not for a modulation (rapid flickering) imposed on the searchlight source. The detecting mechanism is designed to respond only to light fluctuating at the frequency with which the searchlight is modulated. Only when pointed at the spot illuminated by the flickering searchlight does the detector sense this fluctuation in light intensity and thus identify the desired angle.

When it is essential to get a reading on the base of the clouds at airports without ceilometer installations, it is possible to use ceiling balloons to get the answer. A small free balloon of known ascent rate (called *ceiling balloon*) is released and the elapsed time from launch to disappearance in the overcast is measured. The product of the ascent rate and time interval provides the required ceiling value.

F.8   Instrument exposure

Before concluding this discussion of instruments, the problem of the proper exposure of instruments should be mentioned. Returning for the moment to measurement of air temperature, unless properly exposed an accurate thermometer may provide an inaccurate air temperature. To achieve the desired accuracy, the thermometer installation must be such that the sensor assumes the temperature of the air around it. This will not be so if the sensor is exposed to sunlight, in which case it will absorb solar radiation and its temperature will rise above air temperature. At night, if the thermometer is exposed to (within sight of) ground that is colder than the air at thermometer level, it will lose heat to the ground by radiative exchange and take on a temperature lower than the air temperature. Proper exposure of the thermometer requires that it be shielded against such radiative exchange. This is accomplished by installing the thermometer in an instrument shelter, a louvered box having a double roof and painted with highly reflective white paint to reduce radiation effects. Consequently, radiative exchanges with earth and sky are eliminated and those between the inside of the shelter and the thermometer are kept at a minimum. The louvers, of course, permit easy flow of air through the box so that the air in contact with the thermometer is disturbed as little as possible by the shelter.

Another problem involving the exposure of the thermometers arises because in weather map analysis one compares temperatures from station to station. This cannot be done with any accuracy if the temperatures are measured at different heights above the ground at different stations. During a sunny day the temperature decreases very rapidly with increasing height in the first few meters above the ground. If thermometers are not installed at the same level above the ground at all stations, the differences in temperature from station to station may be due to a considerable extent to the height differences. The same is true of wind measurements since air motion also changes rapidly with height in the lowest layer of the atmosphere. The agreed-upon height for thermometers is 1.5 m and for anemometers is 10 m. In the case of wind it is also important that the instrument be remote from trees and buildings that tend to obstruct the movement of air.

Another example involves pressure measurement. The mercurial barometer must be installed in a room that is as free as possible of vertical temperature gradients and in a location in that room that never receives direct sunlight. These precautions are taken so that the instrument has a uniform temperature, the temperature used in determining the density of the mercury.

In installing any meteorological instrument a little common sense usually will avoid the important exposure problems. Would you install the rain gauge under a tree?

G. Beaufort wind scale

Beaufort No.	Description	Specifications L: on land S: at sea far from land	Speed equivalent at height of 10 m	
			m s^{-1}	knots
0	Calm	L: Calm; smoke rises vertically.	0–0.2	1
		S: Sea like a mirror.		
1	Light air	L: Wind direction shown by smoke-drift but not by wind vanes.	0.3–1.5	1–3
		S: Ripples resembling scales are formed, but without foam crests.		
2	Light breeze	L: Wind felt on face; leaves rustle; ordinary vanes moved by wind.	1.6–3.3	4–6
		S: Small wavelets, still short but more pronounced; crests have glassy appearance.		
3	Gentle breeze	L: Leaves and small twigs in constant motion; wind extends small flag.	3.4–5.4	7–10
		S: Large wavelets; crests begin to break; foam of glassy appearance; scattered whitecaps.		
4	Moderate breeze	L: Raises dust and loose paper; small branches are moved.	5.5–7.9	11–16
		S: Small waves, becoming longer; fairly frequent whitecaps.		
5	Fresh breeze	L: Small trees in leaf begin to sway; crested wavelets form on inland waters.	8–10.7	17–21
		S: Moderate waves, taking a more pronounced long form; many whitecaps.		

Beaufort No.	Description	Specifications L: on land S: at sea far from land	Speed equivalent at height of 10 m	
			m s^{-1}	knots
6	Strong breeze	L: Large branches in motion; whistling heard in utility wires; umbrellas used with difficulty. S: Large waves begin to form; the white foam crests are more extensive everywhere (probably some spray).	10.8–13.8	22–27
7	Near gale	L: Whole trees in motion; inconvenience felt when walking against the wind. S: Sea heaps up and white foam from breaking waves begins to be blown in streaks along direction of the wind.	13.9–17.1	28–33
8	Gale	L: Breaks twigs off trees; generally impedes progress. S: Moderately high waves of greater length; edges of crests begin to break into spindrift; foam is blown in well-marked streaks along wind.	17.2–20.7	34–40
9	Strong gale	L: Slight structural damage occurs. S: High waves; dense streaks of foam along wind; crests of waves begin to roll over; spray may affect visibility.	20.8–24.4	41–47

(continued)

G. Beaufort wind scale *(continued)*

Beaufort No.	Description	Specifications L: on land S: at sea far from land	Speed equivalent at height of 10 m	
			m s^{-1}	knots
10	Storm	L: Seldom experienced inland; trees uprooted; considerable structural damage. S: Very high waves with long overhanging crests; foam, in great patches, is blown in dense white streaks along wind; sea takes on a white appearance; tumbling of sea becomes heavy and shocklike; visibility affected.	24.5–28.4	48–55
11	Violent storm	L: Very rarely experienced; accompanied by wide-spread damage. S: Exceptionally high waves (small and medium-sized ships might be for a time lost to view behind waves); the sea is completely covered with long white patches of foam lying along the direction of the wind; everywhere the edges of the wave crests are blown into froth; visibility affected.	28.5–32.6	56–63
12	Hurricane	L: ——— S: The air is filled with foam and spray; sea completely white with driving spray; visibility very seriously affected.	≥ 32.7	≥ 64

Bibliography

Elementary textbooks and popular books

(Recent publications plus a few outstanding earlier ones that may be found in libraries.)

R. A. Anthes et al. *The Atmosphere,* 2nd ed. Columbus, Ohio: Boyd Lane/Merrill, 1978.

R. G. Barry and R. J. Chorley. *Atmosphere, Weather and Climate,* 3rd ed. London: Methuen, 1976.

L. J. Battan. *Fundamentals of Meteorology.* Englewood Cliffs, N.J.: Prentice-Hall, 1979.

D. I. Blumenstock. *The Ocean of Air.* New Brunswick, N.J.: Rutgers University Press, 1959.

S. Bodin. *Weather and Climate.* London: Blandford Colour Series, 1978.

N. Calder. *The Weather Machine.* New York: Viking Press, 1974.

F. Cole. *Introduction to Meteorology,* 3rd ed. New York: Wiley, 1980.

H. S. Critchfield. *General Climatology,* 2nd ed. Englewood Cliffs, N.J.: Prentice-Hall, 1966.

G. M. B. Dobson. *Exploring the Atmosphere.* Oxford: Clarendon Press, 1963.

J. R. Eagleman. *Meteorology: The Atmosphere in Action.* New York: Van Nostrand, 1980.

J. G. Edinger. *Watching for the Wind.* Garden City, N.Y.: Doubleday, 1967.

S. D. Getzelman. *The Science and Wonders of the Atmosphere.* New York: Wiley, 1980.

R. M. Goody and J. C. G. Walker. *Atmospheres.* Englewood Cliffs, N.J.: Prentice-Hall, 1972.

H. Heastic. *A Course in Elementary Meteorology,* 2nd ed. Bracknell, Berks., England: British Meteorological Office. Available in the United States through Pendragon House, Palo Alto, California.

P. J. Hughes. *American Weather Stories*. Washington, D.C.: U.S. Environmental Data Service, 1976. (Order from Government Printing Office, Washington, D.C. 20402.)

E. Linacre and J. Hobbs. *The Australian Climatic Environment*. Brisbane, Australia: Wiley, 1977.

F. K. Lutgens and E. J. Tarbuck. *The Atmosphere*. Englewood Cliffs, N.J.: Prentice-Hall, 1979.

A. Miller and J. C. Thompson. *Elements of Meteorology*, 3rd ed. Columbus, Ohio: Merrill, 1979.

J. G. Navarra. *Atmosphere, Weather and Climate: An Introduction to Meteorology*. Englewood Cliffs, N.J.: Prentice-Hall, 1979.

S. Petterssen. *Introduction to Meteorology*, 4th ed. New York: McGraw-Hill, 1969.

H. Riehl. *Introduction to the Atmosphere*, 3rd ed. New York: McGraw-Hill, 1978.

J. L. Stanford. *Tornado: Accounts of Tornadoes in Iowa*. Ames, Iowa: Iowa State University Press, 1977.

G. T. Trewartha. *An Introduction to Climate*, 4th ed. New York: McGraw-Hill, 1968.

J. E. Weems. *The Tornado*. Garden City, N.Y.: Doubleday, 1977.

Intermediate and advanced general textbooks

H. R. Byers. *General Meteorology*, 4th ed. New York: McGraw-Hill, 1974.

J. A. Dutton. *The Ceaseless Wind: An Introduction to the Theory of Atmospheric Motion*. New York: McGraw-Hill, 1976.

R. G. Fleagle and J. A. Businger. *An Introduction to Atmospheric Physics,* 2nd ed. New York: Academic Press, 1980.

G. J. Haltiner and T. Williams. *Numerical Prediction and Dynamic Meteorology,* 2nd ed. New York: Wiley, 1980.

S. L. Hess. *An Introduction to Theoretical Meteorology*. New York: Henry Holt, 1959.

J. R. Holton. *An Introduction to Dynamic Meteorology*, 2nd ed. New York: Academic Press, 1979.

J. T. Houghton. *The Physics of Atmospheres*. Cambridge, England and New York: Cambridge University Press, 1977. (Paperback published in 1979.)

J. V. Iribarne and H.-R. Cho. *Atmospheric Physics*. Hingham, Mass.: Reidel, 1980.

J. V. Iribarne and W. L. Godson. *Atmospheric Thermodynamics*. Hingham, Mass.: Reidel, 1973.

F. H. Ludlam. *Clouds and Storms*. University Park, Pa.: The Pennsylvania State University Press, 1980.

E. Palmén and C. W. Newton. *Atmospheric Circulation Systems*. New York: Academic Press, 1969.

S. Petterssen. *Weather Analysis and Forecasting. Vol. I: Motion and Motion Systems. Vol. II: Weather and Weather Systems*. New York: McGraw-Hill, 1956.

S. M. Wallace and P. V. Hobbs. *Atmospheric Science: An Introductory Survey*. New York: Academic Press, 1977.

Books, mostly advanced technical, on special topics

Atmospheric Phenomena, Readings from the Scientific American. Introduction by D. K. Lynch. San Francisco: W. H. Freeman and Co., 1980.

W. Bach. *Man's Impact on Climate.* New York: Elsevier, 1979.

E. C. Barrett. *Climatology from Satellites.* London and New York: Methuen, 1974. (Reprinted with paperback edition in 1979.)

B. Bolin, ed. *The Global Carbon Cycle.* New York: Wiley, 1979.

G. Breuer. *Weather Modification—Prospects and Problems.* Cambridge, England and New York: Cambridge University Press, 1979.

M. I. Budyko. *Climatic Change.* Washington, D.C.: American Geophysical Union, 1977.

H. R. Byers and R. R. Braham: *The Thunderstorm.* Washington, D.C.: U.S. Weather Bureau, 1949.

Climate Research Board, NRC/NAS. *Carbon Dioxide and Climate: A Scientific Assessment.* Washington, D.C.: National Academy of Sciences, 1979.

K. L. Coulson. *Solar and Terrestrial Radiation—Methods and Measurements.* New York: Academic Press, 1975.

R. J. Davis and L. O. Grant. *Weather Modification—Technology and Law.* Boulder, Colorado: Westview Press, 1978.

A. Dennis. *Weather Modification by Cloud Seeding.* New York: Academic Press, 1980.

J. R. Eagleton, V. J. Muirhead, and N. Willems. *Thunderstorms, Tornadoes, and Building Damage.* New York: Heath, 1975.

R. H. Eather. *Majestic Lights: The Aurora in Science, History and the Arts.* Washington, D.C.: American Geophysical Union, 1980.

R. H. Golde. *Lightning and Lightning Protection. Vol. 1: Physics of Lightning. Vol. 2: Lightning Protection.* New York: Academic Press, 1977.

W. N. Hess, ed. *Weather and Climate Modification.* New York: Wiley, 1974.

L. F. Hubert and P. E. Lehr. *Weather Satellites.* Waltham, Mass.: Blaisdell, 1967.

W. W. Kellogg. *Effects of Human Activities on Global Climate.* WMO Technical Note No. 156 (WMO No. 486). Geneva, Switzerland: World Meteorological Organization, 1977.

W. W. Kellogg and M. Mead. *The Atmosphere: Endangered and Endangering.* Washington, D.C.: U.S. Department of Health, Education and Welfare, 1977.

W. W. Kellogg and R. Schware. *Climate Change and Society.* Boulder, Colorado: Westview Press, 1981.

G. Kutzbach. *The Thermal Theory of Cyclones. A History of Meteorological Thought in the Nineteenth Century.* Boston: American Meteorological Society, 1979.

H. H. Lamb. *Climate: Present, Past and Future. Vol. 1: Fundamentals and Climate Now, 1972. Vol. 2: Climatic History and the Future, 1977.* London and New York: Chapman and Hall/Methuen.

C. E. Leith and C. G. Little. *Severe Storms: Prediction, Detection, and Warning. Report of the Panel on Short-Range Prediction and the Panel on Severe Storms.* Washington, D.C.: National Academy of Sciences, 1977.

K.-N. Liou. *An Introduction to Atmospheric Radiation*. New York: Academic Press, 1980.

E. N. Lorenz. *The Nature and Theory of the General Circulation of the Atmosphere*. Geneva, Switzerland: World Meteorological Organization, 1967.

C. Magono. *Thunderstorms*. New York: Elsevier, 1980.

Man's Impact on the Global Environment—Assessment and Recommendations for Action. Report of the Study of Critical Environmental Problems. Cambridge, Mass.: MIT Press, 1970.

B. J. Mason. *Clouds, Rain and Rainmaking*, 2nd ed. Cambridge, England and New York: Cambridge University Press, 1975.

————. *The Physics of Clouds*, 2nd ed. Oxford: Clarendon Press, 1971.

W. K. Middleton. *The History of the Barometer*. Baltimore, Md.: Johns Hopkins Press, 1964.

————. *A History of the Thermometer and Its Use in Meteorology*. Baltimore, Md.: Johns Hopkins Press, 1966A.

————. *A History of the Theories of Rain*. New York: Franklin Watts, 1966B.

————. *The Invention of the Meteorological Instruments*. Baltimore, Md.: Johns Hopkins Press, 1968.

S. Niewold. *Tropical Climatology. An Introduction to the Climates of the Low Latitudes*. New York: Wiley, 1977.

Operations of the National Weather Service. Silver Spring, Md.: U.S. Department of Commerce, NOAA, National Weather Service, 1978.

G. W. Paltridge and C. M. R. Platt. *Radiative Processes in Meteorology and Climatology*. New York: Elsevier, 1976.

H. R. Pruppacher and J. D. Klett. *Microphysics of Clouds and Precipitation*. Hingham, Mass.: Reidel, 1978.

H. Riehl. *Climate and Weather in the Tropics*. New York: Academic Press, 1979.

W. O. Roberts and H. Lansford. *The Climate Mandate*. San Francisco: W. H. Freeman and Co., 1979.

R. R. Rogers. *A Short Course in Cloud Physics*, 2nd ed. New York: Pergamon Press, 1979.

R. Scorer. *Clouds of the World*. Harrisburg, Pa.: Stackpole Books, 1972.

R. H. Simpson and H. Riehl. *The Hurricane and Its Impact*. Baton Rouge, La.: Louisiana State University Press, 1981.

A. Stern, ed. *Air Pollution*, 3rd ed. *Vol. 1: Air Pollutants: Their Transformations and Transport; Vol. 2: The Effects of Air Pollution; Vol. 3: Measuring, Monitoring and Surveillance of Air Pollution; Vol. 4: Engineering Control of Air Pollution; Vol. 5: Air Quality Management*. New York: Academic Press, 1976–1977.

U.S. Committee for the GARP, National Research Council. *Understanding Climatic Change*. Washington, D.C.: National Academy of Sciences, 1975.

Weather Modification Advisory Board. *The Management of Weather Resources*. Washington, D.C.: U.S. Department of Commerce, 1978.

O. R. White, ed. *The Solar Output and Its Variation*. Boulder, Colorado: Colorado Associated University Press, 1977.

W. K. Widger. *Meteorological Satellites*. New York: Holt, Rinehart and Winston, 1966.

J. Williams, ed. *Carbon Dioxide, Climate, and Society*. New York: Pergamon Press, 1978.

S. J. Williamson. *Fundamentals of Air Pollution*. Reading, Mass.: Addison-Wesley, 1973.

Workshop on the Global Effects of Carbon Dioxide from Fossil Fuels. Washington, D.C.: U.S. Department of Energy Office of Health and Environmental Research, 1979.

World Meteorological Organization. *Climate Research Programme and the Global Atmospheric Research Programme*. Geneva, Switzerland: World Meteorological Organization, 1980.

———. *World Weather Watch: The Plan and the Implementation Programme 1980–1983*. Geneva, Switzerland: World Meteorological Organization, 1979.

Reference books

F. A. Berry, E. Bollay, and N. R. Beers, eds. *Handbook of Meteorology*. New York: McGraw-Hill, 1945. (Note: While very old, this work contains much standard useful material.)

R. W. Fairbridge, ed. *Encyclopedia of Atmospheric Sciences and Astrogeology*. New York: Academic Press, 1967.

R. E. Huschke, ed. *Glossary of Meteorology*. Boston: American Meteorological Society, 1959, reprinted 1980.

S. Letestu, ed. *International Meteorological Tables*. Geneva, Switzerland: World Meteorological Organization, 1966.

R. J. List. *Smithsonian Meteorological Tables*. Washington, D.C.: Smithsonian Institution, 1951.

D. H. McIntosh (British Meteorological Office). *Meteorological Glossary*. London: H. M. Stationers Office, 1963.

T. F. Malone, ed. *Compendium of Meteorology*. Boston: American Meteorological Society, 1951.

S. P. Parker, ed. *McGraw-Hill Encyclopedia of Ocean and Atmospheric Sciences*. New York: McGraw-Hill, 1980.

J. A. Ruffner and F. E. Blair. *Weather Almanac,* 2nd ed. Detroit, Mich.: Gale Research, 1977.

U. S. Standard Atmosphere, 1976. Published by NOAA, NASA, and U.S. Air Force. (Purchase from U.S. Government Printing Office, Washington, D.C. 20402.)

World Meteorological Organization. *International Cloud Atlas,* Revised ed. Vol. I. Geneva, Switzerland: World Meteorological Organization, 1975.

World Meteorological Organization. *International Cloud Atlas,* Abridged ed. Geneva, Switzerland: World Meteorological Organization, reprinted 1969.

Periodicals and serials

Popular and Semitechnical Journals

Bulletin of the American Meteorological Society
Weather. Bracknell, Berks., England: Royal Meteorological Society.
Weatherwise. Washington, D.C.: Heldref Publications.
WMO Bulletin. Geneva, Switzerland: World Meteorological Organization.

Professional and Technical Journals

Atmosphere-Ocean (A-O). *The Scientific Journal of the Canadian Meteorological and Oceanographic Society*. (Order from University of Toronto Press.)
EOS—Transactions of the American Geophysical Union.
Journal of Applied Meteorology (JAM). (Published by the American Meteorological Society.)
Journal of Atmospheric Sciences (JAS). (Published by the American Meteorological Society.)
Journal of Geophysical Research, Oceans and Atmospheres (JGR). (Published by the American Geophysical Union.)
Monthly Weather Review (MWR). (Published by the American Meteorological Society.)
National Weather Digest (NWD). (Published by the National Weather Association, Clinton, Md.)
Quarterly Journal of the Royal Meteorological Society (QJRMS).

Serials

Meteorological Monographs. (Published by the American Meteorological Society.) Recent issues include No. 33: *Meteorological Observations and Instrumentation, 1970*; No. 35: *Meteorology of the Southern Hemisphere, 1972*; No. 38: *Hail: A Review of Hail Science and Hail Suppression, 1977*; and No. 39: *Solar Radiation and Clouds, 1980*.
Historical Monographs. (Published by the American Meteorological Society.)

Periodical Guides and Abstracting Journals

Chemical Abstracts.
General Science Index.
Meteorological and Geoastrophysical Abstracts.
Physics Abstracts.
Readers Guide to Periodical Literature.

Glossary

This glossary gives brief definitions of most of the specialized terms used in this book. More complete explanations of many of them can be found by looking at the pages referred to in the index. For words not included here, consult any good collegiate or unabridged dictionary, or the *Glossary of Meteorology* published by the American Meteorological Society.

absolute humidity   Mass of water vapor per unit volume; the density of the water vapor in moist air.

absolute instability   A condition of an air column such that a slight vertical displacement of an air parcel would result in further acceleration of the parcel in the direction of the displacement, independent of whether it is saturated or unsaturated. The criterion for absolute instability is that the air column have a lapse rate greater than the unsaturated adiabatic rate of cooling.

absolute stability   A condition of an air column such that a slight vertical displacement would result in an acceleration tending to restore the parcel to its original position, independent of whether the air is saturated or unsaturated. The criterion for absolute stability is that the lapse rate in the air column be less than the saturated adiabatic rate of cooling.

absolute temperature   The temperature in degrees Kelvin or degrees Celsius above absolute zero. The temperature on the Kelvin temperature scale.

absolute vorticity   The vorticity of an air parcel relative to a coordinate system fixed with respect to the stars. It is the sum of the vorticity relative to the earth's surface and the vorticity due to the earth's rotation at the position of the parcel.

absolute zero   The zero point of the Kelvin temperature scale, -273.15 on the Celsius or centigrade scale.

absorption   The process by which radiant energy falling on matter is retained by it. The energy may be converted to heat and raise the temperature of the absorbing substance, or it may ionize or dissociate it.

absorptivity   The fraction of the incident radiant energy that is absorbed by a substance.

acceleration   The rate of change of velocity per unit time. In straight line motion it is the rate of change of speed. If the motion is curved there is an additional part of the acceleration, the *centripetal acceleration* (q.v.), due to the change in direction.

acceleration of gravity   The rate of increase in speed of unit mass falling in a vacuum due to the attractive force of the earth. See *gravity*.

acid rain   Precipitation that carries to earth sulfuric acid and nitric acid accumulated from air pollutants.

adiabat   A curve on a thermodynamic diagram representing the change in state variables (for instance, pressure and temperature) during an adiabatic process.

adiabatic lapse rate   The rate of decrease of temperature with height in an air column when that rate is equal to an adiabatic process rate of cooling. Usually the unsaturated or dry adiabatic rate is meant; when the saturation adiabatic lapse rate is meant it is always specified.

adiabatic process   A change in the value of the state variables (temperature, pressure, density) in which there is no heat added to or withdrawn from the substance under consideration, usually an air parcel in meteorological considerations.

adiabatic rate of cooling   The rate of decrease of temperature of an air parcel when its pressure is decreased in an adiabatic process. If the pressure decrease is due to an upward displacement of an unsaturated parcel, the rate (the unsaturated adiabatic rate) is 0.98 K per 100 m displacement. If the parcel is saturated, the rate (the saturated adiabatic rate) is less by an amount which depends on its temperature and pressure.

advection   The transport of air and its properties by horizontal motion.

advection fog   A fog that forms when moist air moves over a sufficiently colder surface.

aerosol   A suspension of particles in air. By extension, it is sometimes used to mean the particles of dust, etc., that are suspended in it.

air   The mixture of gases comprising the atmosphere.

airglow   A radiant emission from the high atmosphere. Also called *light of the night sky*.

air mass   An extensive body of air with properties (temperature and humidity) approximately homogeneous in the horizontal.

air parcel   A portion of air small enough to have approximately uniform properties, but large enough to contain very many molecules, that can be subjected to imaginary motions and other changes in order to study thermodynamic and dynamic processes.

air pollution   Toxic or otherwise harmful substances introduced into the atmosphere by human activities.

air pollution potential   The degree and frequency of meteorological conditions in a locality favorable to the accumulation and persistence of high concentrations of air pollution, independent of whether there are sources of pollutants in the locality.

albedo   The percentage of the total electromagnetic radiation falling on a surface that is reflected by it.

Aleutian low   The low pressure center, with associated cyclonic circulation, located on mean charts of sea-level pressure, and frequently on individual daily charts, in the vicinity of the Aleutian Islands. It is one of the *centers of action*.

alpha ray A stream of positively charged particles (alpha particles) emitted during radioactive disintegration of the nuclei of certain atoms, such as radon and thoron.

altimeter An instrument used to determine the height above sea level. The pressure altimeter is an aneroid barometer with a scale giving heights corresponding to the pressure in the Standard Atmosphere, and must be corrected for pressure and temperature deviations from the Standard Atmosphere.

altocumulus A middle-level cloud, white or gray in color, consisting of waves, rolls, or small heaplike patches in a layer.

altostratus A middle-level cloud, in the form of a gray or bluish layer or sheet, usually uniform but sometimes striated or fibrous in appearance.

anemometer An instrument for measuring wind speed.

aneroid barometer An instrument for measuring atmospheric pressure. (See Appendix F for description.)

Ångstrom A unit of length, mostly used in expressing wavelengths of short electromagnetic radiation. It is equal to 10^{-10} m.

angular momentum A measure of the amount of rotational motion. Its magnitude is mrv, where m is the mass of the moving body, v its speed, and r the distance from the axis around which it is rotating.

angular velocity The rate of rotation of a body about the axis of rotation.

anticyclone A body of air flowing around a center of high pressure, in a clockwise direction in the Northern Hemisphere or a counterclockwise direction in the Southern Hemisphere.

anticyclonic flow Flow in the direction air flows around an anticyclone—clockwise in the Northern Hemisphere and counterclockwise in the Southern Hemisphere.

aphelion The position in the earth's orbit farthest from the sun.

Arctic front The boundary between arctic air and polar air.

Arctic sea smoke Steam fog forming when frigid air from snow- or ice-covered regions of the Arctic flows over the adjacent open ocean.

atmosphere (1) The envelope of air surrounding the earth and moving with it by virtue of the earth's gravitational attraction. (2) A unit of pressure equal to the pressure at sea level in the Standard Atmosphere—101,325 Pa or 1013.25 mb.

atmospheric pressure The pressure at a point in the atmosphere due to the weight of the air above it.

atmospheric window The infrared spectral region between about 9 and 12 μm wavelength in which the atmosphere is practically transparent except for absorption by ozone at 9.6 μm.

atom The smallest portion into which matter can be divided and still retain its chemical character as an element.

aurora Radiant emissions in the upper atmosphere at high latitudes in the form of rays or curtains of colored light. In the Northern Hemisphere they are called *aurora borealis* or *northern lights;* in the Southern Hemisphere, *aurora australis*.

baroclinic Characterized by density varying along surfaces of constant pressure, so that surfaces of constant pressure and surfaces of constant density intersect.

baroclinicity A baroclinic state of stratification in a region.

barograph   A device for recording the atmospheric pressure. A recording barometer.

barometer   An instrument for measuring atmospheric pressure.

baroswitch   A switching device operated by an aneroid pressure capsule, used in radiosondes to shift from transmission of the temperature signal to a humidity or reference signal and back. The pressure is known each time the switch changes the signal.

barotropic   Characterized by constancy of density on surfaces of constant pressure, so that surfaces of constant density coincide with surfaces of constant pressure.

Beaufort force   A number denoting the strength of the wind as given by the Beaufort Wind Scale.

Beaufort Wind Scale   A system for estimating wind speeds in terms of the character of the sea surface or the effects on trees, people, and structures on land.

Bergeron process   The three-phase process of formation of precipitation by the transfer of water substance from supercooled droplets to ice crystals, because the equilibrium vapor pressure over ice is lower than that of water at the same subfreezing temperature.

bimetallic strip   A thermometer element consisting of a strip formed by welding together two metals having different coefficients of expansion. Because the two metals expand or shrink differently when the temperature changes, the strip curves by an amount that depends on the temperature.

biosphere   The zone at the surface of the earth and in the oceans and other bodies of water in which life occurs. Also, the totality of living organisms that occupy that zone.

black body   A hypothetical substance that can absorb all electromagnetic radiation falling on it, and consequently can radiate the maximum amount of radiation possible at all wavelengths at any given temperature.

black body radiation   The electromagnetic radiation emitted by a black body, that is, the maximum amount of radiant energy at all wavelengths that can be emitted by a substance at a given temperature.

blizzard   An intense snow storm with strong winds, heavy snow, and low temperatures.

boiling point   The temperature at which the equilibrium vapor pressure of a liquid is equal to the atmospheric pressure. The boiling point of pure water at standard pressure is used as a fiducial point for calibrating thermometers, 100°C or 212°F.

bora   A cold northeast wind descending the mountains to the coast of the Adriatic. Being very cold, it descends the mountains like a waterfall. More generally, the term is used in other places where similar fall winds occur.

Buys Ballot's law   A statement of the relation between wind direction and pressure distribution: If you stand with your back to the wind, lower pressure will be to your left in the Northern Hemisphere, to your right in the Southern Hemisphere.

calorie   A unit of heat, defined originally as the amount required to raise the temperature of one gram of pure water from 14.5°C to 15.5°C. It is equivalent to 4.186 joules.

carbon dioxide   A colorless gas, chemical formula CO_2 (its molecule consisting of one atom of carbon and two atoms of oxygen), forming the fourth most plentiful constituent of air.

carbon monoxide   A colorless, odorless gas, chemical formula CO; extremely toxic, it is one of the major harmful constituents of air pollution.

ceiling   The height of the lowest layer of clouds that obscures more than one-half the sky.

ceilometer   A device for measuring and recording the ceiling height. It consists of a searchlight that projects a narrow modulated vertical beam. The angle from the ground to the spot formed on the cloud is observed by a photocell detector and converted automatically to the corresponding cloud height.

Celsius temperature scale   A temperature scale in which the melting point of ice is taken to be 0° and the boiling point of water at standard pressure 100°.

center of action   Any of the semi-permanent highs and lows on charts of mean sea level pressure. The main centers of action in the Northern Hemisphere are the Aleutian low, the Icelandic low, the Pacific high, the Bermuda high, the Siberian high (in winter), and the Asian low (in summer).

centrifugal force   The apparent force, in curving motion, that appears to try to deflect the moving body outward from the center of curvature.

centripetal acceleration   The acceleration involved in the change of direction of a body moving in a curved path. It is equal and opposite to the centrifugal force.

chinook   A foehn-type wind on the eastern side of the Rocky Mountains.

cirrocumulus   A high-level cloud composed of a thin layer of small white elements in the form of grains or ripples.

cirrostratus   A high-level cloud in the form of a whitish fibrous veil which covers part or all of the sky. When it covers the sun it may produce a halo.

cirrus   A high-level cloud composed of detached elements in the form of fibrous white streaks, patches, or narrow bands.

clear-air turbulence (*CAT*)   Turbulence encountered by airplanes when flying in cloud-free air.

climate   The totality of weather over long periods in a region, usually summarized by averages and statistical measures of variability.

cloud   A visible aggregate of minute water drops and/or ice crystals formed by condensation of water vapor at levels above the earth's surface.

cloud seeding   The process of introducing artificially into a cloud particles of substances intended to alter the cloud and its natural development.

coalescence   The merging of two cloud drops into a single larger drop.

cold front   A boundary between two air masses that moves so that the colder air replaces the warmer air.

cold-front-type occlusion   An occluded front in which the air behind the front is colder than the air ahead of it.

cold low   A low pressure center that is completely surrounded by cold air, with warmer air farther from the low center. In this circumstance the low extends upward to great heights.

cold wave   A sudden rapid decrease in temperature to extremely low values for the region.

collision-coalescence process   The process of precipitation formation by growth of larger cloud drops falling more rapidly than small ones and collecting them.

collision efficiency   The fraction of the drops in the path of a larger cloud drop that actually do collide with it.

compound   A chemical substance consisting of molecules composed of atoms of more than one element.

condensation   The process by which a substance changes from its gaseous or vapor phase to its liquid phase. The change from gaseous to solid is sometimes called condensation also, but in meteorology it is called *sublimation* or *deposition*.

condensation nucleus   A particle, usually hygroscopic, on which condensation is initiated in the formation of clouds or fog.

condensation trail   A streamer of cloud formed behind an aircraft flying at high levels in cold moist air.

conditional instability   The condition of an air column such that, if a parcel in it is saturated, a slight vertical displacement would result in further acceleration of the parcel in the direction of the displacement, but if the parcel is unsaturated it would result in an acceleration tending to restore it to its original position. The criterion for conditional instability is that the lapse rate be greater than the saturated adiabatic process rate but less than the unsaturated adiabatic rate of cooling.

conduction   The transfer of energy through an object or from one object to another by molecular vibrations and collisions, without movement or exchange of matter.

coning   The spreading uniformly in all cross-wind directions of a plume from a smoke stack.

constant absolute vorticity trajectory   The path of an air parcel whose motion is governed by the requirement that its absolute vorticity remain constant.

constant-pressure chart   A map summarizing the height, temperature, humidity, and winds on a constant-pressure surface at a given time.

constant-pressure surface   A surface along which the pressure has a prescribed constant value at a particular instant.

contour line (also called *contour*)   A line of constant elevation, used to define the position and configuration of a surface, such as the earth's surface or a constant-pressure surface.

contrail   A condensation trail.

convection   In general, motions of portions of a fluid resulting in transport and mixing of its properties. In meteorology, convection refers to predominantly vertical motions, and advection is used to speak of motions that are predominantly horizontal.

convective condensation level (CCL)   The level at which free convection would be released by heating the bottom of an air column and mixing the heated layer until its lapse rate is dry adiabatic and its mixing ratio is constant at the average of the layer.

convective instability   The condition of an unsaturated air column, the wet-bulb potential temperature of which decreases with height. If such a column is lifted bodily until it is saturated, its lapse rate will be greater than the saturation-adiabatic lapse rate, and it will be unstable.

convergence   Flow into a region. For three-dimensional convergence, the density must increase. Horizontal convergence with no density change results in vertical motions. (Opposite of *divergence*.)

Coriolis force   The force that appears to act on a body moving relative to a rotating system when viewed from that system. Projectiles or air moving relative to the earth's surface behave as though acted on by a force tending to bend their paths to the right in the Northern Hemisphere and to the left in the Southern Hemisphere.

Coriolis parameter   The quantity $2\Omega \sin \phi$, which multiplies the speed v of the moving object to give the magnitude of the horizontal component of the Coriolis force per unit mass.

corpuscular radiation   Transfer of energy by the flow of particles streaming from the sun, to form the *solar wind*.

cross section   A two-dimensional representation of a vertical "slice" through the atmosphere.

cumulus   A cloud of vertical development in the form of individual puffs or heaps, with flat bases and rising domes or towers, the tops of which frequently resemble a cauliflower.

cyclogenesis   The formation, development, or intensification of a cyclone.

cyclone   A closed circulation around a low pressure center, counterclockwise in the Northern Hemisphere and clockwise in the Southern Hemisphere.

cyclone family   A series of wave cyclones along a frontal surface. The more mature, occluded cyclone waves are usually farther poleward, and the young waves are farther away from the pole.

cyclonic   Having the sense of rotation of the circulation around a cyclone.

cyclonic shear   Horizontal variation of wind such that small parcels of air tend to be rotated in a cyclonic sense.

density   Mass of a substance per unit volume.

deposition   Change of phase directly from vapor to solid. Same as *sublimation*.

depression   A center or trough of low pressure.

dew   Water condensed onto a surface, such as grass and other objects on the ground.

dew point (or *dew point temperature*)   The temperature to which air having a particular amount of water vapor must be cooled at constant pressure to produce saturation.

dissociation   The breaking of a molecule of a substance into two or more component parts, usually by absorption of radiation.

diurnal   Varying or occurring during a single day, usually in the sense of recurring every day.

divergence   Flow out of a region. Three-dimensional divergence must be accompanied by a decrease in density. Horizontal divergence with no or little change in density is accompanied by vertical motion. For instance, horizontal divergence in the layers next to the ground results in subsidence of air from higher levels. (Opposite of *convergence*.)

doldrums   The older name for the region of calms or light winds where the trade winds of the two hemispheres meet near the equator. The intertropical convergence zone.

Doppler radar   A radar equipped to detect and interpret the difference in signal frequency due to the velocity of moving targets.

drag   The resisting force air exerts on bodies moving through it, or the resisting force exerted on air by surfaces over which it is moving.

drizzle   Precipitation consisting of numerous very small droplets (less than 0.5 mm in diameter).

drought   A prolonged period during which abnormally little precipitation falls.

dry adiabat   A line on a thermodynamic diagram representing the variation of state variables (e.g., temperature and pressure) in an unsaturated adiabatic process.

dry adiabatic lapse rate   The rate of decrease of temperature with height in an air column when that rate is the same as the rate of cooling in an unsaturated adiabatic process.

dry adiabatic process   Same as unsaturated adiabatic process.

dry air   (1) Air that contains no water vapor. (2) The components of air, omitting its water vapor content.

dry-bulb temperature   The temperature of the air, as determined by the dry-bulb thermometer of a psychrometer.

dust devil   A small vigorous vortex or whirlwind, induced by heating, which lifts dust or sand from the ground to considerable heights.

dynamics   The study of physical processes involving the relationship between forces and motion. In dynamic meteorology these relationships are studied by analyzing and solving special forms of the Navier–Stokes equations of fluid dynamics, the application of Newton's laws of motion to fluids.

dyne   The unit of force in the cgs system. 1 dyne $= 10^{-5}$ newtons.

easterly wave   A wave in the easterly trade wind regions, moving from east to west, usually less rapidly than the wind.

echo-free vault   The region in a super-cell thunderstorm in which the updrafts are so strong that precipitation large enough to reflect radar signals cannot fall into it.

eddy   A swirl or quasi-circular circulation within a larger flow pattern.

electromagnetic radiation   The transfer of energy in the form of a wave disturbance in the electric and magnetic fields through which it passes. Also, the energy so transmitted.

electron   A subatomic particle possessing a very small mass and negative charge.

element   A substance that cannot be broken down by ordinary chemical means into simpler components; a substance having atoms with an identical number of protons in their nuclei.

energy   The capacity for doing work.

entrainment   The addition of environmental air into a rising current or a cloud by mixing.

equation of state   An equation relating the condition of a physical system in terms of its temperature, pressure, and volume or density. For gases such as air, this is represented approximately by the *perfect gas law, pα = RT.*

equilibrium   A balanced state, in which all existing forces are in balance, i.e., the vector sum of the forces is zero.

equinox   One of the two points where the ecliptic—the sun's apparent path among the fixed stars—intersects the plane of the earth's equator. Also, the dates when the sun reaches one of these points, approximately March 21 or September 22.

erg   The unit of energy in the cgs system. 1 erg $= 10^{-7}$ joules.

evaporation   The process of change of phase of a substance from liquid to gas or vapor. The opposite of *condensation.*

extratropical   Occurring at higher latitudes than the Tropics of Cancer and Capricorn.

eye (of the storm)   The region of calms or light winds at the center of a severe tropical storm (hurricane, typhoon, etc.).

Fahrenheit temperature scale   The temperature scale in common use in the United States, in which the ice point (at which ice melts) is at 32° and the boiling point of water is at 212°.

fall wind A strong, cold, downslope wind that descends mountains much like a waterfall. A fall wind differs from a foehn in that the air is so cold initially that it is still relatively cold in spite of being warmed adiabatically because of the descent.

fanning The cross-wind spreading in the horizontal of the plume from a smokestack, with little vertical spreading, in the presence of an inversion.

Ferrel cell The reverse meridional cell that is present at middle latitudes—the latitudes of the prevailing westerlies—in the mean global circulation of the atmosphere.

flux The flow or transport of an atmospheric property. Also, the rate of this transport.

foehn A hot dry wind descending the leeward side of a mountain range.

fog A suspension of minute water drops or ice crystals near the ground, restricting the visibility. By international convention it is termed *fog* if the visibility is 1 km or less, and *mist* if the visibility is larger.

force A physical quantity that produces an acceleration when applied to matter.

fossil fuels Combustible materials resulting from the remains of organisms deposited in past geological eras. They include coal, petroleum, peat, and natural gas.

freezing The transition of a substance from the liquid phase to the solid. The same as *fusion*.

freezing nucleus An ice-forming nucleus that initiates freezing in water. Most freezing nuclei are effective only at temperatures below 0°C.

frequency In describing wave phenomena, the number of oscillations at a given point per unit time, usually per second.

front The sloping boundary between two air masses characterized by different properties, usually temperature, humidity, or density. The boundary is usually regarded as a surface, and the intersection with the ground, a line, but in fact it is ordinarily a zone or layer of relatively rapid transition.

frontal inversion A temperature inversion that is present when the air mass above a sloping front is sufficiently warmer than the air below it.

frontogenesis The process of formation or intensification of a front.

frontolysis The process of weakening and disappearance of a front.

frost A deposit of ice crystals on surfaces when the dew point is below 0°C and the temperature of the surfaces cools below the dew point (or more properly, the *frost point*). Because the saturation vapor pressure over ice is lower than over supercooled water at the same temperature, the frost point temperature is slightly higher than the dew point temperature.

fumigation The transport to the ground of a smoke plume when the lapse rate below the level of the stack exit becomes adiabatic after having been more stable.

fusion (1) The transition from liquid to solid phase of a substance. (Same as *freezing*.) (2) The reaction by which two small atomic nuclei combine to form a larger one, releasing large amounts of energy thereby. Also called *nuclear fusion*.

gale A strong wind, in the range from 14 m s^{-1} to 27 m s^{-1} (28–55 knots).

gas constant The constant factor in the equation of state for perfect gases. The gas constant for any particular gas is the *universal gas constant* divided by the *molecular weight*.

general circulation   The broad aspects of the motions of the earth's atmosphere. Also called the *global circulation*.

geostationary satellite   An artificial satellite circling the earth above the equator at the same speed as the earth rotates, so that it remains above the same spot on earth.

geostrophic wind   The wind that would have the speed and direction for which the Coriolis force would exactly balance the pressure gradient force for the existing pressure distribution.

global circulation   Same as the *general circulation*.

global radiation   The sum of the radiation received directly from the sun and the solar radiation scattered by the atmosphere and received from all parts of the sky.

gradient   The rate of decrease with distance of a quantity, ordinarily in the direction in which it decreases most rapidly.

gradient wind   For the case of curved constant-level isobars or constant-pressure contour lines, the wind velocity that would result in the air following the curved isobar or contour line. For this to be true the wind direction would be tangent to the isobar or contour, and the speed would be such that the sum of the Coriolis force and the pressure gradient force would produce the acceleration required to have the air turn with the curvature of the isobar or contour line.

gram   The unit of mass in the cgs system, originally defined as the mass of 1 cm^3 of water at 4°C.

gravity   The force acting on a mass due to the gravitational attraction of the earth, reduced by the centrifugal force of the earth's rotation.

greenhouse effect   The increase in temperature at and near the earth's surface due to the absorption and re-emission of infrared radiation by water vapor and carbon dioxide in the atmosphere, so that the radiation received at the ground, the combination of incoming solar radiation and the infrared radiation from the atmosphere, is much greater than it would be in the absence of an atmosphere.

gust   A sudden brief increase in wind speed.

gust front   The surge of strong winds and cold air coming out of the forward edge of the downdraft in a severe thunderstorm.

Hadley cell   A direct meridional cell of the mean global circulation, driven by the excess heating near the equator.

hail   Precipitation in the form of quasi-spherical or irregular lumps of ice.

halo   Any of a class of optical phenomena caused by the refraction of light by ice crystals, in the form of rings or arcs around the sun or moon. Halos have relatively large angular distances from the sun, 22° or larger; coronas produced by liquid cloud drops have much smaller radii; and rainbows, produced by raindrops, form circles around the antisolar point.

haze   A suspension of fine particles of dust, salt, droplets of terpene, or other emissions from vegetation, etc., that reduces visibility.

heat   A form of energy, characterized by the fact that it is transferred from bodies at higher temperatures to those at lower temperatures.

Hertz   A unit of frequency, equal to one cycle or oscillation per second.

heterosphere   The portion of the atmosphere above about 80 km in which the concentrations of the major chemical constituents of the air vary with height.

high   A center of high pressure on a constant-height chart or high height of a constant-pressure chart.

homosphere   The portion of the atmosphere below about 80 km in which the concentrations of the major chemical constituents of the air are practically invariant with height.

horizontal convergence or divergence   Horizontal flow in or out of a region. See *convergence* and *divergence*.

humidity   The amount of water vapor in the air. See *absolute humidity, relative humidity, specific humidity, mixing ratio, vapor pressure,* and *dew point* for particular measures of humidity.

hurricane   A severe tropical storm in the North Atlantic Ocean, Caribbean Sea, Gulf of Mexico, and eastern North Pacific Ocean off the coast of Mexico.

hydrologic cycle   The circulation of water substance from atmosphere to earth in the form of precipitation, from land through rivers, lakes, and underground flow to the seas, and from land surfaces and oceans by evaporation and transpiration to the atmosphere.

hydrometeor   Any atmospheric phenomenon associated with water substance, including in particular products of condensation, such as fog, cloud, rain, snow, hail, dew, and frost.

hydrosphere   The water portion of the earth, including rivers, lakes, and oceans.

hydrostatic equation   The equation relating pressure variation with height in the absence of motion in the atmosphere.

hygroscopic   Having an affinity with water, which promotes condensation at relative humidities below 100 percent.

Icelandic low   The low pressure center, with an associated cyclonic circulation, located on mean charts of sea-level pressure, and frequently on individual daily charts, in the vicinity of Iceland. It is one of the *centers of action*.

ice nucleus (also called *ice-forming nucleus*)   A particle that initiates ice crystal formation. If it acts in liquid water, it is a *freezing nucleus*. If it initiates formation of ice directly from the vapor phase, it is a *sublimation nucleus*.

ice point   The temperature at which pure water and pure ice exist in equilibrium at standard atmospheric pressure. This temperature is used as a fiducial point, 0°C and 32°F, in calibrating thermometers. Frequently (incorrectly) called the freezing point.

inertial flow   Flow in the absence of external forces. In the absence of real forces, such as the pressure gradient force or friction, moving air would be acted on by the Coriolis force, producing a centripetal acceleration which turns it in anticyclonically curved paths called *inertia circles*.

inertial period   The time it would take air to go around an inertial circle, namely $D/2 \sin \phi$ seconds, where ϕ is the latitude and D is the length of a sidereal day, namely 86,164.1 seconds.

infrared radiation   Electromagnetic radiation having wavelengths longer than visible light, from about 0.8 μm up to about 0.1 mm.

insolation   Solar radiation received at the earth's surface. Specifically, the rate at which direct solar radiation falls on unit horizontal surface.

instability   A condition in which a small perturbation of a system initially in equilibrium will produce forces that tend to move the system farther from equilibrium.

intertropical convergence zone (ITCZ)   The region near the equator where the northeast trade winds of the Northern Hemisphere and the southeast trade winds of the Southern Hemisphere come together.

inversion   In general the reversal of the usual variation of an atmospheric property with height, and the layer through which the reversal takes place. Specifically and usually the term refers to a *temperature inversion*, in which temperature increases with height instead of the usual decrease.

ion   An electrically charged particle.

ionization   The process by which neutral molecules or other particles in the air become electrically charged, chiefly by radiation or by collisions with high-energy particles.

ionosphere   A layer high in the atmosphere (above about 70 km) having a high density of charged particles.

irradiance   Flux of radiation per unit area.

irrotational motion   Flow in which the vorticity is zero.

isobar   A line along which the pressure has a constant value.

isobaric process   A process—for instance, a temperature change—in which the pressure remains constant.

isobaric surface   A surface at every point of which the pressure has the same value at a particular time. A constant-pressure surface.

isopleth   A line along which a given quantity has a constant value. For instance, an isobar is an isopleth of pressure.

isotach   A line along which wind speed has a constant value.

isotherm   A line along which the temperature has a constant value.

jet stream   A narrow band of very strong winds.

joule   The unit of energy in the SI.

katabatic wind   A downslope wind.

kelvin   The unit of temperature in the Kelvin or absolute temperature scale. It is equal to 1°C.

Kelvin temperature scale   The absolute temperature scale, based on thermodynamic considerations which place its zero at 273.15°C below the ice point. The kelvin (K) equal to the degree Celsius is chosen as the unit, which makes the ice point 273.15 K and the boiling point 373.15 K.

kilogram   The unit of mass in the SI. Originally set at 1000 g (see *gram* for original definition), it is now taken to be the mass of the standard kilogram, which is preserved by the International Bureau of Weights and Measures at Sèvres, France.

kinetic energy   The energy possessed by a body as a result of its motion.

knot   A unit of speed: one nautical mile per hour.

land-and-sea breeze   The cycle of local winds at a coast induced by the heating of the land during the day and its cooling at night.

langley   A unit of energy per unit area: one calorie per square centimeter.

lapse rate   The rate of decrease of temperature with height in an air column.

latent heat   The heat released or absorbed during a change of phase, the temperature and pressure remaining constant during the process.

level of nondivergence   A level in the atmosphere at which there is no horizontal convergence or divergence of the winds.

lidar ("light radar")   An instrument for probing the atmosphere using laser pulses in or near the visible region of the radiation spectrum.

lifting condensation level (LCL)   The level at which a parcel of moist air would become saturated if raised adiabatically.

lightning   The visible effects of electrical discharges in the atmosphere, almost always associated with cumulonimbus clouds in thunderstorms.

lofting   The spreading in upward arcs of a smoke plume emitted into air with an inversion below the stack exit and an unstable lapse rate at and above it.

looping   The spreading in up-and-down arcs of a smoke plume emitted into air with an unstable lapse rate from the ground to above the stack exit.

low   A center of low pressure on a constant-height chart or low height on a constant-pressure chart.

magnetic storm   A severe fluctuation of the earth's magnetic field, usually world-wide and associated with a solar flare or other solar disturbance.

magnetosphere   The configuration of charged particles around the earth arising from the interaction of the solar wind and the earth's magnetic field.

mass   The amount of matter, as measured by its resistance to acceleration.

mean   The average of a group of numbers.

median   The middle value of a distribution of quantities.

melting point   The temperature at which a solid changes to liquid. Ice ordinarily melts at the *ice point*, but freezing usually occurs at a lower temperature.

meridional circulation   An atmospheric circulation in a vertical plane oriented north–south, along a meridian.

meridional flow   Flow along a meridian, i.e., north–south or south–north.

mesohigh   A small high-pressure area associated with a thunderstorm or other severe storm system.

meson   A subatomic particle having mass intermediate between electrons and protons.

mesopause   The boundary between the mesosphere and the thermosphere.

mesoscale   Characterizing the size of atmospheric circulation patterns and other systems intermediate between the small-scale phenomena that are the subject of micrometeorology and large-scale phenomena such as extratropical cyclones. Its range is usually taken to be from about 10 km to 200 km.

mesosphere   The layer above the stratosphere, in which the temperature decreases with height.

meteor   In general, any phenomenon occurring in the atmosphere. This general meaning is not widely used any longer, and usually the term is used to mean a luminous meteor, the streak of light accompanying an object from space passing through the air.

meteorograph   An instrument that automatically makes and records two or more meteorological measurements, e.g., temperature, humidity, and pressure.

micrometeorology   The study of very small-scale atmospheric phenomena, in particular those related to the layer of air in immediate contact with the ground.

microwave radiation   Electromagnetic radiation with wavelengths in the range from 0.1 mm to 1 m.

millibar   A unit of pressure, equal to 100 Pa or 1000 dynes/cm².

mistral   A cold, dry north wind that blows from the Alps down the Rhone Valley in France.

mixing condensation level   The height at which saturation is reached by turbulent stirring of the layer of air from the ground to that height.

mixing ratio   The ratio of the mass of water vapor in moist air to the mass of dry air in it.

moist air   Air that contains water vapor.

moisture   Ordinarily, the water vapor in air. Sometimes the term is used to refer to the total water substance (vapor, droplets, and ice crystals) in the air, and occasionally to precipitation accumulation on the ground.

mole   A unit of mass equal in number of grams to the molecular weight of a substance.

molecule   A group of atoms bound together forming the smallest unit retaining the chemical properties of a compound.

momentum   A measure of the amount of motion of a body, given by the product of its mass and its velocity.

monsoon   A seasonal wind, blowing from sea to heated land in summer and from land to sea in winter.

mountain and valley winds   A diurnal wind pattern, blowing up the valley and up the mountain slopes in the daytime, and in the opposite direction at night.

multicell thunderstorm   A thunderstorm consisting of several convective cells in various stages of development.

neutral equilibrium   A condition of a system initially in equilibrium such that a slight disturbance results in the system remaining in equilibrium, giving rise neither to forces tending to amplify the disturbance nor to forces tending to restore the system to its original state.

neutron   A subatomic particle having no electric charge and mass slightly more than the mass of a proton.

newton   The unit of force in the SI.

nimbostratus   A dark gray layer cloud, ordinarily hanging low, from which rain or snow falls more or less continuously.

numerical weather prediction (NWP)   The forecasting of atmospheric flow patterns and the resulting temperature changes, precipitation, etc., by solving the equations of motion numerically using high-speed computers.

occluded front   The air-mass boundary formed when the cold front of a wave cyclone overtakes the warm front.

occlusion process   The process of the cold air cutting off the warm air at the ground in a wave cyclone. The warm air is lifted as the cold front, moving faster, overtakes the warm front; the warm air is hidden from the ground, or *occluded*. In the process the low center deepens and the wind system intensifies.

orographic   Caused by or pertaining to mountains.

overseeding   Introduction into a cloud of an excess of nucleating material for the effect desired, usually an amount greater than the optimum for maximum precipitation.

ozone   A highly reactive form of oxygen, having three atoms in its molecule.

ozonosphere   The layer, mostly in the stratosphere, in which there is a relatively high concentration of ozone.

pascal   The unit of pressure in the SI, given by 1 newton per square meter.

perihelion   The point on the earth's orbit nearest the sun.

period   In wave motion, the time interval between successive passages of a given phase of the wave, for instance, the crest or trough.

phase   (1) The physical or chemical state of a substance, for example, solid, liquid, or gas. (2) A particular stage in a periodic process or phenomenon, for example, the maximum amplitude in wave motion.

photochemical reaction   A chemical reaction activated by the absorption and/or emission of electromagnetic radiation.

photon   The quantum of electromagnetic radiation.

planetary albedo   The fraction of the solar radiation received by the earth that is reflected by the entire earth and atmosphere as seen from space.

plasma   An ionized gas containing equal numbers of positively and negatively charged particles.

plasmasphere   A portion of the magnetosphere occupied by plasma.

polar easterlies   The easterly winds that on the average are present poleward of the Aleutian and Icelandic lows in the Northern Hemisphere and over the Antarctic Continent in the Southern Hemisphere.

polar front   The front separating air masses originating in polar and subpolar regions from those originating in tropical and subtropical areas.

polar front jet stream   The concentrated narrow band of very strong winds in the upper troposphere associated with the polar front.

potential   In electricity, the potential energy of unit charge at a point in a circuit, the gradient of which is the force acting on such a charge; voltage.

potential energy   The energy of a body due to its position in a field of force. For the atmosphere, the energy of an air parcel due to its position in the field of gravity.

potential temperature   The temperature a parcel of air would attain if brought by an unsaturated adiabatic process to a pressure of 1000 mb (100 kPa).

power   The rate at which energy is released or converted. The unit of power in the SI is the watt, defined as 1 joule per second.

precipitation   Any form of liquid or solid water particles that falls from the atmosphere to the ground. Also, the accumulated amount, expressed in equivalent depth of liquid water, that has fallen.

pressure   The force per unit area exerted normally (perpendicularly) by a fluid on a surface within it or bounding it.

pressure gradient force   The force due to variation of pressure within a fluid.

prevailing westerlies   The predominantly westerly flow showing up on the average in middle latitudes, particularly in the upper troposphere.

prevailing wind   The wind having the direction most frequently observed at a place during a given period.

primary pollutant   An air pollutant that has remained in the air in the form it was emitted.

proton   A positively charged subatomic particle having a mass 1836 times that of the electron.

pseudofront   Same as *gust front*.

psychrometer   An instrument for measuring humidity, consisting of two thermometers on a common back, one of which has its bulb covered with moistened fabric so that when ventilated it cools by an amount depending (inversely) on the amount of water vapor in the air.

quantum   The smallest unit into which energy can be divided; for radiation of frequency v it has the size $\hbar v$, where $\hbar$, Planck's constant, has the value $6.625 \cdot 10^{-27}$ erg sec.

radiant flux density   Same as *irradiance*.

radiation   The process by which energy is propagated through space by high energy particles (corpuscular radiation) or by waves in the electric and magnetic fields in space (electromagnetic radiation). Also, the energy so propagated.

radiation belts   Regions of high radioactivity in the outermost layers of the atmosphere, that is, in the magnetosphere.

radiosonde   An instrument, sent up on a balloon, that measures and transmits by radio the values of temperature, pressure, and humidity at various levels as it rises through the atmosphere.

rainbow   A circular arc of colored bands surrounding the point in the sky opposite the sun or moon, due to the light being refracted and diffracted by raindrops (or other drops of spray). The primary rainbow has an angular radius of about 42°, with red on the inside and blue on the outside. A larger but fainter secondary bow of about 50° angular radius with colors reversed is frequently visible.

rain gauge   An instrument for measuring the amount of rain that has fallen. Sometimes it also measures the rate at which it falls.

rainout   The removal of pollutants from the air by incorporation into cloud drops that grow and fall as rain or snow.

rain shadow   The region in the lee of a mountain range where precipitation is markedly less than on the windward side.

rawinsonde   A system for measuring speed and direction of the wind, in addition to temperature, pressure, and humidity, using a radiosonde and tracking its path by radio direction-finding or radar.

reflectivity   The fraction of radiation of a particular wavelength reflected by a given surface. Reflectivity differs from albedo in that albedo is the fraction of radiation of all wavelengths reflected.

relative advection   The horizontal flow of warmer air over cooler air or cooler air over warmer air, thereby decreasing or increasing the lapse rate.

relative humidity   The ratio, expressed in percent, of the observed vapor pressure to the saturation vapor pressure for the observed temperature.

resultant wind   The vectorial average of all winds for a given height at a particular place for a particular period, say, a month.

ridge   An elongated area of high atmospheric pressure.

Rossby wave   A long wave in the westerly flow of the upper troposphere.

Santa Ana   A hot, dry foehn-type northeast wind that carries air from the high-level desert areas of Nevada and California to the coast of Southern California.

saturation   Strictly, *saturation equilibrium*. The state of moist air such that, when in contact with a plane surface of pure water at the same temperature, the amount of water vapor in the air remains constant. Air containing this amount of water vapor is said to be *saturated*, even though in the absence of particles (nuclei) or surfaces it is possible for it to contain much larger quantities of water vapor without condensation occurring.

saturation adiabat   A curve on a thermodynamic diagram representing the change of state variables (for instance, pressure and temperature) in a saturation adiabatic process.

saturation-adiabatic lapse rate   The lapse rate of an air column that is the same as the rate of cooling in a saturation-adiabatic process.

saturation-adiabatic process   An adiabatic process in which the air is maintained at saturation by condensation or evaporation. When cloudy air ascends without mixing with environmental air, the change of temperature with decreasing pressure follows the saturation-adiabatic process rate of cooling.

saturation mixing ratio   The value of the mixing ratio of air saturated at the observed temperature and pressure. (See *saturation*.)

saturation vapor pressure   The value of the vapor pressure of air saturated at the observed temperature. (See *saturation*.) The saturation vapor pressure is a function of the temperature only, being practically independent of the pressure.

scattering   The process by which small particles intercept electromagnetic radiation and send it in all directions unchanged in wavelength or total amount.

sea breeze   A local wind at a coast that blows from sea to land during the daytime due to the higher temperature on land than at sea.

sea-level pressure   The atmospheric pressure at mean sea level, observed directly where possible, but for most land stations, which are at elevations above sea level, estimated on the basis of the observed station pressure "reduced to sea level" by the weight of an air column of length from sea level to the elevation of the station and temperature based on the temperature observed at the station.

secondary pollutant   An air contaminant that has been converted by chemical reaction or other change from the form in which it was emitted from its source.

shear   With respect to the wind, the variation of wind velocity with distance in a specified direction.

significant figures   The digits, other than zeros adjacent to the decimal point, of a measurement that are known accurately.

skew T–log p diagram   A thermodynamic diagram with temperature and logarithm of pressure as coordinates, but with the isotherms rotated 45° clockwise.

sky radiation   Solar radiation reaching the earth indirectly after having been scattered from the direct solar beam by molecules or larger particles suspended in the air.

sleet   In the United States, precipitation in the form of small quasi-spherical pellets of ice formed by the freezing of raindrops. In British terminology, a mixture of rain and snow.

slope wind   The wind that blows directly up or down the sides of mountains due to the temperature differences between the mountain sides and the free air over the valley or plain at the same levels.

smog   Originally, a mixture of smoke and fog. Currently, any air pollution, whether or not associated with high humidity and fog.

snow   Precipitation of opaque or semi-opaque ice particles in the form of individual crystals, small pellets, or flakes formed by the aggregation of crystals.

solar constant   The rate at which radiation from the sun is received outside the atmosphere on a unit surface perpendicular to the solar beam, at the average distance of the earth from the sun.

solar radiation   (1) The total electromagnetic radiation emitted by the sun. (2) The intensity and spectral distribution of radiation received from the sun by the earth.

solar wind   A stream of particles sweeping out from the sun at high speeds in all directions.

solstice   Either of the two points on the sun's apparent path through the fixed stars (the ecliptic) where it is farthest north or south from the earth's equatorial plane. Also, the times the sun reaches these points, about June 21 for the summer solstice and December 22 for the winter solstice.

sounding   In general, a probing of the environment for scientific observation. In meteorology, a probing of the upper atmosphere by instruments carried aloft by a balloon or rocket.

sounding curve   The representation on a thermodynamic diagram of the state of the various parcels of air in an air column as determined by sounding—for instance, with a radiosonde.

source region   An extensive area of the earth's surface having relatively uniform characteristics, over which bodies of air remain sufficiently long to acquire characteristic air mass properties.

specific heat   The amount of heat required to raise the temperature of unit mass of a substance one degree.

specific humidity   The mass of water vapor in unit mass of moist air.

specific volume   The volume occupied by unit mass of a substance.

spectrum   The distribution of energy of electromagnetic radiation with respect to wavelength or frequency. Also, the display of this distribution when a beam of electromagnetic radiation is spread by a prism or a diffraction grating.

squall line   A nonfrontal line or narrow band of thunderstorms, usually characterized by gusty winds.

stability   The property of a system in equilibrium, such that any small disturbance gives rise to forces tending to restore the system to its initial state.

standard atmosphere   A hypothetical distribution of atmospheric temperature, pressure, and density with height, set by national or international agreement as representative of average conditions in the atmosphere for use in the design of aircraft and missiles and the calibration of instruments used on them.

stationary front   A boundary between two air masses, neither of which is displacing the other, so that the boundary does not move. The air masses can and do move parallel to a stationary front, and the warmer air mass can move normal to it, thereby going up or down the frontal surface over the colder air.

steam   Water vapor, particularly the vapor produced by boiling.

steam fog   Fog formed when water evaporates from a warm surface into much colder air.

storm   An atmospheric disturbance. Usually the term refers to disturbances with strong winds at the ground accompanied by precipitation, ranging in size from a local thunderstorm, of the order of a few kilometers across, to extratropical cyclones thousands of kilometers in diameter.

storm surge   High water driven onto the shore by the strong winds of a storm.

stratocumulus   A low cloud in the form of a gray layer of uneven density or shading, or sometimes of separate rolls or puffs, though not the sharply defined flat-based puffs of cumulus.

stratopause   The boundary between the stratosphere, which it tops, and the mesosphere.

stratosphere   The layer above the troposphere in which the temperature is constant with height or rises as one goes upward. It extends from the tropopause to about 50 km.

stratus   A low, relatively thin cloud in the form of an extensive layer of uniform gray color, with a diffuse base and a sharply defined top.

streamline   A line oriented at every point in the direction of the wind at that point.

sublimation   In general, the transition from solid phase directly to vapor or vice versa. In meteorology sublimation is used exclusively to mean the change directly from water vapor to ice.

sublimation nucleus   A particle that initiates the formation of an ice crystal directly from water vapor.

subsidence   The sinking of an extensive body of air.

subtropical   Pertaining to the region immediately poleward of the Tropic of Cancer and the Tropic of Capricorn, up to about 35° latitude.

subtropical front   A zone of transition in the upper troposphere between the warm air carried poleward by the upper branch of the Hadley cell and the cooler air of middle latitudes.

subtropical high   One of the semi-permanent high pressure areas centered over the oceans at about 30° to 35° latitude.

subtropical inversion   Also called *trade wind inversion*. The layer over the eastern portions of oceans at subtropical latitudes in which the temperature increases with height. It is present almost continuously during the warm half of the year with its base at heights ranging from 300 m or 400 m at the west coasts of continents to 2 km or more in mid-ocean.

supercell thunderstorm   A thunderstorm consisting of a single large intense cell that persists for as much as several hours.

supercooled water drops   Liquid drops at temperatures below 0°C.

supersaturation   When the relative humidity is above 100 percent, the amount of its excess, in percent.

surface inversion   A layer of air with base at the earth's surface in which the temperature increases with height.

surface observation   A weather observation made using instruments that remain at the ground, as opposed to upper-air observations using sounding devices.

surface weather map   A chart summarizing the weather distribution near the ground, including sea-level pressure, temperature, humidity, clouds as observed from the earth's surface, and precipitation.

swell   A long ocean wave that has moved from the region where it was generated by the wind.

synoptic   In general, giving an overall view; in meteorology, summarizing the weather situation at a particular time over a large area.

synoptic meteorology   The branch of meteorology concerned with the study of atmospheric phenomena by using synoptic charts.

synoptic observations   Observations taken for use on synoptic charts; consequently, observations taken at the same time throughout a region or the entire world.

temperature   The property of a body that determines whether heat will flow to or from it when placed in contact with another body. If body A is hotter, that is, at a higher temperature, than body B, heat will flow from A to B. Temperature is measured on one of several temperature scales (Kelvin, Celsius, or Fahrenheit) with a thermometer.

terminal velocity   The speed attained by an object falling through a fluid when the frictional drag exerted by the fluid on the object exactly equals the force exerted by gravity, at which time the object is no longer accelerated but falls at constant speed.

terrestrial radiation   The total radiation emitted by the earth's surface. It is in the far infrared region of the spectrum.

theodolite   An optical instrument for measuring the angular position of a distant object, for instance, a balloon used to measure upper winds.

thermal   A rising current of air, caused by the air near the ground being rendered unstable by heating during the day.

thermal low   An area of low pressure due to high temperatures caused by intense heating in such regions as the deserts of subtropical latitudes in summer.

thermal wind   The change of geostrophic wind with height due to horizontal temperature variation.

thermistor   A device in which the electrical resistance varies relatively rapidly with temperature. It is used as the sensor in thermometers, particularly in radiosondes.

thermodynamic diagram   A chart on which the variation of state variables, for instance, temperature and pressure, during thermodynamic processes can be represented. Usually in meteorological thermodynamic diagrams the coordinates are temperature and a function of pressure that is roughly proportional to height. Another frequently used diagram has as coordinates temperature and entropy, a quantity that is constant in an adiabatic process. In addition to the basic coordinate lines, usually isobars and isotherms, curves representing unsaturated and saturated adiabatic processes and constant values of saturation mixing ratio are shown.

thermograph   An instrument for continuously recording the temperature.

thermometer   An instrument for measuring temperature.

thermosphere   The region of the atmosphere, extending upward from the mesopause, in which the temperature increases with height. From very low values at about 80 km, the temperature attains about one thousand kelvins at a height of several hundred kilometers.

three-phase process   The process of formation of precipitation by the growth of ice crystals in the presence of supercooled liquid water drops in moist air.

thunder   The sound caused by the sudden expansion of air heated by lightning discharges.

thunderhead   The popular name for the anvil-shaped top of a cumulonimbus cloud.

thunderstorm   A rainstorm accompanied by lightning, thunder, strong wind gusts, and sometimes hail. It is produced by intense convection in one or more cumulonimbus clouds.

tornado   A violent windstorm in the form of a rotating column of air extending downward from a cumulonimbus cloud, with a funnel cloud at its center.

torque   The moment of force about a point, that is, the action of a force in such a way as to produce a rotational motion.

trade wind cumulus   The characteristic cumulus cloud of the trade winds over the ocean, having bases at about 600 m and tops penetrating slightly into the subtropical inversion.

trade wind inversion   The subtropical inversion.

trade winds   The persistent easterly winds blowing around the southern part of the subtropical highs and converging toward the intertropical convergence zone.

trajectory   A curve showing the positions in space successively occupied by a moving body (a parcel of air, in meteorology).

tropical cyclone   A cyclonic storm originating at low latitudes.

tropopause   The boundary between the troposphere and stratosphere, sloping upward from 8 to 10 km in polar regions to 16 or 17 km at low latitudes.

troposphere   The layer of the atmosphere extending from the earth's surface to the tropopause in which the temperature usually decreases with height.

trough   An elongated area of relatively low pressure.

turbidity   The reduction of atmospheric transparency due to particles suspended in the atmosphere.

turbulence   Small-scale irregular fluctuations of motion.

typhoon   An intense tropical cyclone over the western Pacific Ocean.

ultraviolet radiation   Electromagnetic radiation with wavelengths shorter than the visible region of the spectrum, that is, shorter than about 0.4 μm, but longer than x rays.

unstable equilibrium   A state of a system that is in equilibrium such that any disturbance gives rise to forces that tend to bring the system farther from equilibrium.

upper-air observation   An observation of conditions in the atmosphere above the earth's surface, made by radiosonde or other sounding device.

upslope fog   Fog caused by the flow of air upward over sloping terrain. The upward motion produces adiabatic cooling of the air to its dew-point.

urban heat island   The area affected by the warmth of a city in comparison to the cooler surrounding rural areas.

valley fog   Fog produced by radiation from the ground of valleys and adjoining mountain slopes, with drainage of the cool air from the slopes allowing the fog to form in the valley but leaving the upper slopes of the mountains clear.

valley wind   A wind blowing upward along a valley floor due to daytime heating of the higher land.

vapor line   A line on a thermodynamic diagram passing through points with a constant value of saturation mixing ratio.

vapor pressure   The partial pressure due to the molecules of the water vapor in moist air.

vector   A quantity, such as force, velocity, or acceleration, that has both magnitude and direction. From the strict mathematical standpoint there are additional requirements in terms of the laws by which they combine.

vertical wind shear   The variation of wind velocity with height.

viscous stress   The tangential force exerted on a parcel of moving fluid by the surrounding fluid.

visibility   The greatest distance in a given direction that a dark object can be distinguished from the sky in the daytime, or a moderately intense unfocused light source can be seen at night by the unaided eye. Also called *visual range*.

visible radiation   Electromagnetic radiation of wavelengths between about 0.4 μm and 0.7 μm, to which the eye is sensitive. The same as *light*.

vortex   A pattern of fluid flow involving rotation about an axis.

vorticity   A measure of the turning of a small parcel of fluid.

warm front   A boundary between two air masses that are moving in such a fashion that the warmer air mass is displacing the cooler one at the ground.

warm front fog   Fog in the cold air just in advance of a warm front, produced by the evaporation of rain falling from the warm air above the frontal surface.

warm-front-type occlusion   An occluded front formed when the cold air behind the cold front that overtakes the warm front is not as cold as the cold air ahead of the warm front.

warm sector   The warm air protruding into the cold air in a frontal wave cyclone.

washout   The process of collection of pollutants by rain or snow falling through polluted air. Compare *rainout*.

watt   The unit of power or rate of conversion of energy in the SI. It is equal to 1 joule per second.

wave cyclone   A cyclone occurring at the crest of a wave on a front.

wavelength   The distance between corresponding parts of successive waves, for instance, from crest to crest.

wave speed   The speed with which an identifiable part of the wave moves, for instance, the trough.

weather   The state of the atmosphere, particularly with respect to its effects on human activities and in terms of its short-term variability.

westerlies   The winds with predominantly a west-to-east component that are present most of the time in the lower troposphere in middle latitudes and over most of the earth in the upper troposphere. Same as *prevailing westerlies*.

westerly wave   A wave disturbance in the westerly flow of the upper troposphere.

wet adiabat   Same as *saturation adiabat*.

wet-bulb depression   The difference in temperature between the dry-bulb thermometer and the wet-bulb thermometer of a psychrometer, assuming they are properly ventilated.

wet-bulb potential temperature   The temperature an air parcel would attain by being raised unsaturated adiabatically from its initial state until it is saturated, and then brought down saturated adiabatically until its pressure is 1000 mb.

wind   The horizontal motion of air relative to the earth's surface.

wind shear   The variation of wind velocity with distance.

wind vane   An instrument used to indicate wind direction.

x ray Electromagnetic radiation having a very short wavelength, in the range from 0.01 nm to 10 nm.

zodiacal light A faint cone of light sometimes visible extending upward from the horizon in the west just after sunset or in the east just before sunrise.

zonal Along parallels of latitude, that is, from west to east or from east to west.

Abbreviations and Acronyms

See Appendix A for abbreviations of units of measurement and Table 1.1 for abbreviations of cloud names.

A   Arctic air

AFOS   Automation of Field Operations and Services

AIDS   Aircraft Integrated Data System

AIREP   Aircraft Report

ALPEX   Alpine Experiment

AMTEX   Air Mass Transformation Experiment

APT   Automatic Picture Transmission

ASDR   Aircraft to Satellite Data Relay

ATS   Applied Technology Satellite

BOMEX   Barbados Oceanographic and Meteorological Experiment

BTU   British Thermal Unit

CAV   Constant Absolute Vorticity

CCL   Convective Condensation Level

CCN   Cloud Condensation Nuclei

cP   Continental polar air

cT   Continental tropical air

E   Equatorial air

EPA   Environmental Protection Agency

ESSA   Environmental Survey Satellite; also Environmental Science Services Administration (predecessor to NOAA)

EST   Eastern Standard Time

FGGE   First GARP Global Experiment

FIDO   Fog Investigation and Dispersal Operation

GARP   Global Atmospheric Research Program

GATE   GARP Atlantic Tropical Experiment

GDS   Global Data-Processing System

GMT   Greenwich Meridian Time

GOES   Geostationary Operational Environmental Satellite

GOS   Global Observing System

GTS   Global Telecommunication System

ICSU   International Council of Scientific Unions

IN   Ice Nucleus

IR   Infrared (Radiation)

ITCZ   Intertropical Convergence Zone

ITOS   Improved TIROS Operational System

JASIN   Joint Air–Sea Interaction Experiment

LIE   Line Islands Experiment

LND   Level of nondivergence

METROMEX   Metropolitan Meteorological Experiment

MONEX   Monsoon Experiment

mP   Maritime polar air

mT   Maritime tropical air

NASA   National Aeronautics and Space Administration

NESS   National Earth Satellite Service (formerly National Environmental Satellite Service)

NHRE   National Hail Research Experiment

NMC   National Meteorological Center

NOAA   National Oceanic and Atmospheric Administration

NWP   Numerical Weather Prediction

NWS   National Weather Service

OPEC   Organization of Petroleum Exporting Countries

PACE   Precipitation Augmentation for Crops Experiment

PEP   Precipitation Enhancement Project

PFJS   Polar front jet stream

PIBAL   Pilot Balloon Observation

PIREP   Pilot Report

POLEX   Polar Experiment

ppm   parts per million

RAOB   Radiosonde Observation

RAWIN   Rawinsonde

SI   International System of Units

SL   Streamline

SMS   Synchronous Meteorological Satellite

SOP   Special Observational Period

STJ   Subtropical jet stream

TIROS   Television Infrared Observational Satellite

TOS   TIROS Operational System

TOVS   TIROS-N Operational Vertical Sounder

WMO   World Meteorological Organization

WWW   World Weather Watch

Index